CISM COURSES AND LECTURES

Series Editors:

The Rectors of CISM
Sandor Kaliszky - Budapest
Mahir Sayir - Zurich
Wilhelm Schneider - Wien

The Secretary General of CISM
Giovanni Bianchi - Milan

Executive Editor
Carlo Tasso - Udine

The series presents lecture notes, monographs, edited works and proceedings in the field of Mechanics, Engineering, Computer Science and Applied Mathematics.
Purpose of the series is to make known in the international scientific and technical community results obtained in some of the activities organized by CISM, the International Centre for Mechanical Sciences.

INTERNATIONAL CENTRE FOR MECHANICAL SCIENCES

COURSES AND LECTURES - No. 403

OPTICAL METHODS IN EXPERIMENTAL SOLID MECHANICS

EDITED BY

KARL-HANS LAERMANN
BERGISCHE UNIVERSITY OF WUPPERTAL

This volume contains 266 illustrations

This work is subject to copyright.
All rights are reserved,
whether the whole or part of the material is concerned
specifically those of translation, reprinting, re-use of illustrations,
broadcasting, reproduction by photocopying machine
or similar means, and storage in data banks.
© 2000 by CISM, Udine
Printed in Italy
SPIN 10763422

In order to make this volume available as economically and as
rapidly as possible the authors' typescripts have been
reproduced in their original forms. This method unfortunately
has its typographical limitations but it is hoped that they in no
way distract the reader.

ISBN 3-211-83325-0 Springer-Verlag Wien New York

PREFACE

Recent developments and achievements in measurement techniques especially in optical and electro-optical methods like holography, shearography, speckle interferometry, grid-methods, modern modification of photoelasticity and photoviscoelasticity have opened new fields of applying experimental analysis in solid mechanics, in research as well as in industrial practice. Combined with computer techniques in digital image processing, data recording and data evaluation complex measuring systems can be realised nowadays with the possibility of far-reaching automation of the entire analysing processes.

The sensitivity and the high resolving power of recording equipment requires thorough investigation of any effects on the precision and reliability of the measurements. Therefore it is necessary, to deal with the physics and the theories of the measurement methods in order to interpret the observed phenomena correctly. Furthermore proper algorithms must be available to evaluate the measured data to get the finally wanted information, which generally do not coincide with the quantities taken from the measurements originally. This fact requires complementary scientific considerations of the theories concerning the problems to be analysed. At least the combination of experimental and mathematical/numerical procedures for evaluation and interpretation, known as "hybrid technique", demands additional contemplation.

The lectures cover
i.   the theory and the physics of advanced optical measuring methods and problems of experimental performance, recent achievements in 2- and 3-dimensional linear/non-linear photoelasticity including photovisco-

elasticity, Moiré- and grid-techniques, interferometric methods (holography, speckle interferometry, shearography);

ii. the theory of digital image processing and its performance, data-recording, -compression, -processing, -visualization;

iii. mathematical and numerical procedures, informatics for evaluation of measured, digitized data.

As the co-ordinator of the CISM-course "Modern Optical Methods in Experimental Solid Mechanics" and the editor of the lecture notes I gratefully thank the lecturers/authors for the effective co-operation and thorough revising the manuscripts, the resident Rector, the General Secretary, the Chief-Editor,- and explicitly the staff-members of CISM Secretariat for support and excellent organisation of the course. Many thanks to the Springer-Verlag for publishing the lecture notes.

Karl-Hans Laermann

CONTENTS

Preface .... Page

I. HYBRID TECHNIQUES IN EXPERIMENTAL SOLID MECHANICS
by K-H. Laermann 1

    1. INTRODUCTION 2
        1.1   Roll and importance of optical methods in experimental Mechanics
        1.2   Influence and effect on safety and reliability of structures
        1.3   Presumption of applicability in industrial practice

    2. PRINCIPLE OF HYBRID TECHNIQUES 7

    3. NON-LINEAR PHOTOELASTICITY 13
        3.1   Non-linear stress-strain relations
        3.2   Non-linear relation between stress state and birefringence
        3.3   Iterative discrete solution
        3.4   Determination of material response
        3.5   Examples of application

    4. PHOTOVISCOELASTICITY 25
        4.1   Preliminary remarks
        4.2   Basic relations between the stress tensor and the refraction tensor
        4.3   Determination of the optical relaxation function
        4.4   Solution of the VOLTERRA's integral equation

    5. EVALUATION OF PHOTOELASTIC/VISCOELASTIC MEASUREMENTS BY MEANS OF BOUNDARY-ELEMENT METHOD 34
        5.1   Mathematical basis
        5.2   Plane plates with inclusions
        5.3   Application to two-dimensional photoelasticity
        5.4   Elastic/viscoelastic response of material

    6. ANALYSIS OF 3-D-STRESS-STRAIN STATES BY MEANS OF DISCRETE BOUNDARY-INTEGRAL 47
        6.1   Determination of the surface displacement
        6.2   Mathematical basis
        6.3   The reduction procedure
        6.4   Example of application

    7. ANALYSIS OF 3-D-STRESS-STRAIN STATES BY COMBINING **ESPI** WITH **FEM** 56
        7.1   Determination of surface displacements by ESPI

    7.2   Evaluation of the experimental data by means of FEM
    7.3   Example of application

8. REMARKS GENERALLY TO CONSIDER IN PERFORMING
   EXPERIMENTAL MECHANICS 64
    8.1   Multidisciplinary character of experimental mechanics
    8.2   Necessity of advanced theories of mechanical problems
    8.3   Inverse problems in final evaluation
    8.4   Transfer problems related to measuring systems

9. REFERENCES 70

II. RECENT DEVELOPMENTS IN 3-D-PHOTOELASTICITY AND GRATING STRAIN MEASUREMENT
by A. Lagarde 73

    1. THREE DIMENSIONAL PHOTOELASTICITY 74
       1.1   Present practice and new possible way
       1.2   Propagation of the light wave through photoelastic medium
             1.2.1 The classical scheme
             1.2.2 ABEN schematisation
             1.2.3 Hypothesis for a thin slice
             1.2.4 Discrete analysis into thin slices
       1.3   Whole-field analysis with a plane polariscope
       1.4   Whole-field optical slicing method
             1.4.1 Method on the contrast measurement of one recording intensity field
             1.4.2 Method based on the variance measurement of the combination of three recording intensity fields
       1.5   Separation of isoclinic and isochromatic patterns of the slice. Isostatics plotting
       1.6   Conclusion and perspectives
       1.7   References

    2. GRATING STRAIN MEASUREMENT 87
       2.1   Introduction
       2.2   Local strain measurement
             2.2.1 Recall: Description on the move of a continuous medium
             2.2.2 Principle of the method
             2.2.3 Grating realisation
             2.2.4 Measurement by optical diffraction
             2.2.5 Measurement by spectral analysis
       2.3   Improvement of the accuracy
             2.3.1 The tools of the accuracy
                  2.3.1.1 The phase shifting method
                  2.3.1.2 The spectral interpolation method
             2.3.2 The device
       2.4   Holo-grating analysis

        2.4.1 Holo-grating recording
        2.4.2 Holo-grating reconstruction
        2.4.3 Application to the ductile fracture
  2.5 Local strain measurement in dynamic
        2.5.1 Diffraction in oblique incidence
        2.5.2 Principle of the method
        2.5.3 Implementation
        2.5.4 Experimental tests
        2.5.5 Hopkinson bar investigation
  2.6 Local strain measurement on cylindrical specimen
  2.7 Conclusion and perspectives
  2.8 References

III. AUTOMATED IN-PLANE MOIRÉ TECHNIQUES AND GRATING INTERFEROMETRY
by M. Kujawinska 123

1. THE BASIC RULES IN FULL-FIELD GRID TECHNIQUES 124
  1.1 Basic grid moiré technique
  1.2 Fixed grid method
  1.3 Principles of grid theory
        1.3.1 Formation of moiré in incoherent light
        1.3.2 Interferometric techniques

2. GRID (GRATING) TECHNOLOGY 132
  2.1 Amplitude grids
  2.2 Phase gratings

3. PRINCIPLES OF AUTOMATIC ANALYSIS OF RESULTS 136
  3.1 Introduction
  3.2 Fringe pattern analysis
        3.2.1 Phase shifting methods
        3.2.2 Fourier transform method
  3.3 Phase unwrapping
  3.4 Calculation of strain fields

4. THE MOIRÉ FRINGE METHOD 144
  4.1 Physical superimposition of gratings
  4.2 Projected superimposition
  4.3 Double exposures and moiré photography

5. GRATING (MOIRÉ) INTERFEROMETRY 148
  5.1 Introduction
  5.2 Principle of grating interferometer
  5.3 Grating interferometer systems
        5.3.1 The laboratory system (LGI)
        5.3.2 The workshop portable system
        5.3.3 Fibre optics grating interferometer sensor (FOS)
        5.3.4 Waveguide grating microinterferometer (WGI)
  5.4 Engineering features of automated grating interterometers

6. MATERIAL ENGINEERING AND MICRO-MECHANICS 158
  6.1 Introduction
  6.2 Local approach to material engineering
  6.3 Microelements testing
  6.4 Electronic chips and packages studies
  6.5 Mechanical/material joint testing

7. HYBRID METHODS OF RESIDUAL STRESS ANALYSIS 175
  7.1 Introduction
  7.2 Objects of measurements
      7.2.1 Railway rail
      7.2.2 Laser beam weldment
  7.3 Experimental set-up and procedure
  7.4 Experimental results and discussion
      7.4.1 Residual strain determination for railway rails
      7.4.2 Residual stress determination in laser beam weld
  7.5 Conclusions

8. CONCLUSIONS AND FUTURE POTENTIALS 187
  8.1 Moiré in relation to other methods
  8.2 Direction of development and future potentials

9. REFERENCES 189

IV. INTERFEROMETRIC METHODS
by W. Jüptner 197

1. INTRODUCTION 198
  1.1 Historical remarks
  1.2 Properties of electromagnetic waves
  1.3 Interference of waves
  1.4 Diffraction of light
  1.5 References

2. HOLOGRAPHIC INTERFEROMETRY 209
  2.1 Holography
      2.1.1 Fundamentals of holography
      2.1.2 Amplitude and phase holograms
      2.1.3 Types of holograms
      2.1.4 Stability requirements for recording holograms
  2.2 Holographic interferometry
      2.2.1 Fundamentals of holographic interferometry
      2.2.2 Evaluation of interference pattern
  2.3 Application of holographic interferometry
      2.3.1 Non-destructive evaluation by holographic interferometry
      2.3.2 Fracture mechanics
      2.3.3 Vibration analysis
  2.4 References

3. DIGITAL HOLOGRAPHY ... 235
  3.1 Fundamentals of digital holography
    3.1.1 Principle of digital holography
    3.1.2 Recording of digital holograms
    3.1.3 Reconstruction of digital holograms
    3.1.4 Elimination of the zero-order diffraction wave
  3.2 Application of digital holography
    3.2.1 Light-in-flight measurements
    3.2.2 Deformation measurement
    3.2.3 Shape measurement
    3.2.4 Non-destructive evaluation
  3.4 References

4. SPECKLE PHOTOGRAPHY ... 257
  4.1 Fundamentals of speckles
    4.1.1 Origin of speckles
    4.1.2 Statistics of speckles
    4.1.3 Size of speckles
  4.2 Speckle photography
    4.2.1 Principles of speckle photography
    4.2.2 Speckle photography with spatial filtering
  4.3 Digital speckle photography (DSP)
    4.3.1 Principle of digital speckle photography
    4.3.2 Application of DSP: Deformation of a weld seam
  4.4 References

5. ELECTRONIC (DIGITAL) SPECKLE INTERFEROMETRY ... 275
  5.1 Fundamentals of ESPI
  5.2 Application of ESPI methods
    5.2.1 Deformation analysis of small components
    5.2.2 Vibration analysis of small membranes
  5.3 References

6. DIGITAL SPECKLE SHEARING INTERFEROMETRY ... 282
  6.1 Fundamentals of shearography
  6.2 Applications of shearography
    6.2.1 Deformation measurements
    6.2.2 Non-destructive testing
  6.3 References

V. **DIGITAL PROCESSING AND EVALUATION OF FRINGE PATTERNS IN OPTICAL METROLOGY**
by W. Osten ... 289

  1. INTRODUCTION ... 290

  2. INTENSITY MODELS; DISTURBANCES AND SIMULATIONS ... 293
    2.1 Intensity relations in optical metrology

2.2 Modelling of the image formation process in holographic interferometry
2.3 Computer simulation of holographic interference pattern

3. TECHNIQUES FOR DIGITAL PHASE RECONSTRUCTION 308
3.1 Methods for pre-processing of fringe pattern
    3.1.1 Smoothing of time dependent electronic noise
    3.1.2 Smoothing of speckle noise
    3.1.3 Shading correction
3.2 Methods for automatic phase measurement
    3.2.1 Fringe tracking or skeleton method
    3.2.2 Fourier-transform method
    3.2.3 Carrier-frequency method
    3.2.4 Phase-sampling method
3.3 Method for post-processing of fringe pattern
    3.3.1 Segmentation of fringe pattern
    3.3.2 The numbering of fringe pattern
        3.3.2.1 The fringe counting problem
        3.3.2.2 Methods of fringe numbering
    3.3.3 Unwrapping of mod2π - phase distributions
    3.3.4 Absolute phase measurement

4. MEASUREMENT OF 3D-DISPLACEMENT FIELDS 363
4.1 Planning of the experiment
    4.1.1 Basic relations
    4.1.2 Valuable a-priori knowledge
    4.1.3 Interferometer design
    4.1.4 The influence of sensitivity
4.2 Acquisition of the data
    4.2.1 Shape measurement by projected fringe technique
    4.2.2 Shape measurement by holography
    4.2.3 Shape measurement by digital holography
4.3 Evaluation of the data
4.4 Displacement calculation and data presentation

5. TECHNIQUES FOR THE QUALITATIVE EVALUATION
   OF FRINGE PATTERN 396
5.1 The technology of HNDT
5.2 Evaluation with neural networks
5.3 Evaluation with knowledge based systems
5.4 Material fault recognition in HNDT using recognition by synthesis

6. MODERN SOFTWARE SYSTEMS FOR DIGITAL PROCESS
   FRINGE PATTERN 409
6.1 The simulation mode
6.2 The processing mode
6.3 The tool box
6.4 The Fringe Processor^{TM} shell

7. LITERATURE 412

CHAPTER I

HYBRID TECHNIQUES IN EXPERIMENTAL SOLID MECHANICS

K-H. Laermann
Bergische University of Wuppertal, Wuppertal, Germany

ABSTRACT

Optical methods in experimental solid mechanics, yielding field information, combined with digital image processing and on-line evaluation of the experimentally obtained data by means of numerical procedures enable the stress-strain analysis of many problems, which couldn't be analysed satisfactorily as yet. Thus the effects of non-linear elastic, of viscoelastic material response and of any combination of such materials on the stress-strain state can be considered. Hybrid techniques, i.e. the combination of measurement techniques with numerical methods for data evaluation based on advanced mathematical algorithms yield reliable knowledge on the actual state and the real reactions of any kind of structures.. As the originally obtained experimental data generally do not meet the finally wanted information these pre-processed and digitized data are to evaluate by numerical procedures like the boundary-integral method and its discrete modification, the boundary-element method or the finite-element method.

1. INTRODUCTION IN THE SCOPE AND THE OBJECTIVES OF THE COURSE

1.1. Roll and importance of optical methods in experimental mechanics

At present methods of electrical measurement of mechanical quantities are still the mostly applied experimental methods in industrial practice, whereas optical methods are mainly used in basic and applied research. However one can predict, that advanced optical methods will be introduced more and more into the industrial practice, especially in inspection of manufacturing processes, quality control and supervising technical systems. The main advantages of modern achievements in grid-methods, holographic- and speckle-interferometry, shearography are to be seen in

1.) the whole field information, not restricted to a finite reference length and a point to point information as e.g. by electrical strain gauges;
2.) the possibilty of non-contact gauging;
3.) the possibility to analyse statc, dynamic processes, vibration and impact effects;
4.) the far-reaching automation of recording the observed optical phenomena the digital image processing and on-line evaluation as well as further hybrid analysis.

In Fig.1.1 the main optical experimental methods are shown together with the most substantial modification. However there still exists a much larger variety of more or less different modified arrangements in experimental set-ups, which cannot be listed up all.

This course will impart the recent achievements in photoelasticity, moire- and grid-techniques interferometric methods as well as in data evaluation procedures.

Experiments have been used to analyse deformations, stress and stability of solids and structures long before numerical and computational methods were known or could be used. Already Galileo Galilei (1564 – 1642) had expressed more than fourhundred years ago: *"It is necessary to measure everything that can be measured and to try making measurable what isn't as yet."*

But recently the variety of new scientific perceptions and developments in computer techniques seemed to displace experimental analysis. According to the fast developments in the past and still ongoing at present in hardware and especially in software with advanced numerical procedures as e. g. finite-element- and boundary-element-methods the question has been discussed world-wide whether methods of experimental analysis are nevertheless necessary and useful. However with regard to the modern achievements in techniques of measurement, the increasing resolving power of modern measuring devices, especially such based on optical methods, recording systems combined with automatic data acquisition and data evaluation including digital image processing experimental analysis of solid as well as of fluid mechanics problems has gained increasing importance. Nowadays optical methods in experimental mechanics are introduced and applied to a much larger extend than at any time before in research institute as well as in industry. They are becoming more and more important just because of the tremendous extension and involvement of computer analysis.

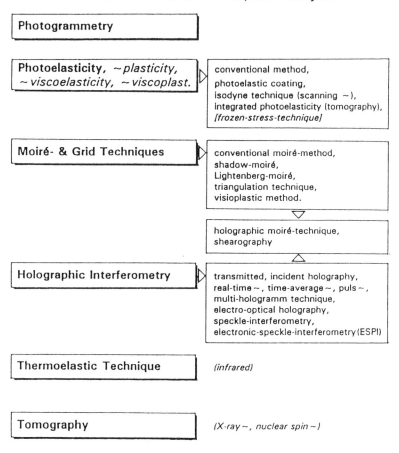

Fig.1.1. Survey over basic optical methods in experimental solid mechanics

Experimental techniques are no longer used only to determine material response in destructive and non-destructive material testing and to predict the state of displacement and strain and finally the stress-state in structures, part of structures and in structural elements, but they have to undertake new functions as to confirm theoretical perceptions and to verify results of numerical analysis, because such results are strongly depending on the validity of assumptions and suppositions which generally are necessary to enable mathematical modelling and mathematical/numerical analysis of complex problems. Methods of experimental mechanics have become inalienable to design structures and products of a large variety to improve safety against failure and to guarantee reliability of products, structures and even complex technical systems. Therefore the principles of experimental mechanics and measuring methods are involved in quality management and controlling production processes.

Experimental methods are applied in system identification to analyse the dynamic response of structures; they are developed especially as tools for supervising operating systems, machines and installations in order to guarantee a higher degree of safety and to minimise still existing risks. In the future therefore methods of experimental mechanics will become very important and inalienable in strategies of risk-management. Furthermore measurement techniques and results of measurements, or let them be called "experiences", are necessary to establish expert systems. Because methods of "artificial intelligence" for computer-oriented automatic decision processes demand as cogent supposition the availability of a tremendous amount of "experiences"; the development of expert systems can be based only on data, the reference to reality of which can be verified exclusively by measurements.

As a matter of fact one has to recognise, that new fields are open for application of experimental techniques. The principal domains of application are shown in Fig.1.2.

1.2. Influence and effect on safety and reliability of structures.

According to an English dictionary "safety" has been defined as "avoiding of and/or protection against risks". However this definition seems to be too strictly, possibly misleading because one may draw the conclusion, that the notion "safety" might be valid absolutely and every risk could be excluded in contradiction to all experiences, that there still exists a residual risk always no matter how large the residue might be. Restricting the considerations on anthropogeneous risks, especially on technical risks in the sense of safety science, then "safety" may be defined as follows: *"A technical state or process can be considered as safe, if the risk is smaller than a prescribed acceptable limiting value.*

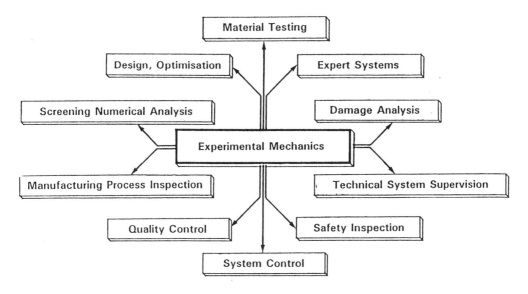

Fig.1.2. Principal application domains of experimental solid mechanics

Such a quantitative classification at any rate demands a quantitative conception of risk despite the uncertainty, who sets the limiting values and in which intention and what is the meaning of "acceptable". Therefore risk will be defined as the product of damage and the probability the damage can occur. This risk concept is materialised by the notion "damage", which anew will be determined by the failure mode of a technical system and the consequences of failure, e.g. whether and if, how many human beings might be afflicted by a failure. With concern to complex and large-scale structures of high risk potentials the question must be answered, whether deterministic methods only for predicting strain and stress in the design phase are sufficient to yield reliable information on the safety against failure and/or damage, ultimate load capacity and on the life span of such structures. Moreover it must be checked whether changes in utilisation, time-depending response of material, fatigue, environmental conditions etc. might influence the factor of safety of structures and their reliability respectively. Furthermore not only because of safety measures and risk-minimising but also because of economical reasons assessment of the structural conditions and their monitoring must be considered as of outmost importance. The infrastructure in many countries of the world is ageing. And there is an increasing awareness of the need to assess the severity of the damage occurring to infrastructure. Limited resources preclude the replacement of all structures, which need to be repaired or which have exceeded their life-time.

1.3. Presumptions of applicability in industrial practice.

It is of substantial importance not to look at the whole measurement configuration, i.e. the measurement system as a "black box". To assure reliability of the finally obtained information the relations between the originally measured input signal and the output signal of complete analysis including data evaluation, i.e. the flow of energy in the system, must be known. The real transfer functions, the impedances, the different signal-noise ratios as well as the measuring range of the single elements in the configuration and their relation to one another must be considered very carefully. This is necessary to avoid uncontrolled changes or distortions of the original input signals. To follow the course of signal transmission through the measuring system thoroughly step-by-step is an extremely important principle in modern experimental mechanics.

A tendency is to observe towards an ever increasing complexity of the measuring systems. But it must be considered very carefully, whether such often highly sophisticated configurations still fit the requirements of practical application outside the laboratories, in which they have been developed. It must be apprehended, that more or less "academic orchids" may withdraw themselves from actual practicability. Therefore it is a challenge to all those, who are engaged in the field of experimental mechanics, to make scientific perceptions and new experimental procedures easy and reliable to manage in practical industrial application.

This statement includes the requirement to intensify the activities in education and training. Because of the fact, that experimental mechanics and measurement methods are applied more and more in industry to control production- and manufacturing –processes, to supervise and monitor machines, installations and structures attention must be focussed not only on educating engineers and laboratory assistants but also on vocational training of those people, who have to perform and to operate the measurement techniques in practice; their qualification must be ensured too.

2. PRINCIPLE OF HYBRID TECHNIQUES.

To analyse any "event" it is necessary to design a logic model of the real "event" at first. Then it must be decided whether the further analysis shall be done in mathematical/numerical procedures or by experimental methods. In the first case the logic model must be described by an advanced mathematical model, which should be based on advanced theoretical perceptions. Generally however assumptions and suppositions are to introduce formulating a simplified mathematical model for quantitative solution, e.g. by means of discrete numerical methods like finite-difference- (FDM), finite-element- (FEM) and boundary- element-method (BEM). This leads to an approach of the reality only. The reliability of thus obtained results is often unknown or hardly to estimate unless they are proved in reality, i.e. mainly by measurements. Such simplified mathematical models might be characterised as "heuristic models", where "heuristic" means *the provisional supposition for the purpose of better understanding the events.*

On the other hand the real events can be modelled much closer to reality by "iconic models", which might be the real objects themselves, prototypes of products and structures, scaled-down replicas or even analogies.

It might be argued such iconic modelling, considering prototypes especially, to be too expensive, time-consuming and sometimes too risky, therefore computer-simulation should be given preference. But in order to develop simulation-models and to proof their reference to reality it is necessary to introduce "experience" and at least numerous sets of data, which are to be taken by measurement, ergo by experiments in the broadest sense. On the other hand simulation-techniques can be used advantageously to help for instance setting up the test- and measurement-installations, choosing the best set of sensor locations, determining the minimum number of sensors and their optimal location, indicating e.g. the best locations of excitation devices for modal tests.

As already mentioned the developments in methods as well as in equipment and in measuring systems warrant high resolution of the data. Simultaneously the amount of data to be handled increases rapidly. Therefore controlling the measurement system and the measuring process itself, data acquisition and evaluation requires powerful computer equipment. On-line procedures, i.e. the connection of the experiment and the measurement system with the computer system has become necessary and inalienable. It must be regarded further-on, that generally the observed phenomena are not identical with the finally wanted information. By means of optical methods in experimental mechanics light-intensities and their changes caused by mechanical interactions, depending on mechanical and optical responses of material of the specimens and/or objects under consideration are measured in principle. These original input-signals are then to transform by means of the respective physical relations into mechanical quantities like displacements, strains, deflections, depending on the used optical method (photoelasticity, moire-techniques, holographic-/speckle-interferometry, shearography).The accuracy and reliability of the transformed data don't depend on the exactness of measuring, converting and recording of the optical signals only, but also strongly on the physical theory of the respective method taken as basis of the transformation algorithms. However as generally internal forces, the stress state and internal parameters like the material properties are looked for, the obtained data of displacements and their derivatives respectively are to introduce into advanced theories, e.g. of elasticity, plasticity and viscoelasticity in order to calculate the finally wanted results. Because of the high sensitivity and resolving power of the modern methods it doesn't make sense to evaluate the metered data according to simplified classical algorithms of elasticity, this would be a back set in modern achievements. In modern experimental analysis therefore proper theories and advanced mathematical models must be introduced of the experimental methods as well as of the "event" to be investigated.

These considerations consequently lead towards the "principle of hybrid techniques" (Fig.2.1), i.e. the combination of theory and experiment, of mathematical/numerical procedures and measurement, a symbiosis of "heuristic" and "iconic models, which yields more realistic knowledge on the actual stress- and strain-state as well as of the reactions in structural elements, structures and structural systems. Different phases of combining mathematical/numerical and experimental procedures in analysing mechanical problems are possible (Fig.2.2), depending on different conditions like for instance the considered problem itself, the purpose and objectives of the investigation, the demanded accuracy, the availability of experimental equipment and computer capacity, the expertise of the investigator and his staff. It must be pointed out, that it is of outmost importance to look at the whole analysing process as a cybernetic process. Because in modern engineering the results of any analysis are to feed back and to relate always to the real event considered, no matter whether these results are taken by mathematical/numerical, experimental or hybrid analysis.But no matter, which type of modelling will be used in analysing any event, either "heuristic" or "iconic" modelling, attention should be paid to the *Five Don'ts* formulated by GOLOMB (Fig.2.3)

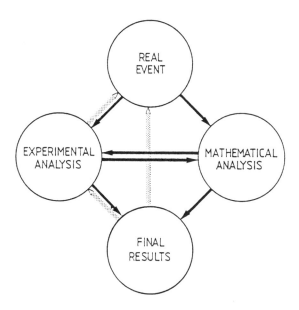

Fig.2.1. Principle of hybrid technique

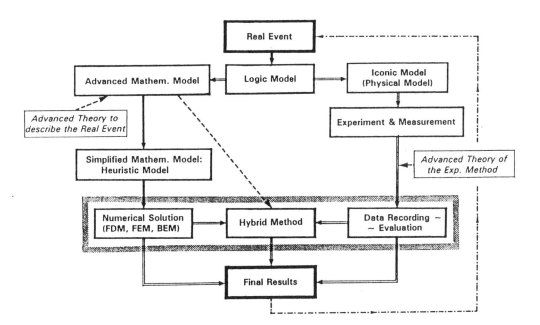

Fig.2.2. Bloc-diagram of hybrid procedures

> **THE FIVE "DON'TS" OF MODELING**
>
> 1. Don't believe that the model is the reality.
> 2. Don't extrapolate beyond the region of fit.
> 3. Don't distort reality to fit the model.
> 4. Don't retain a discredited model.
> 5. Don't fall in love with your model.
>
> S.W. GOLOMB, Simulation 14 (1970)

Fig.2.3. The *Five Don'ts* of GOLOMB

As an example of application a circular plate, simply supported and centrally loaded by a concentrated load, will be considered [2.1]. According to the classical KIRCHHOFF-LOVE plate theory it is assumed, that bending stresses occur only. However as proved by experiments the bending stress state is always superimposed by a membrane stress state. This real state can be described by two coupled differential equations:

$$N \nabla^2 \nabla^2 w = p + h \cdot L(w,F) ; \quad \nabla^2 \nabla^2 F = \frac{1}{2} E \cdot L(w,w) \tag{2.1}$$

where N denotes the bending stiffness, E the YOUNG's modulus, h the plate thickness, F denotes AIRY's stress function and L a differential operator

$$L(w,F) = w_{,11} F_{,22} + w_{,22} F_{,11} - 2 w_{,12} F_{,12} \tag{2.2}$$

In such a plate the bending stresses σ_{ij}^M are superimposed to membrane stresses σ_{ij}^N, the principal directions of which do not coincide.

For experimental analysis a two-layered plate model is needed, which consists of a thin photoelastic sheet of thickness d and a second layer of different material, e.g. aluminium-alloy, the surface of which is mirrored by a galvanic process; the two layers are glued together by a reflective adhesive. By means of LIGTENBERG's moire-technique the gradient of the deflection w is determined in discrete points (ι,κ) by digital image processing, enabling the calculation of the bending stresses and their principal direction $\psi^M(\iota,\kappa)$. By means of the photoelastic reflection method the isochromatic fringe order $\delta^{M+N}(\iota,\kappa)$ and so-called "characteristic directions" $\varphi^{M+N}(\iota,\kappa)$ [2.2] of the superimposed stress state are obtained. The principal direction of this stress state is function of the coordinate perpendicular to the central plane of the plate; the boundary values $\sigma_{ij0}{}^M$ are apparently much larger than the membrane stresses, these are constant over the plate thickness; the photoelastic layer can be considered as an optical symmetrical medium. Therefore the characteristic direktion can be expressed by the angle bisector of the principal directions of the superimposed stress state at the interface of the two layers and at the surface of the photoelastic layer. The relation between the isochromatic fringe order of the superimposed stress state can be expressed by WERTHEIM's law as an integral over the thickness of the photoelastic layer:

$$\delta^{M+N} = \frac{2}{S}\int_{a-d}^{a}[\Delta\sigma_0^M \frac{x_3}{a} + (\sigma_{11}^N - \sigma_{22}^N)]\frac{dx_3}{\cos 2\psi(x_3)} \qquad (2.3)$$

with

$$\Delta\sigma_0^M = \sigma_{11_0}^M - \sigma_{22_0}^M$$

The characteristic direction can be described as

$$\varphi^{M+N} = \frac{1}{2}[\psi(a) + \psi(a-d)] \qquad (2.4)$$

Finally the explicit non-linear relations are obtained:

$$\delta^{M+N} = \frac{a}{2S}\Delta\sigma_0^M\{(1+Q)^2[1+(Q\eta)^2/(1+Q)^2]^{1/2} - (1-d/a+Q)^2[1+(Q\eta)^2/(1-d/a+Q)^2]^{1/2} +$$

$$+(Q\eta)^2 \ln\frac{(1+Q)(1+[1+(Q\eta)^2/(1+Q)^2]^{1/2})}{(1-d/a+Q)(1+[1+(Q\eta)^2/(1-d/a+Q)^2]^{1/2})}\} \qquad (2.5)$$

$$\varphi^{M+N} = \frac{1}{4}\arctan\frac{Q\eta(2-d/a+2Q)}{(1+Q)(1-d/a+Q)-(Q\eta)^2} \qquad (2.6)$$

For abbreviation it has been introduced

$$Q = (\sigma_{11}^N - \sigma_{22}^N)/\Delta\sigma_0^M; \quad \eta = \tan 2\psi^N \qquad (2.7)$$

The eqn.s 2.5 and 2.6 can be solved intelligently by computer only. The results Q and φ^N and furthermore the difference of the principal membrane stresses are related to the direction of the bending stress state in each discrete point (ι,κ); finally these values must be transformed onto the reference coordinate-system (x_1,x_2). The bloc-diagram demonstrates the principle of the described hybrid method (Fig.2.4).

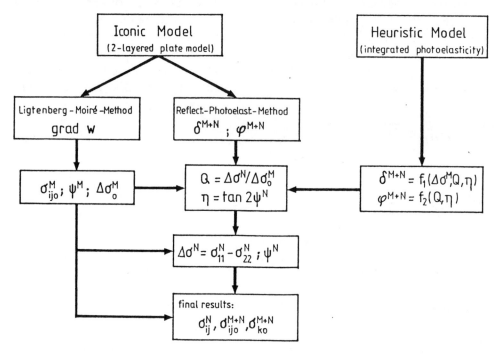

Fig.2.4. Bloc-diagram of the hybrid analysis of plates in bending

3. NON-LINEAR PHOTOELASTICITY

3.1. Non-linear stress-strain relations

As yet in photoelasticity it has been supposed mainly HOOKE's law of elasticity to be valid for all model materials and consequently linear relations between birefringence effects and stresses to be existing [3.1]. However in areas of high stress concentration, e.g. in the vicinity of crack tips, notches and inclusions, and with regard to some of the mainly used photoelastic model materials considerable uncertainties may result. Therefore non-linear stress-strain relations will be introduced. On the supposition the strains still to be small such non-linear relations may be formulated according to Kauderer [3.2] for an isothermal state:

$$\varepsilon_{ij} = [\frac{1}{3K}\varkappa(s) - \frac{1}{2G}g(\tau_0^2)]s\,\delta_{ij} + \frac{1}{2G}g(\tau_0^2)\sigma_{ij} \tag{3.1}$$

where $\varkappa(s)$ denotes a compression function, formulated as a potential series

$$\varkappa(s) = 1 + \sum_{\nu=1}^{n} \varkappa_\nu \frac{1}{(3K)^\nu} s^\nu \tag{3.2}$$

and similarly $g(\tau_0^2)$ denotes a shear function

$$g(\tau_0^2) = 1 + \sum_{\nu=1}^{n} g_{2\nu} \frac{1}{(2G)^{2\nu}} \tau_0^{2\nu} \tag{3.3}$$

With the initial values of the compression modulus and the shear modulus respectively

$$K = \frac{E_0}{3(1-2\nu)}\,;\; G = \frac{E_0}{2(1+\nu)} \tag{3.4}$$

the coefficients $\varkappa_\nu$ and $g_{2\nu}$ describe the non-linear material response. The moduli K and G as well as the coefficients are to determine by material testing procedures (see sect. 3.4). Furthermore in eqn's (3.2) and (3.3) s denotes the mean tension and τ_0 the reference stress.

$$s = \frac{1}{3}(\sigma_{11} + \sigma_{22} + \sigma_{33})$$

$$\tau_0^2 = \frac{2}{3}[\frac{1}{3}(\sigma_{11}^2 + \sigma_{22}^2 + \sigma_{33}^2 - \sigma_{11}\sigma_{22} - \sigma_{22}\sigma_{33} - \sigma_{33}\sigma_{11}) + (\sigma_{12}^2 + \sigma_{23}^2 + \sigma_{31}^2)] \tag{3.5}$$

According to Neumann [3.3] (see also Coker/Filon [3.4] and Mindlin [3.5]) in amorphous material the birefringent effect caused by mechanically induced strains is assumed to be linear. And of course there are some polymeres, which meet this assumption as has been proved [3.6].

3.2. The non-linear relation between stress-state and birefringence.

With n_0, the refraction index of the unstrained material, n_{ij} the refraction tensor and the strain-optical coefficients d_1 and d_2 the relation between birefringence and strain holds

$$n_{ij}^{-2} - n_0^{-2}\delta_{ij} = d_1 e_{ij} + \frac{1}{3} d_2 e \delta_{ij} \tag{3.6}$$

In eq.(3.6) e denotes the volume change and e_{ij} the strain deviation. Introducing eq.(3.1) into eq.(3.6) yields the non-linear relation between the stress tensor and the refraction tensor:

$$n_{ij}^{-2} - n_0^{-2}\delta_{ij} = A_1 s \delta_{ij} + A_2 \sigma_{ij} \tag{3.7}$$

with the denotations

$$A_1 = d_2 \frac{\varkappa(s)}{3K} - d_1 \frac{g(\tau_0^2)}{2G}; \quad A_2 = d_1 \frac{g(\tau_0^2)}{2G} \tag{3.8}$$

Next a plane stress-state in plane (x_1, x_2) will be considered. The direction of the incident light rays shall be parallel to the x_3-axis. Assuming a rectilinear light path through the two-dimensional model the component n_{33} of the refraction tensor doesn't have any influence on the birefringence effects, and because of $\sigma_{13} = \sigma_{23} = 0$ the components n_{13} and n_{23} are also equal to zero. Thus the refraction tensor is reduced to $[n_{\alpha\beta}^{-2}]$, $\alpha,\beta \in [1,2]$. The principal axes ψ_N of the index ellipse with reference to the x_1-axis are given by the proper vectors of the tensor $[n_{\alpha\beta}^{-2}]$. From eq.(3.7) it follows:

$$\begin{aligned} n_{11}^{-2} - n_0^{-2} &= A_1 s + A_2 \sigma_{11} \\ n_{22}^{-2} - n_0^{-2} &= A_1 s + A_2 \sigma_{22} \\ n_{12}^{-2} &= A_2 \sigma_{12} \end{aligned} \tag{3.9}$$

and furthermore

$$\tan 2\psi_N = \frac{2n_{12}^{-2}}{n_{11}^{-2} - n_{22}^{-2}} = \frac{2A_2 \sigma_{12}}{A_2(\sigma_{11} - \sigma_{22})} = \tan 2\psi_S \tag{3.10}$$

Thus the principal axes of the index ellipse coincide with the principal axes of the stress tensor. The eigen values of the tensor $[n_{\alpha\beta}^{-2}]$ are the principal values n_α^{-2} which are given as

$$n_1^{-2} = n_0^{-2} + (A_1 + \frac{3}{2} A_2) s + \frac{1}{2} A_2[(\sigma_{11} - \sigma_{22}) \cos 2\psi_N + 2\sigma_{12} \sin 2\psi_N]$$

$$n_2^{-2} = n_0^{-2} + (A_1 + \frac{3}{2} A_2) s - \frac{1}{2} A_2[(\sigma_{11} - \sigma_{22}) \cos 2\psi_N + 2\sigma_{12} \sin 2\psi_N] \tag{3.11a}$$

Eqn.s (3.11a) are transformed yielding the refraction indices

$$n_1 = n_0\{1+n_0^2(B\cdot s + \frac{1}{2} A_2[(\sigma_{11}-\sigma_{22})\cdot\cos 2\psi_N + 2\sigma_{12}\sin 2\psi_N])\}^{-1/2}$$

$$n_2 = n_0\{1+n_0^2(B\cdot s - \frac{1}{2} A_2[(\sigma_{11}-\sigma_{22})\cdot\cos 2\psi_N + 2\sigma_{12}\sin 2\psi_N])\}^{-1/2} \quad (3.11b)$$

$$B = A_1 + \frac{3}{2}A_2$$

In these equations the term $\{\cdots\}^{-1/2}$ is expanded in a series, which can be truncated after the second term because of the weak birefringence.

$$\{\cdots\}^{-1/2} \approx 1 - \frac{1}{2}n_0^2(B\cdot s \pm \frac{1}{2}A_2[\cdots]) + \text{trunc} \quad (3.12)$$

Then the difference of the refraction indices holds

$$n_1 - n_2 \approx -\frac{1}{2}n_0^3 A_2[(\sigma_{11}-\sigma_{22})\cos 2\psi_N + 2\sigma_{12}\sin 2\psi_N] \quad (3.13)$$

confirming the relation

$$n_2^{-2} - n_1^{-2} \approx \frac{2}{n_0^3}(n_1 - n_2) \quad (3.14)$$

On one hand the birefringence is defined as the difference of the refraction indices

$$\Delta := n_1 - n_2 \quad (3.15)$$

and on the other hand as

$$\Delta = \frac{\lambda}{d} \delta \quad (3.16)$$

related to the wave length λ of the monochromatic light, to the actual thickness of the two-dimensional model

$$d = d_0 [1 - \nu(\varepsilon_{11} + \varepsilon_{22})] \quad (3.17)$$

and to δ, the order of the isochromatic fringes, which as well as the principal directions ψ_S (= ψ_N, see eq.(3.10)) are to be taken in photoelastic experiments by digital image processing.

With A_2 according to eq.(3.8) and the shear-function $g(\tau_0^2)$, eq.(3.3), taking into account

$$\frac{2\sigma_{12}}{\sin 2\psi_N} = (\sigma_{11}-\sigma_{22})\cos 2\psi_N + 2\sigma_{12}\sin 2\psi_N \quad (3.18)$$

the relation between the stress state and the birefringence runs

$$\left[1 + \sum_{\nu=1}^{n} g_{2\nu} \frac{1}{(2G)^{2\nu}} \tau_0^{2\nu}\right] \sigma_{12} = \hat{S} \frac{\delta}{d} \sin 2\psi_N \qquad (3.19)$$

where $\hat{S}$ denotes the stress-optical coefficient

$$\hat{S} = -\frac{4G}{d_1} \cdot \frac{\lambda}{n_0^3} \qquad (3.20)$$

which is to determine in material testing (see sect. 3.4).

In the expression of the reference stress τ_0 the component σ_{22} of the stress tensor will be eliminated, then eq.(3.3) holds

$$\tau_0^2 = \frac{2}{9}[\sigma_{11}^2 - 2\sigma_{11} \cdot \sigma_{12} \tan^{-1} 2\psi_N + (3 + 4\tan^2 2\psi_N)\sigma_{12}] \qquad (3.21)$$

The stress component σ_{11} will be expressed by integrating the equilibrium condition:

$$\sigma_{11} = C(x_2) - \int \sigma_{12,2} \, d\bar{x}_1 \qquad (3.22)$$

In this equation $C(x_2)$ denotes the value $^\Gamma\sigma_{11}$ at the boundary Γ of the object. Along a load-free boundary the principal direction coincides with the tangent at the boundary and the normal to it respectively. The principal stress $^\Gamma\sigma_2$ perpendicular to the boundary equals zero, whereas $^\Gamma\sigma_1$ parallel to the boundary follows from eq.(3.19).

$$\left[1 + g_2 \frac{1}{18 G^2} \, ^\Gamma\sigma_1^2\right] {}^\Gamma\sigma_1 = \hat{S} \frac{d}{{}^\Gamma\delta} \qquad (3.23)$$

The boundary values of the components $^\Gamma\sigma_{\alpha\beta}$ hold

$$\begin{aligned}
{}^\Gamma\sigma_{11} &= {}^\Gamma\sigma_1 \cdot \cos^2 {}^\Gamma\psi_N = C(x_2) \\
{}^\Gamma\sigma_{22} &= {}^\Gamma\sigma_1 \cdot \sin^2 {}^\Gamma\psi_N \\
{}^\Gamma\sigma_{12} &= \tfrac{1}{2} {}^\Gamma\sigma_1 \cdot \sin(2\, {}^\Gamma\psi_N)
\end{aligned} \qquad (3.24)$$

The three eqn,s (3.19),(3.21),(3.22) are the basic relations to calculate the components of the stress tensor under the supposition the stress state to be two-dimensional even in areas of high stress concentration!

3.3. Iterative discrete solution.

To start the iterative procedure as a first approach values of $\sigma_{12}^{(0)}$ are calculated from the photoelastic data δ and ψ_N on the basis of the relations describing linear photoelasticity.

$$\sigma_{12}^{(0)}(j,k) \approx \frac{1}{2} \cdot \frac{\hat{S}}{d_0} \delta(j,k) \sin 2\psi_N(j,k) \qquad (3.25)$$

and inserted into eq.(3.22). To integrate this equation the predictor-corrector method of Heun [3.7] will be applied. In discrete points (j,k) the predictor is determined according to the recurrent formula of EULER-CAUCHY, Fig.3.1.

$$\sigma_{11}^{(0)}(j,k) = \sigma_{11}^{(0)}(j-1,k) - [\sigma_{12}^{(0)}(j-1,k+1) - \sigma_{12}^{(0)}(j-1,k-1)] \frac{\Delta x_1}{2\Delta x_2} \qquad (3.26)$$

Considering the definition of $\tau_0^{2(0)}(j,k)$, eq.(3.21), the results of the eqn.s(3.22) for each point (j,k) are inserted into eqn.s(3.19), thus algebraic equations of grade 3 in $\sigma_{12}^{(1)}(j,k)$ will be obtained, because two terms only of the shear-function, eq.(3.3), are considered. These algebraic equations can be solved e.g. by means of the PEGASUS-method [3.8], giving improved values of the shear stresses, which then are inserted into the so-called corrector (Fig.3.2) in consecutive steps μ, $\mu \in [0/m]$.

$$\sigma_{11}^{(\mu+1)}(j,k) = \sigma_{11}^{(\mu+1)}(j-1,k) - [\sigma_{12}^{(\mu+1)}(j-1,k+1) + \sigma_{12}^{(\mu)}(j,k+1) - \sigma_{12}^{(\mu+1)}(j-1,k-1) - \sigma_{12}^{(\mu)}(j,k-1)] \frac{\Delta x_1}{4\Delta x_2} \qquad (3.27)$$

The results are used to calculate the shear stresses $\sigma_{12}^{[\mu+1]}$ in the next step of the iteration process and so on.

The described iterative procedure is to continue until the difference $\sigma_{12}^{(\mu+1)} - \sigma_{12}^{(\mu)}$ between two successive steps is smaller then a given limiting value. The local order of error will be reached already after a few steps of iteration, if the interval Δx_1 is sufficiently small, i.e. if it is valid for the numerical code of steps K: $0.05 \leq K \leq L \times \Delta x_1 = 0.20$, where L denotes the LIPSCHITZ-constant of the function $\sigma_{12,2}$. Note, that the iteration must not reach a standstill! With the final results of the iteration process

$$\sigma_{11}(j,k) = \sigma_{11}^{(m+1)}(j,k); \quad \sigma_{12}(j,k) = \sigma_{12}^{(m+1)}(j,k) \qquad (3.28)$$

the component $\sigma_{22}(j,k)$ is given as well:

$$\sigma_{22}(j,k) = \sigma_{11}(j,k) - 2\sigma_{12}(j,k) \cdot \tan^{-1} 2\psi_N \qquad (3.29)$$

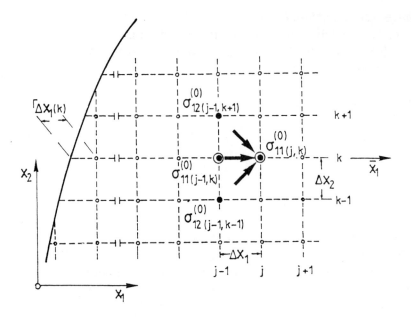

Fig.3.1. Discrete integration, first step: Predictor.

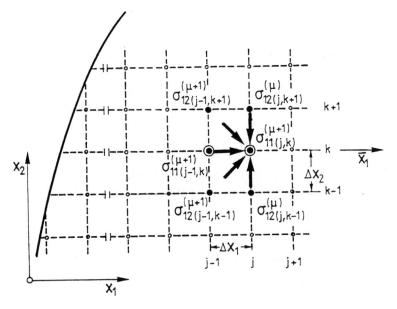

Fig.3.2. Discrete integration, step $\mu \in [1/m]$: Corrector.

3.4. Determination of material response.

The material response, i.e. the mechanical as well as the optical coefficients can be determined in a uniaxial tensile test. At first the initial value of the YOUNG's modulus E_0 and the POISSON's ratio v are determined by means of the well-known classical methods. Generally it can be assumed v to be constant or at least nearly constant Then the compression- as well as the shear-modulus are given according to eq.(3.4). Regarding eq.(3.1) the stress-strain relation in a uniaxial test holds

$$\varepsilon_1 = [\frac{1}{3K}\varkappa + \frac{1}{G}g]\frac{1}{3}\sigma_1 \tag{3.30}$$

Introducing the moduli K and G, the compression-function $\varkappa(s)$, eq.(3.2), and the shear-function $g(\tau_0^2)$, eq.(3.3), the following equation will be obtained:

$$E_0\varepsilon_1 = \sigma_1 + c_1\sigma_1^2 + c_2\sigma_1^3 \tag{3.31}$$

with the coefficients

$$c_1 = \frac{1}{9}\varkappa_1\frac{(1-2v)^2}{E_0} \; ; \; c_2 = \frac{4}{27}g_2\frac{(1+v)^3}{E_0^2} \tag{3.32}$$

The tensile stress amounts to

$$\sigma_1 = \frac{P}{b_0 d_0} \cdot \frac{1}{(1-v\varepsilon_1)^2} \tag{3.33}$$

where $b_0 d_0$ denotes the cross-section of the specimen in the unstrained state and P the test load. Introducing eq.(3.33) into eq.(3.31) yields the relation between the test load and the measured strain ε_1.

$$E_0\varepsilon_1 = \frac{P}{b_0 d_0}\left[\frac{1}{(1-v\varepsilon_1)^2} + c_1\frac{P}{b_0 d_0}\cdot\frac{1}{(1-v\varepsilon_1)^4} + c_2\frac{P^2}{(b_0 d_0)^2}\cdot\frac{1}{(1-v\varepsilon_1)^6}\right] \tag{3.34}$$

Regarding $v\varepsilon \ll 1$ the respective terms in the aforesaid relation are expanded in TAYLOR-series, which are truncated after the second term, thus yielding

$$E_0\varepsilon_1 \approx \frac{P}{b_0 d_0}\left[(1+2v\varepsilon_1) + c_1\frac{P}{b_0 d_0}(1+4v\varepsilon_1) + c_2\frac{P^2}{(b_0 d_0)^2}(1+6v\varepsilon_1)\right] \tag{3.35}$$

The terms with the unknown coefficients c_1 and c_2 are separated from the terms, which only contain given quantities and measured data respectively.

A set of equations will be obtained for different test loads P_i, $i \in [1/N]$.

$$\underbrace{\frac{b_0 d_0}{P_i}(E_0 \varepsilon_{1i}) - (1+2\nu\varepsilon_{1i})}_{f(P_i, \varepsilon_{1i})} = \underbrace{c_1 \frac{P_i}{b_0 d_0}(1+4\nu\varepsilon_{1i}) + c_2 \frac{P_i^2}{(b_0 d_0)^2}(1+6\nu\varepsilon_{1i})}_{\phi(P_i, \varepsilon_{1i}, c_j)} \qquad (3.36)$$

which will be solved in a discrete linear approximation by means of the GAUSSIAN least-square method

$$\min \underbrace{\sum_{i=1}^{N}(f(P_i, \varepsilon_{1i}) - \phi(P_i, \varepsilon_{1i}, c_j))^2}_{D(c_j)^2} ; \; j \in [1,2] \qquad \frac{\partial D^2}{\partial c_j} = 0 \qquad (3.37)$$

With the partial derivatives

$$\frac{\partial \phi}{\partial c_j} = \varphi_j : \quad \begin{aligned} \varphi_1 &= \frac{P_i}{b_0 d_0}(1+4\nu\varepsilon_{1i}) \\ \varphi_2 &= \frac{P_i^2}{(b_0 d_0)^2}(1+6\nu\varepsilon_{1i}) \end{aligned} \qquad (3.38)$$

a linear system of equations results, the GAUSSIAN normal-equations.

$$\begin{bmatrix} \sum_{i=1}^{N} \varphi_1 \cdot \varphi_1 & \sum_{i=1}^{N} \varphi_1 \cdot \varphi_2 \\ \sum_{i=1}^{N} \varphi_2 \cdot \varphi_1 & \sum_{i=1}^{N} \varphi_2 \cdot \varphi_2 \end{bmatrix} \times \begin{bmatrix} c_1 \\ c_2 \end{bmatrix} = \begin{bmatrix} \sum_{i=1}^{N} f \cdot \varphi_1 \\ \sum_{i=1}^{N} f \cdot \varphi_2 \end{bmatrix} \qquad (3.39)$$

These equations yield the solutions c_j and according to the definition (3.32) the coefficients of the compression-function as well as of the shear-function.

$$\varkappa_1 = c_1 \frac{9 E_0}{(1-2\nu)^2} ; \quad g_2 = c_2 \frac{3}{4} \cdot \frac{9 E_0^2}{(1-2\nu)^3} \qquad (3.40)$$

Considering the uniaxial stress-state (see eq.(3.23)) and eq.(3.33) the stress-optical coefficient is given to

$$\hat{S} = \frac{P}{b_0}(1+\nu\varepsilon_1)\left[1 + \frac{3}{2} c_2 \frac{(1+\nu)^2}{(1-2\nu)^3} \cdot \frac{P^2}{(b_0 d_0)^2}(1+4\nu\varepsilon_1)\right]\frac{1}{\delta} \qquad (3.41)$$

Regarding the definition of the stress-optical coefficient, eq.(3.20), this coefficient must be independent of the applied stress. And indeed this has been confirmed by respective experiments. As an example the material testing results of an often used model-material VP 1527, a polyester-resin, are presented in Fig.3.3.

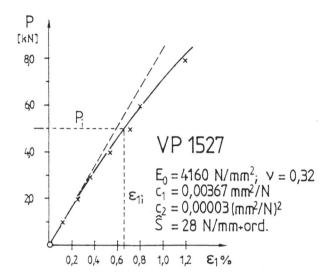

Fig.3.3. Results of material testing.

3.5. Examples of application.

To convey an impression on the effects of non-linear material response of photoelastic model material and on the results of the evaluation of the experimental data two examples are presented. These are simple ones to promote the understanding of the described method.

Fig. 3.4 shows the isochromatic fringe-pattern of a half-plane under a concentrated load and the course of the principal stress along the central axis x_1 for linear as well as non-linear material response. The difference between $\sigma_{non-linear}$ and σ_{linear} is noticeable, in a depth $x_1 = 4$ mm for instance amounting to an error f of about 8%.

As the second example the corner of a frame loaded by two concentrated loads will be considered. Fig. 3.5 exhibits the isochromatic fringe-pattern. The results of the fringe evaluation along the free boundary, these are the stresses $^\Gamma\sigma_1$, and the stress difference $\sigma_{11} - \sigma_{22}$ along the axis x_2 (Fig.3.6), confirm the remarkable effects of non-linearity on the stress-state, the error f_{max} amounts to about 16%.

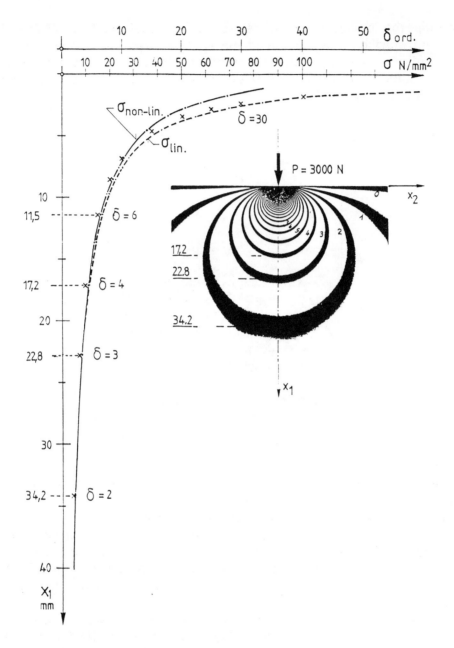

Fig.3.4. Half-plane under concentrated load, material: polyester VP1527 (see Fig.3.3)

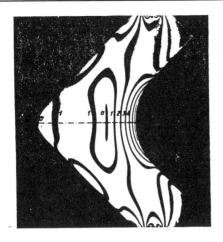
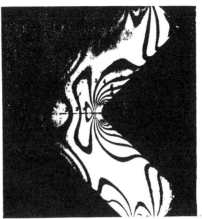

Fig. 3.5. Isochromatic fringe-pattern of a frame corner.

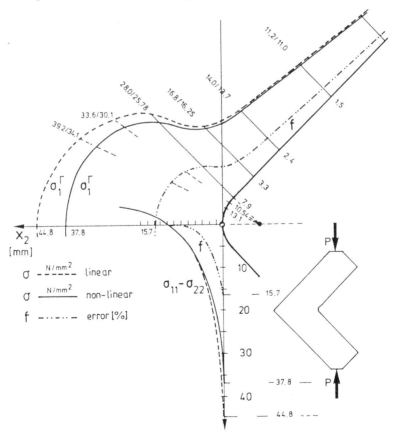

Fig. 3.6. Comparison of stresses in a frame corner, material: VP1527

Both these examples prove the necessity to take into account non-linear response of model material in areas of high stress concentration. (Remind the supposition of a two-dimensional stress-state! Therefore the calculated stresses are mean values over the thickness of the specimen and the object respectively!)

As a matter-of-fact the high arithmetical expenditure demands proper computing capacity for digital image processing as well as for the evaluation of the measured and recorded data. However it is not necessary to analyse the object entirely according to the presented method. The application of the non-linear evaluation-process can be restricted to areas of apparently high stress concentration only, whereas in the remaining area of the object the methods of conventional linear photoelasticity may be applied. The whole domain of the object to be analysed may be denoted by Ω, the domain, in which non-linearity should be taken into account, by $\Omega_n \subset \Omega$, the remaining domain, where linear material response can be taken approximately as basis for data evaluation, by Ω_l, $\Omega_l \cup \Omega_n = \Omega$; Γ_n marks the boundary between both the domains, furthermore $\Delta\Gamma_n \subset \Gamma$ the external boundary of domain Ω_n (Fig.3.7). Then $\sigma_{\alpha\beta}$ along Γ_n will be given by means of linear evaluation in domain Ω_l; along $\Delta\Gamma_n$ however non-linear material response is to consider in order to calculate $^\Gamma\sigma_1$. With the thus determined boundary values the evaluation in domain Ω_n is to perform as described in the preceding sections.

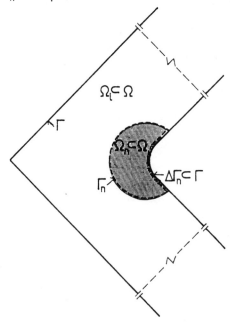

Fig. 3.7. Domains for separate evaluation procedures.

4. PHOTOVISCOELASTICITY

4.1. Preliminary remarks.

Rational design of structures and structural components exhibiting viscoelastic response demands studying the complete history of the stress distribution in such objects. This can be done by numerical as well as by experimental analysis, however the latter in hybrid technique only. If the experimental analysis is carried out by means of photoviscoelastic methods, not only the mechanical response of proper model material depending on time but also their time-depending optical response must be considered. To study the problem and effects of photoviscoelasticity different high polymeres can be used. Pindera [4.1] had performed basic research in this field, but mainly to obtain information on the reliability of photoelastic experiments. Theocaris [4.2] however investigated the essential rheological properties, the mechanical as well as the optical, of pure and plasticised cold-setting epoxy polymeres. He came to the conclusion, that such modified polymeres show a linear viscoelastic behaviour to a satisfactory degree of accuracy. Dill, Fowlkes and Coleman, [4.3] to [4.6], have described a photoviscoelastic theory, which is based on MAXWELL's electro-magnetic equations, regarding the dielectric tensor and the refraction tensor as linear functions of the stress tensor history, assuming a "fading memory".

The following considerations are restricted to the theory of linear visco- and photoviscoelastic response of material. The linearity of viscoelastic response is manifested by the fact, that at any time the time-dependent functions are approximately linear-proportional to the applied constant stresses and/or strains, provided

i. stresses and strains are limited in such a range, that BOLTZMANN's principle of superposition is still valid,
ii. thermodynamic effects can be neglected and isothermal processes are guaranteed,
iii. displacements are small corresponding to the LAGRANGE-formulation of the equilibrium conditions.

4.2. Basic relations between the stress tensor and the refraction tensor.

In the LAPLACE-transform the relation between both this tensors holds according to Coleman/Dill [4.7]

$$\tilde{\sigma}_{ij}(p) = p \cdot \tilde{C}^*(p) \cdot \tilde{n}_{ij}(p) + \frac{1}{3} p [\tilde{G}^*(p) - \tilde{C}^*(p)] \cdot \tilde{N}(p) \cdot \delta_{ij} \qquad (4.1)$$

In this equation $C^*(p)$ denotes the optical relaxation function for shear, $G^*(p)$ the optical relaxation function for dilatation, $N(p)$ the trace of the refraction tensor and δ_{ij} the Kronecker-delta. For a two-dimensional stress state the components of the stress tensor hold

$$\tilde{\sigma}_{11}(p) = p \cdot \tilde{C}^*(p) \cdot \tilde{n}_{11}(p) + p[\tilde{G}^*(p) - \tilde{C}^*(p)] \cdot \tilde{N}(p)/3$$
$$\tilde{\sigma}_{22}(p) = p \cdot \tilde{C}^*(p) \cdot \tilde{n}_{22}(p) + p[\tilde{G}^*(p) - \tilde{C}^*(p)] \cdot \tilde{N}(p)/3 \qquad (4.2)$$
$$\tilde{\sigma}_{12}(p) = p \cdot \tilde{C}^*(p) \cdot \tilde{n}_{12}(p)$$

The sum of the normal stresses is given to

$$\tilde{\sigma}_{11}(p) + \tilde{\sigma}_{22}(p) = \frac{1}{3} p [2\tilde{G}^*(p) + \tilde{C}^*(p)] \cdot [\tilde{n}_{11}(p) + \tilde{n}_{22}(p)] \qquad (4.3)$$

and the difference respectively to

$$\tilde{\sigma}_{11}(p) - \tilde{\sigma}_{22}(p) = p \cdot \tilde{C}^*(p) \cdot [\tilde{n}_{11}(p) - \tilde{n}_{22}(p)] \qquad (4.4)$$

At any time the direction of the principal axes of the index ellipse to the axis x_1 of a reference co-ordinate system are given by the proper vectors of the refraction tensor

$$\tan 2\psi_N = 2n_{12}[n_{11} - n_{22}]^{-1} \qquad (4.5)$$

The eigen-values of $[n_{\alpha\beta}]$ are the principal values, the refraction indices n_1 and n_2

$$n_{1,2} = \frac{1}{2}(n_{11} + n_{22}) \pm \sqrt{(n_{11} - n_{22})^2 + 4n_{12}^2} \qquad (4.6)$$

Then the difference of the refraction indices holds

$$n_1 - n_2 = \sqrt{(n_{11} - n_{22})^2 + 4n_{12}^2} \qquad (4.7)$$

and analogously the difference of the principal stresses

$$\sigma_1 - \sigma_2 = \sqrt{(\sigma_{11} - \sigma_{22})^2 + 4\sigma_{12}^2} \qquad (4.8)$$

Considering eq.s (4.2), (4.4) and (4.7) it follows from eq.(4.8)

$$\tilde{\sigma}_1(p) - \tilde{\sigma}_2(p) = p \cdot \tilde{C}^*(p) \cdot [\tilde{n}_1(p) - \tilde{n}_2(p)] \qquad (4.9)$$

Considering the convolution in the LAPLACE-transform to being commutative eq.(4.9) reads after retransformation

$$\sigma_1(t) - \sigma_2(t) = C^*(0^+)[n_1(t) - n_2(t)] + \int_{0^+}^{t} C^*(t-\tau) \cdot \frac{\partial}{\partial \tau}[n_1(\tau) - n_2(\tau)] d\tau \tag{4.10}$$

The stress-induced retardation $\Delta(t)$ is defined as the difference of the refraction indices perpendicular to the light path, assuming the light path to be rectilinear and not influenced by the stress gradient:

$$\Delta(t) := n_1(t) - n_2(t) \tag{4.11}$$

On the other hand $\Delta(t)$ is experimentally given by the order $\delta(t)$ of the isochromatic fringes

$$\Delta(t) = \lambda \cdot \delta(t) \cdot d(t)^{-1} \tag{4.12}$$

where λ denotes the wave-length of the monochromatic light and $d(t)$ the geometrical length of the light path through the plane model, i.e. the thickness of the model perpendicular to the light path, considering the change of this thickness over time t caused by lateral contraction

$$d(t) = d_0[1 - \nu(t) \cdot \varepsilon_{jj}(t)] = d_0[1 - \nu(t) \cdot [\varepsilon_1(t) + \varepsilon_2(t)]] \tag{4.13}$$

It has been proved, that the influence of rheological material response on POISSON's ratio is weak and therefore can be neglected.

$$\nu(t) \approx \nu(0^+) = \text{const} \tag{4.14}$$

Since $\varepsilon_{jj}(t)$ is unkwown as yet as a first approach $d(t)$ will be set equal to d_0, the thickness of the undeformed model. Having calculated $\sigma_{jj}(0^+)$ at time $t = 0^+$ the deformed thickness $d(t)$ will be approximated by $d(0^+)$, thus considering the elastic displacements only; the error caused by this approach is negligible small. Finally eq.(4.10) runs

$$\sigma_1(t) - \sigma_2(t) = \frac{\lambda}{d(0^+)} \{C^*(0^+) \cdot \delta(t) + \int_{0^+}^{t} C^*(t-\tau) \frac{\partial \delta(\tau)}{\partial \tau} d\tau\} \tag{4.15}$$

which is a VOLTERRA-integral equation of the second kind. With regard to eq.s(4.2) it will be obtained

$$\sigma_{11}(t) - \sigma_{22}(t) = \frac{\lambda}{d(0^+)} \{C^*(0^+) \cdot \delta(t) \cos 2\psi_N(t) + \int_{0^+}^{t} C^*(t-\tau) \frac{\partial}{\partial \tau} \delta(\tau) \cos 2\psi_N(\tau) d\tau\} \tag{4.16.1}$$

$$\sigma_{12}(t) = \frac{\lambda}{d(0^+)}\{C^*(0^+)\cdot\delta(t)\sin 2\psi_N(t) + \int_{0^+}^{t} C^*(t-\tau)\frac{\partial}{\partial\tau}\delta(\tau)\sin 2\psi_N(\tau)d\tau\} \qquad (4.16.2)$$

The directions of the principal axes of the refraction tensor, eq.(4.5), are given as the isoclinic pattern in the photoviscoelastic measurements. It must be pointed out, that generally the principal axes of the refraction tensor do not coincide with the principal axes of the stress tensor. The axes coincide only, if the stress state is uniaxial or a pure shear stress state exists or if load-free boundaries or axes of symmetry are considered. But as also for time-dependent stress states the equilibrium conditions must be satisfied at any time, the principal directions of the stress tensor related to the reference co-ordinate system can be determined.

$$\tan 2\psi_s(t) = 2\sigma_{12}(t)\cdot[\sigma_{11}(t) - \sigma_{22}(t)]^{-1} \qquad (4.17)$$

4.3. Determination of the optical relaxation function.

For photoviscoelastic analysis it is necessary to determine the optical relaxation function $C^*(t)$. However obviously it seems to be easier in performing the respective material tests to determine the optical creep compliance $C(t)$ and then to transform the creep compliance into the relaxation function. With the equations

$$\tilde{n}_{11}(p) - \tilde{n}_{22}(p) = p\,\tilde{C}(p)\cdot[\tilde{\sigma}_{11}(p) - \tilde{\sigma}_{22}(p)]$$
$$\tilde{\sigma}_{11}(p) - \tilde{\sigma}_{22}(p) = p\,\tilde{C}^*(p)\cdot[\tilde{n}_{11}(p) - \tilde{n}_{22}(p)] \qquad (4.18)$$

the relation between $C^*(t)$ and $C(t)$ reads in the LAPLACE-transform

$$\tilde{C}(p)\cdot\tilde{C}^*(p) = \frac{1}{p^2} \qquad (4.19)$$

and re-transformation yields

$$C^*(t) = \frac{1}{C(0^+)}\left[1 + \int_{0^+}^{t} C^*(\tau)\frac{\partial}{\partial\tau} C(t-\tau)d\tau\right] \qquad (4.20)$$

This is a VOLTERRA-integral equation of the second kind, which can be solved by discrete integration as the creep compliance is experimentally given as a set of measured data at different points of time t_i, $i \in [0/n]$.

$$C^*(t_n) = \frac{1}{C(0^+)} \left[1 + \sum_{i=1}^{n} \int_{t_{i-1}}^{t_i} C^*(\tau) \frac{\partial}{\partial \tau} C(t_n - \tau) d\tau \right] \tag{4.21}$$

The integral over time-interval t_{i-1} to t_i will be expressed by the finite formulation

$$\int_{t_{i-1}}^{t_i} \cdots = \frac{1}{2} [C^*(t_{i-1}) + C^*(t_i)] \cdot [C(t_n - t_i) - C(t_n - t_{i-1})] \tag{4.22}$$

which is to introduce into eq.(4.21). This equation then reads

$$C^*(t_n) = \frac{1}{C(0^+)} \left[1 + \frac{1}{2} \sum_{i=1}^{n} [C^*(t_{i-1}) + C^*(t_i)] \cdot [C(t_n - t_i) - C(t_n - t_{i-1})] \right] \tag{4.23}$$

With regard to error propagation it can be anticipated, that the above discrete representation is extremely stable. Because of the fading memory of the polymeres used as model material the relation between the optical creep compliance and the optical relaxation function can be expressed approximately as

$$C^*(t) \approx 1/C(t) \tag{4.24}$$

The material test can be performed either by tensile tests or by shear tests. It has been proved, that the shear tests should be given preference. Assuming POISSON's ratio to be approximately constant over time t the stress state in a shear test specimen is independent of time and equal to an equivalent elastic stress state according to the *correspondence principle* (see Flügge [4.8]). Considering a pure shear stress state the thickness of the specimen in the centre of the test specimen is independent of time, then. $d(t) = d_0$, and the angles of the principal directions of the refraction tensor are given to $\psi_N = \pm 45°$. Then the relation is valid

$$C(t) = \frac{1}{2} \cdot \frac{\lambda}{d_0} \cdot \frac{1}{\sigma_{12}(0^+)} \cdot \delta(t) \tag{4.25}$$

which means, that the creep compliance is proportional to the measured quantities of the isochromatic fringe order. The shear stress $\sigma_{12}(0^+)$ can be calculated on the basis of the theory of elasticity, e.g. by means of the finite-element method. Care must be taken in the loading process, because applying the test load according to HEAVISIDE's step function as assumed is not possible in the reality. To avoid dynamic effects the load must be applied continuously to the specimen from zero to the wanted test load. In practice this takes about one to two seconds and therefore it is impossible, to take exact data of the quantities to be measured exact at time t = 0, however correction is possible by backwards extrapolation. It must be guaranteed furthermore, that during the measurements the temperature is kept constant absolutely; even the influence of radiation must be avoided, because a temperature change of about 1°C falsifies the results considerably.

To calculate the creep compliance $C(t_i)$, (eq.(4.25)), the data of the fringe order δ are taken at different points of time t_i, $i \in [0/n]$, Then the results of the calculation are inserted into eq.(4.23) to determine the values of the relaxation function $C^*(t_i)$.

As an example Fig.4.1 and 4.2 show the results of the described shear test and its evaluation for one of the different model materials, which have been tested. To give in impression at least on the photoviscoelastic effects the isochromatic fringe pattern of a simply supported beam under a concentrated load taken at different points of time are represented in Fig.4.3.

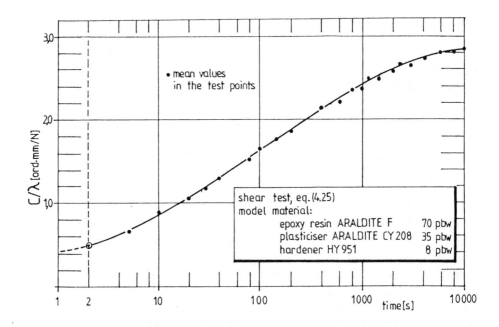

Fig.4.1. Optical creep compliance C(t), test results.

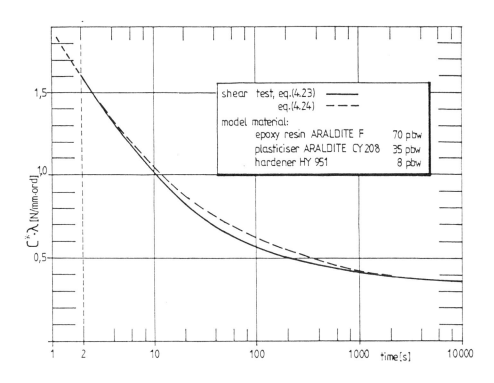

Fig.4.2. Optical relaxation function $C^*(t)$.

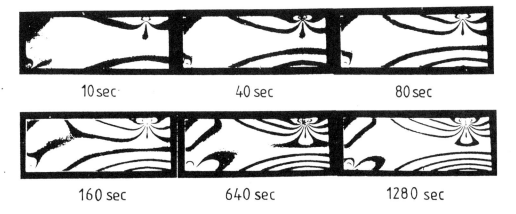

Fig.4.3. Isochromatic fringe pattern, beam under concentrated load.

4.4. Solution of the VOLTERRA's integral equation.

To evaluate the basic relations (eq.s 4.15 and 4.16) it must be regarded, that the relaxation function is given at different discrete points of time. Therefore it is recommended to approximate this function by a series of e-functions, so-called PRONY-DIRICHLET-series.

$$C^*(t) = C^*(0^+) \sum_{\mu=1}^{m} a_\mu \cdot \exp(-\frac{t}{b_\mu}) \qquad (4.26)$$

The coefficients a_μ and b_μ are determined by the non-linear least-square method

$$\min C^*(0^+) \underbrace{\sum_{j=1}^{k} [f_j - \phi_\mu(t_j)]^2}_{D^2} \qquad (4.27)$$

with

$$\phi_\mu(t_j) = \sum_{\mu=1}^{m} a_\mu \cdot \exp(-\frac{t_j}{b_\mu}); \quad f_j = C_j^*/C^*(0^+); \quad C_j^* \equiv C^*(t_j) \qquad (4.28)$$

C^*_j denotes the data according to eq. (4.23).
The conditions must be accomplished

$$\frac{\partial D^2}{\partial a_\mu} = 0: \quad \sum [f_j - \phi_\mu(t_j)] \frac{\partial \phi_\mu(t_j)}{\partial a_\mu} = 0$$
$$\frac{\partial D^2}{\partial b_\mu} = 0: \quad \sum [f_j - \phi_\mu(t_j)] \frac{\partial \phi_\mu(t_j)}{\partial b_\mu} = 0 \qquad (4.29)$$

which lead to a system of 2m non-linear transcendent equations. This system can be solved for instance by means of the NEWTON-method or different modifications respectively (see Engeln-Müllges/Reutter [4.9].

As the convolution in the LAPLACE-transform is commutable eq.(4.15) for instance can be rewritten as follows.

$$\sigma_1(t) - \sigma_2(t) = \frac{\lambda}{d(0^+)} \{ C^*(0^+) \cdot \delta(t) - \int_{0^+}^{t} \delta(\tau) \frac{\partial}{\partial \tau} C^*(t-\tau) \, d\tau \} \qquad (4.15a)$$

The partial derivative of $C^*(t-\tau)$, eq.(4.26), runs

$$\frac{\partial}{\partial \tau} C^*(t-\tau) = C^*(0^+) \sum_{\mu=1}^{m} \frac{a_\mu}{b_\mu} \exp(-\frac{t}{b_\mu}) \exp(\frac{\tau}{b_u}) \tag{4.30}$$

Inserting this expression into eq.(4.15a) yields

$$\sigma_1(t) - \sigma_2(t) = \frac{\lambda}{d(0^+)} C^*(0^+) \{\delta(t) - \int_{0^+}^{t} \delta(\tau) \sum_{\mu=1}^{m} \frac{a_\mu}{b_\mu} \exp(-\frac{t}{b_\mu}) \exp(\frac{\tau}{b_\mu}) d\tau \} \tag{4.31}$$

To optimise the numerical evaluation process the values of the difference of the principal stresses are calculated at discrete points of time. To save processing time and to reduce the necessary computer capacity a recurrent formula is recommended, because it must be pointed out, that for each arbitrary time t_n the calculation of $\sigma_1(t_n) \cdot \sigma_2(t_n)$ must be started at time $t(0^+)$ always.

$$\sigma_1(t_n) - \sigma_2(t_n) = \frac{\lambda}{d(0^+)} C^*(0^+) \{\delta(t_n) - \sum_{\mu=1}^{m} \frac{a_\mu}{b_\mu} \exp(-\frac{t_n}{b_\mu}) \cdot J(t_n)\} \tag{4.32}$$

$$J(t_n) = J(t_{n-1}) + \int_{t_{n-1}}^{t_n} \delta(\tau) \cdot \exp(\frac{\tau}{b_\mu}) d\tau \; ; \; J(0^+) = 0 \tag{4.33}$$

As an acceptable approach linear course of the isochromatic fringe order in the time interval Δt_i may be assumed to solve the integral in eq.(4.33) strictly, otherwise a discrete EULER-CAUCHY solution is possible.

Analogously the eq.s (4.16) are solved, taking into account the directions ψ_N of the principal axes of the refraction tensor. To receive finally the components of the stress tensor further evaluation can performed according to the procedures in linear photoelasticity or by advanced discrete evaluation methods as presented later.

5. EVALUATION OF PHOTOELASTIC/-VISCOELASTIC MEASUREMENTS BY MEANS OF BOUNDARY-ELEMENT METHOD.

5.1. Mathematical basis.

Let Ω be a finite domain with the boundary Γ, u and v two functions, which are steady and which have steady derivatives up to the second order in Ω including Γ, i.e. the functions u and v are harmonic in Ω. The problem is given to determine the harmonic function u for given values u^Γ on the boundary Γ. This problem is usually denoted as DIRICHLET's problem, described by GREEN's formula [5.1]

$$\iint_\Omega (u \cdot \nabla^2 v - v \cdot \nabla^2 u) d\Omega = \int_\Gamma (u \cdot v_{,n} - v \cdot u_{,n}) d\Gamma \tag{5.1}$$

where v denotes a fundamental solution of the LAPLACE-differential equation, $u_{,n}$ and $v_{,n}$ are the derivatives in direction of the external normal to boundary Γ. For a two-dimensional problem as a fundamental solution the function v is chosen:

$$v = \ln \frac{1}{r} \tag{5.2}$$

where r denotes the distance between a definite point in Ω and a variable point on Γ.

Based on eq.(5.1) the values u_M of the harmonic function in arbitrary points M inside the domain Ω and on the boundary Γ can be expressed by the values of this function and its derivative normal to the boundary

$$u_M = \frac{1}{2\pi} \int_\Gamma [\ln \frac{1}{r} q^\Gamma - (\ln \frac{1}{r})_{,n} u^\Gamma] d\Gamma \tag{5.3}$$

where q^Γ denotes the normal derivative of u^Γ.

To solve eq.(5.3) according to the Boundary-Element Method [5.2] the boundary Γ is divided into N discrete elements μ, $\mu \in [1/N]$, with the nodal points v, $v \in [1/N]$ in the centre of the elements (Fig.5.1). With the interpolation functions ψ and φ, which may be e.g. constant, linear or quadratic functions, the boundary value of the harmonic function u in a nodal point κ holds

$$u_\kappa^\Gamma = \frac{1}{2\pi} \sum_{v=1}^{N} [q_v^\Gamma \cdot \ln \frac{1}{r_v} \int_{\Gamma_\mu} \psi(\underline{x}) d\Gamma - u_v^\Gamma (\ln \frac{1}{r_v})_{,n} \int_{\Gamma_\mu} \varphi(\underline{x}) d\Gamma] \tag{5.4}$$

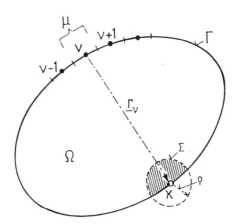

Fig.5.1. Boundary element division; singularity in point $v = \kappa$.

If $v = \kappa$ the first term of eq.(5.4) is going to infinity, the second term is equal to zero, because the vector r is orthogonal to the normal to the boundary in κ. In order to take the influence of u_κ^Γ into account a circle with the small radius ρ (Fig.5.1) is considered. If the boundary is smooth in κ, integration of the second term in eq.(5.3) yields

$$\int_\Sigma u^\Sigma \cdot (\ln \tfrac{1}{\rho})_{,n} d\Sigma = -\tfrac{1}{\rho} \int_\Sigma u^\Sigma d\Sigma = -\pi u_\kappa^\Gamma \tag{5.5a}$$

It is to consider, that the integral is to take over πr. The integration of the first term leads with passage to the limit $\rho \to 0$ to

$$\lim_{\rho \to 0} \int_\Sigma q^\Sigma \cdot \ln \tfrac{1}{\rho} d\Gamma = \lim_{\rho \to 0} (\ln \tfrac{1}{\rho} \int_\Sigma q^\Sigma d\Gamma) = \lim_{\rho \to 0} (\ln \tfrac{1}{\rho} \cdot q^\Sigma \pi \rho) = 0 \tag{5.5b}$$

with that eq.(5.4) runs

$$\tfrac{1}{2} u_\kappa^\Gamma = \tfrac{1}{2\pi} \sum_{\substack{v=1 \\ v \neq \kappa}}^N [q_v^\Gamma \cdot \ln \tfrac{1}{r_v} \int_{\Gamma_\mu} \psi(\underline{\hat{x}}) d\Gamma - u_v^\Gamma (\ln \tfrac{1}{r_v})_{,n} \int_{\Gamma_\mu} \varphi(\underline{\hat{x}}) d\Gamma] \tag{5.6}$$

and the normal derivative q^Γ for given boundary values u^Γ can be calculated in all discrete points v including κ on the boundary. In matrix notation eq.(5.6) reads

$$(q^\Gamma) = [G^\Gamma]^{-1} \cdot [H^\Gamma] \cdot (u^\Gamma) \tag{5.7}$$

with the elements of the matrices $[G^\Gamma]$ and $[H^\Gamma]$

$$h^\Gamma_{\kappa v} = \tfrac{1}{2}\delta_{\kappa v} + \tfrac{1}{2\pi}\left(\ln\tfrac{1}{r_v}\right)_{,n}\int_{\Gamma_\mu}\varphi(\hat{x})\,d\Gamma \quad (\delta_{\kappa v} = \text{KRONECKER-Delta})$$

(5.8)

$$g^\Gamma_{\kappa v} = \tfrac{1}{2\pi}\ln\tfrac{1}{r_v}\int_{\Gamma_\mu}\psi(\hat{x})\,d\Gamma$$

Now the values of the harmonic function u_M in arbitrary discrete points M inside the domain Ω can be calculated according to

$$u_M = (g_{Mv})^T \cdot (q^\Gamma_v) - (h_{Mv})^T \cdot (u^\Gamma_v)$$

(5.9)

5.2. Plane plates with inclusions.

A plane plate with several inclusions consisting of materials with different mechanical properties will be analysed (Fig.5.2).

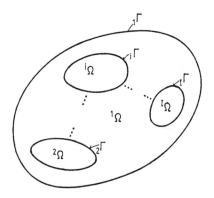

Fig.5.2. Plate with inclusions of different materials

The plate itself is denoted by $^1\Omega$, whereas the inclusions are denoted as sub-domains $^i\Omega$, $i\in[2/I]$ with the boundaries $_i\Gamma$. The boundary values ^1_iu of the harmonic function are given in the discrete points v of the outer boundary, the values along the internal boundaries $_i\Gamma$, i.e. the interfaces between the domain $^1\Omega$ and the sub-domains $^i\Omega$, are unknown. Then according to eq.5.7 the relations are valid

i. for the domain $^1\Omega$:
$$\sum_{i=1}^{I} [^1_iH](^1_iu) = \sum_{i=1}^{I} [^1_iG](^1_iq); \quad i\in[1/I]$$
(5.10a)

ii. for the sub-domains $^i\Omega$:
$$[^i_iH](^i_iu) = [^i_iG](^i_iq); \quad i\in[2/I]$$
(5.10b)

The contact conditions, assumed to be constant along the internal boundaries $_i\Gamma$, may be described generally by

$$(_i^i u) = {_i}a\,(_i^1 u),\quad (_i^i q) = {_i}b\,(_i^1 q),\quad i \ne 1 \tag{5.11}$$

Inserting these conditions into eq.(5.10b) yields the following equation,

$$(_i^1 q) = \frac{{_i}a}{{_i}b}\,[_i^i G]^{-1}[_i^i H]\,(_i^1 u),\quad i \ne 1 \tag{5.12}$$

which is to insert into eq.(5.10a) yielding a set of linear equations.

$$\underbrace{\sum_i \left[\,[_1^1 G];\,[_2 K];\,\cdots\,;[_i K];\,\cdots\,;[_I K]\,\right]}_{[K]} \cdot \underbrace{\begin{bmatrix}(_1^1 q)\\(_2^1 u)\\\vdots\\(_i^1 u)\\\vdots\\(_I^1 u)\end{bmatrix}}_{(x)} = [_1^1 H]\cdot(_1^1 u)\;\sum_i \tag{5.13}$$

The matrices $[_i K]$, $i \ne 1$ are defined as

$$[_i K] = \frac{{_i}a}{{_i}b}\,[_1^i G][_i^i G]^{-1}[_i^i H] - [_1^i H] \tag{5.14}$$

and Σ_i is the sum of all nodal points on the boundaries $_i\Gamma$.
Introducing the super-vector (x), which includes all values $(_i^1 q)$ as well as the values $(_i^1 u)$ and the matrix [K] according to the definition given in eq.(5.13), this equation holds

$$(x) = [K]^{-1}\cdot[_1^1 H]\cdot(_1^1 u) \tag{5.13a}$$

The solutions are set into eq.s.(5.11) and (5.10) to calculate the values of the harmonic function and its normal derivative in the discrete boundary points of the sub-domains. Thus all boundary values in the discrete points on the external and internal boundaries of domain $^1\Omega$ as well as on the boundaries of the sub-domains $^i\Omega$ are given and the values u_M of the harmonic function in any discrete arbitrary point M inside the different domains can be calculated on the basis of eq.(5.9).

5.3. Application to two-dimensional photoelasticity.

The first invariant S of the two-dimensional stress tensor, i.e. the sum of the principal stresses, satisfies the LAPLACE-differential equation, therefore S is a harmonic function and in all previous equations u and q are substituted by S and $S_{,n}$ respectively, e.g. in eq.s.(5.6) and (5.7):

$$\frac{1}{2}S_x^\Gamma = \frac{1}{2\pi} \sum_{\substack{v=1 \\ v \neq \kappa}}^{N} [S_{,nv}^\Gamma \ln\frac{1}{\Gamma_v}\int_{\Gamma_\mu}\psi(\tilde{x})\,d\Gamma - S_v^\Gamma (\ln\frac{1}{\Gamma_v})_{,n}\int_{\Gamma_\mu}\varphi(\tilde{x})\,d\Gamma] \qquad (5.15)$$

$$(S_m^\Gamma) = [G^\Gamma]^{-1} \cdot [H^\Gamma] \cdot (S^\Gamma) \qquad (5.16)$$

From photoelastic measurements the isochromatic fringe orders $_1^1\delta$ in the discrete points v along load-free parts of $_1\Gamma$ are taken. With the so-called photoelastic constant $^1\hat{s}$ of the material of domain $^1\Omega$ and with the thickness d of the plate the boundary values hold

$$(_1^1 u) \triangleq (_1^1 S) \triangleq (_1^1 \sigma) = \frac{^1\hat{s}}{d}(_1^1\delta) \qquad (5.17)$$

which are to insert into eq.(5.13).

The plate may be loaded by a concentrated load P_m, applied to the external boundary $_1\Gamma$ in point m in the direction normal to the boundary. According to the principle of DE ST.VENANT a circle with the small radius ρ, being tangent to the boundary in m, will be considered, (Fig.5.3). The BOUSSINESQ-solution yields the principal stresses along the circle:

$$\sigma_{rr}^\Sigma = -\frac{P_m}{\pi} \cdot \frac{1}{d_o} \cdot \frac{1}{\rho} \qquad \sigma_{\varphi\varphi}^\Sigma = 0 \qquad (5.18)$$

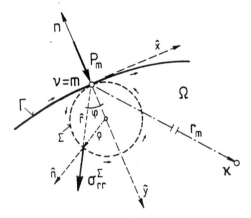

Fig.5.3. Concentrated load P_m normal to the boundary.

Then the sum S^Σ of the principal stresses along Σ is given to

$$S^\Sigma \equiv \bar{\sigma}^\Sigma_{rr} = -\frac{P^{(m)}}{\pi} \cdot \frac{1}{d_0} \cdot \frac{1}{\rho} \tag{5.19}$$

and the normal derivative

$$S^\Sigma_{,\hat{n}} = S^\Sigma_{,\hat{x}} \frac{d\hat{x}}{d\hat{n}} + S^\Sigma_{,\hat{y}} \frac{d\hat{y}}{d\hat{n}} \tag{5.20}$$

runs after some transformation regarding eq.(5.18)

$$S^\Sigma_{,\hat{n}} = \frac{P_m}{2\pi\rho} \cdot \frac{1}{d_0} \cdot \frac{1}{\cos^2\varphi} \tag{5.21}$$

With reference to eq.(5.4) the influence of P_m on the boundary value of the harmonic function in the discrete point κ is given to

$$^mS^\Gamma_\kappa = \frac{1}{2\pi} [\ln\frac{1}{r_m} \int_0^{2\pi} S^\Sigma_{,\hat{n}} \rho\, d\varphi - (\ln\frac{1}{r_m})_{,n} \int_0^{2\pi} S^\Sigma \rho\, d\varphi] \tag{5.22}$$

where r_m denotes the distance between points m and κ. Introducing eq.s.(5.18) and (5.21) the influence of each load P_m on S^Γ amounts to

$$^mS^\Gamma_\kappa = \frac{P_m}{\pi d_0} (\ln\frac{1}{r_m})_{,n} \tag{5.23}$$

This term is to add to eq.(5.16), which then runs

$$(S^\Gamma_m) = [G^\Gamma]^{-1} \cdot \{[H^\Gamma] \cdot (S^\Gamma) + [^mH^\Gamma] \cdot (^mS^\Gamma)\} \tag{5.24}$$

with the elements of the matrix $[^mH^\Gamma]$

$$^mh^\Gamma_{\kappa\nu} = \begin{cases} \frac{1}{2\pi}(\ln\frac{1}{r_\nu})_{,n} & \text{für } \nu = m \\ 0 & \text{für } \nu \neq m \end{cases} \tag{5.25}$$

and the elements of the vector $(^mS^\Gamma)$

$$^ms^\Gamma_{\kappa\nu} = \begin{cases} 2P_m/d_0 & \text{für } \nu = m \\ 0 & \text{für } \nu \neq m \end{cases} \tag{5.26}$$

Consequently eq.(5.13a) holds with regard to the denotations in Fig.5.2:

$$(x) = [K]^{-1} \cdot \{[{}_1^1H] \cdot ({}_1^1S) + [{}_1^mH] \cdot ({}_1^mS)\} \tag{5.27}$$

With the solution of this equation and after calculation of ${}_i^1S$ and ${}_i^iS_{,n}$ the sum of the principal stresses in arbitrary points M in the domain ${}^1\Omega$ as well as in the sub-domains ${}^i\Omega$ can be determined.

$$ {}^1S_M = \sum_{i=1}^{I}\{({}_i^1h_{Mv})^T \cdot ({}_i^1S) - ({}_i^1g_{Mv})^T \cdot ({}_i^1S_{,n})\} + ({}_1^mh_{Mv})^T \cdot ({}_1^mS); \; i \in [1/I] \tag{5.28}$$

$$ {}^iS_M = ({}_i^ih_{Mv})^T \cdot ({}_i^iS) - ({}_i^ig_{Mv})^T \cdot ({}_i^iS_{,n}); \; i \in [2/I] \tag{5.29}$$

Based on the principal relations in linear photoelasticity the difference of the principal stresses in arbitrary points M is given by the measured isochromatic fringe order in these points:

$$ {}^i\Delta_M = ({}^i\sigma_1 - {}^i\sigma_2)_M = \frac{{}^i\hat{S}}{d} {}^i\delta_M; \; i \in [1/I] \tag{5.30}$$

The principal stresses are obtained by adding or subtracting the results of eq.s (5.28) or (5.29) respectively and (5.30).

$$ {}^i\sigma_{1M} = \tfrac{1}{2}({}^iS_M + {}^i\Delta_M); \; {}^i\sigma_{2M} = \tfrac{1}{2}({}^iS_M - {}^i\Delta_M) \tag{5.31}$$

It must be pointed out as a special advantage of the described method, that in contradiction to the classical method to evaluate photoelastic data the isoclinics are not necessary to determine the principal stresses, thus improving the accuracy of results and saving time in measurement and evaluation.

As an instructive example the stress state in circular disc with a central circular inclusion of different material under two diametrical concentrated loads has been analysed (Fig.5.4). The result of evaluation of the isochromatic fringe pattern (Fig.5.5a) are shown in Fig.5.5b/c. To formulate the boundary conditions between ${}^1\Omega$ and ${}^i\Omega$, eq.s (5.11), as an approach it has been assumed, that the sum of the principal strain in both the adjacent domains are equal and the normal derivatives of this sum differ by the factor (-1) only:

$$ {}_i^1(\varepsilon_1 + \varepsilon_2) = {}_i^i(\varepsilon_1 + \varepsilon_2); \; {}_i^1(\varepsilon_1 + \varepsilon_2)_{,n} = - {}_i^i(\varepsilon_1 + \varepsilon_2)_{,n} \; i \in [2/I] \tag{5.32}$$

It must be mentioned, that with this approach the effects of different lateral contraction of the inclusion and the plate has been neglected

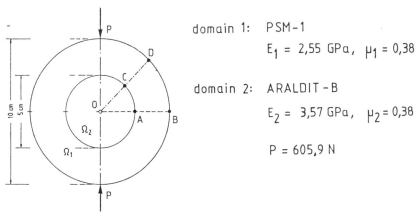

Fig.5.4. Circular disc with central circular inclusion.

domain 1: PSM-1
$E_1 = 2{,}55$ GPa, $\mu_1 = 0{,}38$

domain 2: ARALDIT-B
$E_2 = 3{,}57$ GPa, $\mu_2 = 0{,}38$

$P = 605{,}9$ N

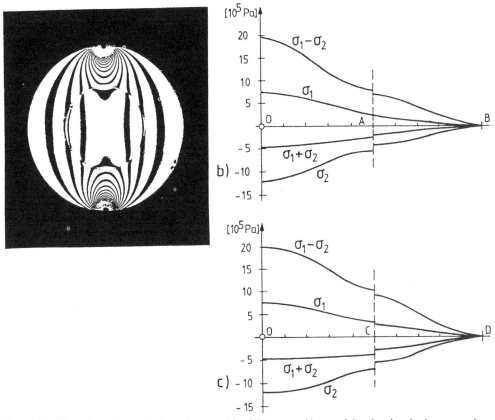

Fig.5.5. Circular disc; a) isochromatic fringe pattern; b) principal stresses in sect. O-A-B, c) in sect. O-C-D

5.4. Elastic/viscoelastic response of material.

Supposed the materials of the inclusions showing viscoelastic response, then beside the mechanical response the optical rheological response of the materials must also be taken into account(see chapter 4).The boundary conditions along the interfaces between the domain $^1\Omega$, consisting of linear-elastic material, and the inclusions, i.e. the subdomains $^i\Omega$, $i \neq 1$, may be defined as

$$^1_i\varepsilon_\eta(t_m) = {^i_i\varepsilon_\eta(t_m)}; \quad {^1_i\varepsilon_\eta(t_m)}_{,n} = -{^i_i\varepsilon_\eta(t_m)}_{,n}; \tag{5.33}$$

where ε_η denotes the sum of the principal strains. This formulation of the boundary conditions must be regarded as an approach only as in the interface the stress-strain state must be three-dimensional because of the different lateral contraction in $^1\Omega$ and $^i\Omega$.

Obviously these boundary conditions are depending on time t. The stress-strain relations for linear viscoelastic materials can be described by a VOLTERRA's integral equation of the second kind [5.3].

$$^i_i\varepsilon_\eta(t_m) = (1-{^i\mu})\{\frac{1}{^iE(t_m)}{^i}S(0^+) + \int_{0^+}^{t_m}\frac{1}{^iE(t_m-\tau)}\cdot\frac{\partial}{\partial\tau}[{^i}S(\tau)]d\tau\} \tag{5.34}$$

The discrete solution of this equation yields with reference to the subdomains $^i\Omega$, $i \neq 1$, [5.4]:

$$^i_i\varepsilon_\eta(t_m) = (1-{^i\mu})\{\frac{1}{^iE(t_m)}{^i}S(0^+) + \frac{1}{2}\sum_{j=1}^{m}[\frac{1}{^iE(t_m-t_j)} + \frac{1}{^iE(t_m-t_{j-1})}]\cdot[{^i}S(t_j)-{^i}S(t_{j-1})]\} \tag{5.35}$$

Although the material of domain $^1\sigma$ exhibits elastic response the strain in this domain depends on time t also, because the stress-strain state will be influenced by the time-depending stress-strain state in the inclusions.

$$^1_i\varepsilon_\eta(t_m) = \frac{(1-{^1\mu})}{^1E}{^i}S(t_m) \tag{5.36}$$

(Note: To avoid confusion with the denotation n of the normal, the final point of time is denoted by t_m.)

Considering eq.s.(5.35) and (5.36) it follows from eq.s (5.33)

$$^i S(t_m) = \frac{1}{^iB(t_m)}\{\frac{1}{^iC}{^i}S(t_m) - {^i}A(t_{m-1})\}; \quad i \neq 1 \tag{5.37}$$

$$_i^i S(t_m)_{,n} = -\frac{1}{_i B(t_m)} \{\frac{1}{_i^i C} {}_i^1 S(t_m)_{,n} - {}_i^i A(t_{m-1})_{,n}\} ; \quad i \neq 1 \tag{5.38}$$

with the definitions

$$_i^i A(t_{m-1}) = \frac{1}{_i^i E(t_m)} {}_i^i S(0^+) + \frac{1}{2}\sum_{j=1}^{m-1}[\frac{1}{E(t_m-t_j)} + \frac{1}{E(t_m-t_{j-1})}]\cdot[{}_i^i S(t_j)-{}_i^i S(t_{j-1})] - {}^i B(t_m) \, {}_i^i S(t_{m-1})$$

$$_i B(t_m) = \frac{1}{2}[\frac{1}{_i^i E(0^+)} + \frac{1}{_i^i E(t_m-t_{m-1})}] \tag{5.39}$$

$$_i^i C = \frac{1-{}^i\mu}{1-{}^1\mu} \, {}^1 E$$

The expressions for the sum of the normal stresses along the interfaces and for the normal derivatives according to eq.s.(5.37) and (5.38) are introduced into eq.s.(5.10 a/b), (see section 5.2.), yielding a system of linear equations

$$[{}_1^1 G]\cdot({}_1^1 S_{,n}) + \sum_{i=2}^{I}\{[{}_i K]\cdot({}_1^1 S) + {}^i C \, [{}_1^1 G]\cdot({}_i^i A_{,n})\} = [{}_1^1 H]\cdot({}_1^1 S) + \sum_{i=2}^{I} {}^i C[{}_i L]\cdot({}_i^i A) \tag{5.40}$$

which runs in matrix denotation:

$$(x) = [K]^{-1}\cdot[H_L]\cdot(U) \tag{5.41}$$

The matrices and vectors are defined as follows:

$$[{}_i L] = [{}_1^1 G]\cdot[{}_1^1 G]^{-1}\cdot[{}_1^i H]$$

$$[{}_i K] = [{}_1^1 G][{}_1^1 G]^{-1}[{}_1^i H] - [{}_1^i H] = [{}_i L] - [{}_1^i H]$$

$$[K] = \left[[{}_1^1 G],[{}_2 K],\cdots,[{}_i K],\cdots,[{}_I K],{}^2 C[{}_2^1 G],\cdots,{}^i C[{}_i^1 G],\cdots,{}^I C[{}_I^1 G]\right] \tag{5.42}$$

$$[H_L] = \left[[{}_1^1 H],{}^2 C[{}_2 L],\cdots,{}^i C[{}_i L],\cdots,{}^I C[{}_I L]\right]$$

$$(U)^T = \{({}_1^1 S);({}_2^2 A);\cdots;({}_i^i A);\cdots;({}_I^I A)\}$$

$$(x)^T = \{({}_1^1 S_m);({}_2^1 S);\cdots;({}_i^1 S);\cdots;({}_2^2 A_m);\cdots;({}_i^i A_m);\cdots;({}_I^I A_m)\}$$

External loads applied to the outer boundary $_1\Gamma$ of the domain $^1\Omega$ are taken into account analogously to eq.s.(5.24 to (5.27). However this is true only if the values of all external loads are given and do not change over time, because in such cases, where e.g. the reactions on support are depending on the state of deformation, these reactions are depending on time also and of course this dependency must be taken into consideration. To analyse for instance the stress state in an object as shown in Fig.5.6 the reactions on the supports in points a, b and c are to consider as external loads, which are depending on time, whereas the applied external load P remains independent of time. Therefore during measuring the optical phenomena in the object the reactions A, B, C must be measured simultaneously.

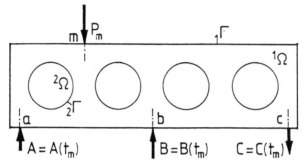

Fig.5.6. Example of external forces depending on time.

Because of the rheological response of the inclusions materials the system of linear equations (5.41) can be solved in a recurrent procedure assuming fading memory of the material. After each time interval Δt_m the elements of the supervector (x) are to calculate, the results are to introduce into eq.s.(5.37) and (5,38) considering the actual values of the elements of the matrices $[_i^i A]$ and $[^i B]$; the solutions of these equations then enable calculating the finally wanted values of the harmonic function S in arbitrary points M or in nodal points of an evaluation grid in $^1\Omega$ and in $^i\Omega$ as well (see section 5.3. eq.s.(5.28, 5.29).

$$^1S_M = \sum_{i=1}^{I} \{(^1_i h_{Mv})^T \cdot (^1_i S) - (^1_i g_{Mv})^T \cdot (^1_i S_{/n})\} + (^m_1 h_{Mv})^T \cdot (^m_1 S); \ i \in [1/I] \tag{5.28}$$

$$^i S_M = (^i_i h_{Mv})^T \cdot (^i_i S) - (^i_i g_{Mv})^T \cdot (^i_i S_{/n}); \ i \in [2/I] \tag{5.29}$$

The difference of the principal stresses in the inclusions, the material of which exhibits viscoelastic response, will be determined according to chapt.4.

In the domains $^i\Omega$, i.e. in the inclusions consisting of viscoelastic material, the difference of the principal stresses in arbitrary points M holds

$$^i\Delta_M = [^i\sigma_1(t_m) - ^i\sigma_2(t_m)]_M = \frac{\lambda}{d(0^+)} \cdot ^iC^*(0^+) \cdot \{^i\delta_M(t_m) - \sum_{\mu=1}^{\hat{m}} \frac{^ia_\mu}{^ib_\mu} \exp(-\frac{t_m}{^ib_\mu}) \cdot ^iJ(t_m)\} \qquad (5.43)$$

where the integral $^iJ(t_m)$ is defined as

$$^iJ(t_m) = ^iJ(t_{m-1}) + \int_{t_{m-1}}^{t_m} {^i\delta_M(\tau)} \exp(\frac{\tau}{^ib_\mu}) \, d\tau; \quad ^iJ(0^+) = 0 \qquad (5.44)$$

In the domain $^1\Omega$ the difference of the principal stresses in arbitrary points M is given as

$$^1\Delta_M = [^1\sigma_1(t_m) - ^1\sigma_2(t_m)]_M = \frac{^1\hat{S}}{d(t_m)} {^1\delta_M(t_m)} \qquad (5.45)$$

Having determined the sum as well as the difference of the principal stresses these can be calculated according to eq.s.(5.31).

$$^i\sigma_{1M} = \frac{1}{2}(^iS_M + ^i\Delta_M); \quad ^i\sigma_{2M} = \frac{1}{2}(^iS_M - ^i\Delta_M) \qquad (5.31)$$

In the preceding described procedure it has been supposed the material in the domain $^1\Omega$ to exhibit linear elastic response, in the subdomains $^i\Omega$ however linear viscoelastic response. But similarly the contrary case can be considered, that domain $^1\Omega$ consists of viscoelastic, the subdomains $^i\Omega$ however consisting of elastic material. Furthermore respective relations can be derived, if the material of different inclusions exhibit different mechanical as well as optical response, either elastic or viscoelastic.

As an example of application the problem of a circular ring, already dealt with (Fig.s.5.4 and 5.5), will be analysed, where the outer ring consists of the elastic material **VP1527**, the subdomain $^2\Omega$ however now consists of viscoelastic material (ARALDITE **F** / ARALDITE **CY208** /Hardener **HY951**). The isochromatic fringe pattern taken at different times after loading (Fig.5.7) compared with the fringe pattern as shown in Fig.5.5. and the course of the principal stresses over the diameter O-A-B at time $t \approx 0$ and at time $t = 120$ sec after loading (Fig.5.8) demonstrate the influence of the rheological material response very instructively.

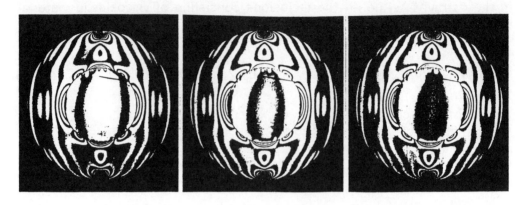

Fig.5.7.: Isochromatic fringe pattern at time t = 0, 30, 120 sec.

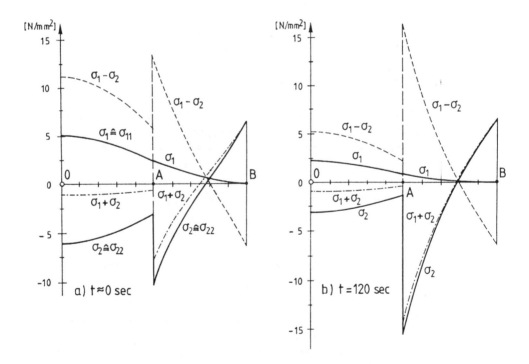

Fig.5.8.: Principal stresses along diameter O-A-B;
a) at t ≈ 0 sec; b) at t = 120 sec.

6. ANALYSIS OF 3-D-STRESS-STRAIN STATES BY MEANS OF A DISCRETE BOUNDARY-INTEGRAL METHOD

6.1. Determination of the surface displacements.

Modern optical methods like e.g. holographic interferometry and speckle-interferometry combined with digital image processing enable the determination of the spatial displacements on the surface of three-dimensional solids. Based on the boundary-integral method a reduction procedure will be derived to calculate the displacements and subsequently the strain- and stress-state in arbitrary points inside of three-dimensional simply-connected solids proceeding from the measured boundary values.

The measurements are carried out by means of multi-hologram analysis. This method yields the spatial displacement vector $(u_i^\Gamma)_{\nu,K}$ in discrete points (ν,K) according to eq.(6.1), [6.1], [6.2].

$$(u_i^\Gamma)_{\nu,K} = \pm \lambda [A_{i\rho}]_{\nu,K}^{-1} (N_\rho)_{\nu,K} \tag{6.1}$$

where λ denotes the wave-length of the laser-light. The elements of the vector $(N_\rho)_{\nu,K}$ are the interference fringe order in points (ν,K) with reference to at least three non-complanar holograms H_ρ; the elements of the matrix of geometry for each (ν,K) and each H_ρ hold

$$A_{i\rho} = a_{i\rho} + a_{iq}\,; \quad i \in [1/3]\,;\ \rho \in [1/3] \tag{6.2}$$

where a_{ip} and a_{iq} are the cosines of the angles between the axes of the reference co-ordinate system and the position-vectors $(r_{p(\nu,K)})$ and $(r_{q(\nu,K)})$ respectively (Fig.6.1).

In order to get the measured data of the whole surface the object including the loading frame is revolved stepwise around the axis x_3 of the reference co-ordinate system. For allotting the different sets of holograms cursor points (I,K) are marked on the surface Γ of the object.

Having determined the components u_i^Γ in all nodal points (ν,K) including the cursor points (I,K) they are transformed into a local co-ordinate system (x_3) related to the tangential plane at the surface in each nodal point (Fig.6.2). The derivatives $u_{i,2}^\Gamma$ and $u_{i,3}^\Gamma$ are given by the measured data. Considering a load-free surface and linear elastic response of the material of the object it follows from the equilibrium conditions

$$\bar{u}_{1,1}^\Gamma = -\frac{\mu}{1-\mu}(\bar{u}_{2,2}^\Gamma + \bar{u}_{3,3}^\Gamma)\,;\quad \bar{u}_{2,1}^\Gamma = -\bar{u}_{1,2}^\Gamma\,;\quad \bar{u}_{3,1}^\Gamma = -\bar{u}_{1,3}^\Gamma\,. \tag{6.3}$$

The derivatives are then to transform onto the local co-ordinate system $(\tilde{x}_i)$ yielding $\tilde{u}_{i,j}^r$.

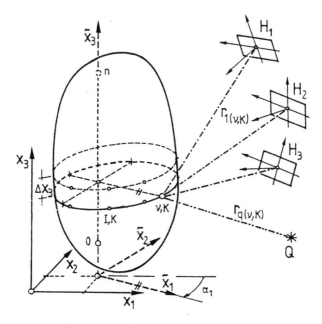

Fig. 6.1.: Multi-hologram set-up.

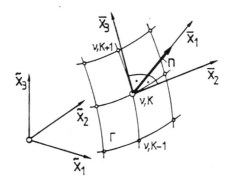

Fig. 6.2.: Orientation of the boundary co-ordinate system.

6.2. Mathematical basis.

Assuming elastic, isotropic and homogeneous response of material the state of displacement in a three-dimensional object can be described by

LAMÉ-NAVIER's differential equation, which runs with regard to EINSTEIN's summation convention

$$\nabla^2 u_i + \frac{1}{1-2\mu}\phi_{,i} = 0; \quad \phi \equiv u_{k,k}; \quad i,k \in [1/3] \tag{6.4}$$

As Φ is a harmonic function it can be proved, that (see [6.3])

$$\nabla^2 (x_i \cdot \phi) \equiv 2\phi_{,i} \tag{6.5}$$

Introducing eq.(6.5) into eq.(6.4) yields a LAPLACE-differential equation

$$\nabla^2 [u_i + \frac{1}{2}\frac{1}{1-2\mu}(x_i \phi)] = \nabla^2 f_i = 0 \tag{6.6}$$

i.e. f_i denotes a harmonic function, the boundary values f_i^Γ of which are given by the preceding described measurements.

Let two harmonic functions U and V be given, which are steady and show steady derivatives up to the 2nd order in a three-dimensional domain O including the boundary Γ. Then the values of the function U can be calculated in arbitrary points M inside Ω according to GREEN's formula [6.4]

$$U_{(M)} = \frac{1}{4\pi} \iint_\Gamma [V \cdot U_{,n}^\Gamma - U^\Gamma \cdot V_{,n}] \, d\Gamma \tag{6.7}$$

With the fundamental solution $V = 1/r$ and the harmonic function $U = f_i$ eq.(6.7) runs

$$f_{i(M)} = \frac{1}{4\pi} \iint_\Gamma [\frac{1}{r} \cdot f_{i,n}^\Gamma - f_i^\Gamma (\frac{1}{r})_{,n}] \, d\Gamma \tag{6.8}$$

As the values f_i^Γ of the harmonic function f_i are given DIRICHLET's problem of the 1st kind must be solved. Eliminating the normal derivatives f_i^Γ in the points M^Γ and assuming, that a GREEN's function $G_1(M^\Gamma,M)$ with the boundary values $1/r$ on Γ exists, eq.(6.8) yields

$$f_{i(M)} = \iint_\Gamma f_{i(M)}^\Gamma G(M^\Gamma,M) \, d\Gamma \tag{6.9}$$

with the correlation function

$$G(M^\Gamma,M) = -\frac{1}{4\pi}[\frac{1}{r} - G_1(M^\Gamma,M)]_{,n} \tag{6.10}$$

To determine this correlation function an algorithm for discrete evaluation will be derived.

In an arbitrary point $M(0,0,0)$ in Ω corresponding to the point of origin of a local co-ordinate system $(\tilde{x}_j)$ the harmonic function f_i is expanded in a TAYLOR-series in the positive and negative direction of the co-ordinate axes

$$f_i(\delta \tilde{x}_j, 0, 0) = f_i(0,0,0) + \sum \frac{1}{n!} \frac{\partial^n f_i(0,0,0)}{\partial \tilde{x}_j^n} (\pm \delta \tilde{x}_j)^n \qquad (6.11)$$

and these series are truncated after the fourth term. For simplicity the value of f_i in point $M(0,0,0)$ is denoted by f_{i0}, the values $f_i(\pm \delta \tilde{x}_j, 0, 0)$ are denoted by $^{\pm}f_{ij}$.

$$^+f_{ij} = f_{i0} + \frac{\partial f_{i0}}{\partial \tilde{x}_j}|^+\delta \tilde{x}_j| + \frac{1}{2!}\frac{\partial^2 f_{i0}}{\partial \tilde{x}_j^2}|^+\delta \tilde{x}_j|^2 + \frac{1}{3!}\frac{\partial^3 f_{i0}}{\partial \tilde{x}_j^3}|^+\delta \tilde{x}_j|^3 + \text{trunc} \qquad (6.11a)$$

$$^-f_{ij} = f_{i0} - \frac{\partial f_{i0}}{\partial \tilde{x}_j}|^-\delta \tilde{x}_j| + \frac{1}{2!}\frac{\partial^2 f_{i0}}{\partial \tilde{x}_j^2}|^-\delta \tilde{x}_j|^2 - \frac{1}{3!}\frac{\partial^3 f_{i0}}{\partial \tilde{x}_j^3}|^-\delta \tilde{x}_j|^3 \pm \text{trunc} \qquad (6.11b)$$

Multiplying eq.(6.11a) by $|^-\delta \tilde{x}_j|$ and eq.(6.11b) by $|^+\delta \tilde{x}_j|$, adding the results and suppressing those terms, which are small of higher order it will be obtained finally with regard to the LAPLACE-differential equation (6.6), considering EINSTEIN's summation convention:

$$f_{i(M)} W_{jj} = W_j (^+f_{ij} \cdot ^+h_k + ^-f_{ij} \cdot ^-h_k) \cdot \delta_{jk}; \; j,k \in [1/3] \qquad (6.12)$$

with the KRONECKKER-delta δ_{ik} and the denotations

$$W_j = [|^+\delta \tilde{x}_j| \cdot |^-\delta \tilde{x}_j|]^{-1}$$
$$^{\pm}h_j = |^{\mp}\delta \tilde{x}_j| \cdot l_j^{-1}; \; l_j = |^+\delta \tilde{x}_j| + |^-\delta \tilde{x}_j| \qquad (6.13)$$

Introducing the distances $^{\pm}\tilde{\xi}_j$ between the origin of the local co-ordinate system $(\tilde{x}_j)$ and the intersection points of the co-ordinate axes $\tilde{x}_j$ with the boundary Γ instead of $\pm \delta \tilde{x}_j$ and substituting f_{ij} by the boundary values f_{ij}^Γ, which are given by measurement regarding eq.(6.6), then eq.(6.12) yields an approximate value of $f_{i(M)}$ and thus a rough approach of the correlation function $G(M^\Gamma,M)$ only. The expression w_j adopts the character of a weighting function and $^\pm h_j$ can be defined as a linear form-function (Fig.6.3).

$$w_j = [|^+\tilde{\xi}_j|\cdot|^-\tilde{\xi}_j|]^{-1}$$
$$^\pm h_j = |^\mp\tilde{\xi}_j|\cdot l_j^{-1}; \quad l_j = |^+\tilde{\xi}_j| + |^-\tilde{\xi}_j| \tag{6.14}$$

These expressions are to insert into eq.(6.12)

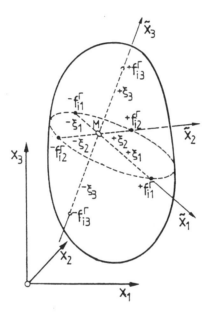

Fig.6.3.: Determination of weighted values $f_{i(M)}$.

Considering the definition of the harmonic function f_i (see eq.(6.6)) the components of the displacement vector in points M inside domain Ω hold with reference to the respective local co-ordinate system

$$\tilde{u}_i \cdot [w_{j,j} + m\, w_i] \approx w_j \cdot (^+\tilde{u}^\Gamma_{i,j} h_k + ^-\tilde{u}^\Gamma_{i,j} h_k)\, \delta_{jk} + m\, w_i \cdot (^+\tilde{u}^\Gamma_{i,i} h_i + ^-\tilde{u}^\Gamma_{i,i} h_i) +$$
$$+ m \cdot (^+\tilde{u}^\Gamma_{ji,j} - ^-\tilde{u}^\Gamma_{ji,j}) \frac{1-\delta_{ij}}{l_j} \tag{6.15}$$
(not summing up over i)

where for abbreviation m denotes

$$m = \frac{1}{2} \cdot \frac{1}{1-2\mu} \tag{6.16}$$

Eq.(6.15) can be written in the form

$$\tilde{u}_i \cdot W_i \approx \tilde{K}_i \tag{6.15a}$$

The meaning of $\tilde{K}_i$ and W_i follows from eq.(6.15)

Transformation of the components $\tilde{u}_i$ into the reference co-ordinate system (x_i) and taking the weighted mean over a sufficient number of these values of a triple, which can be regarded as a numerical integration in the $\mathbf{R}^3$-space, yields the final values of the components of the displacement vector $u_{i(M)}$ with satisfactory accuracy.

$$(u_{k(M)}) = \sum_{(\tilde{x}_j)} [a_{ki}] \cdot (\tilde{K}_i) \cdot \left(\sum_{(\tilde{x}_j)} W_i\right)^{-1} \tag{6.17}$$

The accuracy is the better the more local co-ordinate systems are taken into account in eq.(6.17). To solve this equation a reduction-procedure will be applied.

6.3. The reduction-procedure.

The object will be divided into parallel cross-sections t, $t \in [0/n]$, of equal distances $\Delta \tilde{x}_3$ perpendicular to the axis $\tilde{x}_3$ (Fig.6.4).

For further evaluation the process is split up into two phases. In the first phase the components u_β, $\beta \in [1,2]$, in cross-section t, i.e. in the plane (x_1, x_2) are determined.

$$(u_\beta) = [\sum_{(v,K)} [a_{\beta\alpha}](\tilde{K}_\alpha) + \frac{1}{2} w_3 \, ({}^+\tilde{u}_{\alpha 3} + {}^-\tilde{u}_{\alpha 3})] \cdot \left(\sum_{(v,K)} W_\beta\right)^{-1}; \quad \alpha, \beta \in [1,2] \tag{6.18}$$

with the denotations

$$\tilde{K}_\alpha = w_\beta ({}^+\tilde{u}^\Gamma_{\alpha\beta} h_\gamma + \tilde{u}^\Gamma_{\alpha\beta} h_\gamma) \delta_{\beta\gamma} + mw_\alpha ({}^+\tilde{u}^\Gamma_{\alpha\alpha} h_\alpha + \tilde{u}^\Gamma_{\alpha\alpha} h_\alpha) + m({}^+\tilde{u}^\Gamma_{\beta\alpha,\beta} - \tilde{u}^\Gamma_{\beta\alpha,\beta}) \frac{1-\delta_{\alpha\beta}}{l_\alpha} \tag{6.19}$$

(not summing up over α)

$$W_\beta = w_{jj} + m \, w_\beta; \quad j \in [1/3], \; \beta \in [1,2] \tag{6.20}$$

Introducing for abbreviation

$$(K^*_{\alpha t}) = \sum_{(v,K)} [a_{\alpha\beta}](\tilde{K}_{\beta t}) \cdot (\sum_{(v,K)} W_{\alpha t})^{-1}$$

$$R_{\alpha t} = \frac{1}{2} \sum_{(v,K)} \Delta x_3^{-2} \cdot (\sum_{(v,K)} W_{\alpha t})^{-1} \qquad (6.21)$$

and regarding

$$^-\tilde{u}_{\alpha 3} = \tilde{u}_\alpha(t-1) = u_\alpha(t-1); \quad ^+\tilde{u}_{\alpha 3} = \tilde{u}_\alpha(t+1) = u_\alpha(t+1) \qquad (6.22)$$

the following equation is obtained to calculate the components u_1 and u_2 along the respective axis of evaluation.

$$u_{\alpha t} = K^*_{\alpha t} + R_{\alpha t}(u_\alpha(t-1) + u_\alpha(t+1)) \qquad (6.23)$$

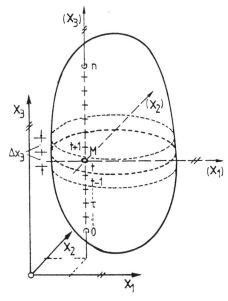

Fig.6.4: Performance scheme of the reduction process.

As in this equation the values $u_\alpha(t+1)$ are unknown the reduction process starts at cross-section $t = 0$ proceeding along the axis of evaluation parallel to axis x_3 to cross-section $t = n$; in both these intersection points the values $u^r_\alpha(0)$ and $u^r_\alpha(n)$ are given as measured data. The reduction process is described by

$$(u_{\alpha t}^\Gamma) = (y_{\alpha t}) + \sum_{\tau=t}^{n-2}[T_{\alpha\tau}](y_{\alpha(\tau+1)}) + [T_{\alpha(n-1)}](u_{\alpha n}^\Gamma) \qquad (6.24)$$

$$(y_{\alpha t}) = \sum_{\tau=1}^{t}[T_{\alpha\tau}][R_{\alpha\tau}]^{-1}(K_{\alpha\tau}^*) + [T_{\alpha 1}](u_{\alpha 0}^\Gamma) \qquad (6.25)$$

The elements of the transmission matrix $[T_{\alpha\tau}]$ are given to

$$T_{\alpha\tau} = T_\alpha(t) \times \cdots \cdots \times T_\alpha(\tau) \qquad (6.26)$$

where $T_\alpha(\tau)$ is calculated according to the recurrent formula

$$T_\alpha(\tau) = R_{\alpha\tau}Q_{\alpha\tau}; \quad Q_{\alpha\tau} = [1 - R_{\alpha\tau}R_{\alpha(\tau-1)}Q_{\alpha(\tau-1)}]^{-1} \qquad (6.27)$$

Having determined the components u_α, $\alpha \in [1,2]$, of the displacement vector in a sufficient large number of points $M(x_\alpha)$ in all sections t the derivatives $u_{\alpha 3,\beta}$ are determined by cubic spline approximation.

The second phase to evaluate eq.(6.17), i.e. calculating u_3, runs analogously to the first phase, taking into consideration, that the components related to the axis (x_3) must not be transformed on the reference co-ordinate system, because (x_3) is parallel to x_3. Thus the component u_3 of the displacement vector holds

$$u_3 = [\sum_{(\nu,K)} K_3 + \frac{1}{2}(1+m)\Delta x_3^{-2}(^+u_{33} + ^-u_{33})] \cdot (\sum_{(\nu,K)} W_3)^{-1} \qquad (6.28)$$

with the denotations

$$K_3 = w_\alpha(^+u_{3\alpha}^\Gamma h_\beta + ^-u_{3\alpha}^\Gamma h_\beta)\delta_{\alpha\beta} + m(2\Delta x_3)^{-1} \cdot (^+u_{\alpha 3,\alpha} - ^-u_{\alpha 3,\alpha}) \qquad (6.29)$$

$$W_3 = w_{jj} + m\Delta x_3^{-2} \qquad (6.30)$$

Introducing for abbreviation the expressions

$$K_{3t}^* = \sum_{(\nu,K)} K_{3t}(\sum_{(\nu,K)} W_{3t})^{-1} \quad R_{3t} = \frac{1}{2}\sum_{(\nu,K)}(1+m)\Delta x_3^{-2}(\sum_{(\nu,K)} W_{3t})^{-1} \qquad (6.31)$$

and taking into account

$$^-u_{33} = u_3(t-1); \quad ^+u_{33} = u_3(t+1) \qquad (6.32)$$

yields u_{3t} to be calculated in analogy to eq.s.(6.24) to (6.27).

$$u_{3t} = K_{3t}^* + R_{3t}(u_3(t-1) + u_3(t+1)) \qquad (6.33)$$

6.4. Example of application.

For an axisymmetric cylindrical solid on a rigid support and axially loaded by a concentrated load P, two holograms are necessary only to determine the components u_r^Γ and u_3^Γ of the displacement vector along the generatrix of the surface of revolution (Fig.6.5).

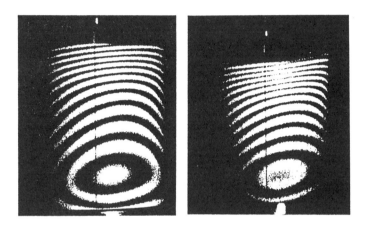

Fig.6.5: Interferograms H1, H2 of a cylinder under axial load.

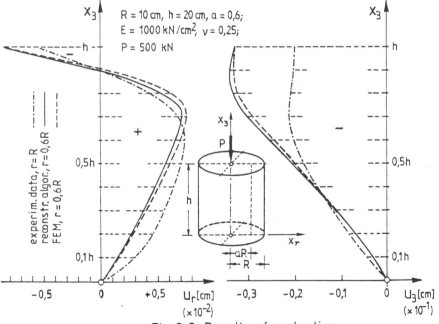

Fig.6.6: Results of evaluation.

The results of evaluation according to the preceding described reconstruction algorithm are shown in Fig.6.6. To prove the reliability and the accuracy of the method, the results of FEM-analysis are given also, demonstrating a sufficient correspondence.

7. ANALYSIS OF 3-D STRESS-STRAIN STATES BY *ESPI* AND *FEM*.

7.1. Determination of surface displacements by ESPI.

A method will be described, based on the principle of „hybrid technique", which enables analysing the stress-strain state of three-dimensional objects like shell-structures, pipe-branching etc. This method combines electronic speckle interferometry (ESPI) with the finite-element method (FEM) to evaluate the experimentally given data of the spatial displacements in discrete points on the object surface. These data are inserted as input-data into a respective FE-program to calculate the remaining nodal displacements of the elements of a three-dimensional FE-grid and finally the strain and stress state. It will be demonstrated, that it is possible to analyse only a section or a cut-out of the object under consideration, which e.g. is difficult to describe by analytical modelling, and then to insert the results into the merely numerical analysis of the object altogether.

The displacements of object surfaces can be determined by means of the field method of electronic speckle interferometry [7.1],[7.2]. The displacement vector $d_s(\iota,\kappa)$ in the pixel(ι,κ) on the surface in the direction of the sensitivity vector is related to the phase difference $\delta(\iota,\kappa)$:

$$\delta(\iota,\kappa) = \frac{2\pi}{\lambda}(\underline{e}_Q + \underline{e}_B)^T \cdot \underline{d}_s(\iota,\kappa) \tag{7.1}$$

which will be calculated from the light intensity in each pixel (ι,κ) recorded by a CCD-camera

$$I_r(\iota,\kappa) = I_0(\iota,\kappa) \cdot [1 + k(\iota,\kappa) \cdot \cos(\delta(\iota,\kappa) + \varphi_r)] \tag{7.2}$$

As the values of I_0, k and δ are unknown the phase-shift method will be applied, which means the phase of the impinging laser-light to be changed, e.g. in four steps $\varphi_r = \pi \, r/4$; $r = -3, -1, +1, +3$, thus yielding four interferograms. Then the phase difference caused by the surface displacements holds

$$\delta'(\iota,\varkappa) = \arctan \frac{I_{+3}(\iota,\varkappa) - I_{-1}(\iota,\varkappa)}{I_{-3}(\iota,\varkappa) - I_{+1}(\iota,\varkappa)} \mod \pi \qquad (7.3)$$

Because of the unsteadiness of the arctan-function rapid phase changes occur at π. Therefore the originally obtained "saw-tooth" images must be unwrapped by adding integer multiples of π to the calculated value δ'(ι,κ) in order to obtain continuous run of the phase difference δ(ι,κ).

$$\delta(\iota,\varkappa) = \delta'(\iota,\varkappa) + 2\pi \left[\sum_{k=2}^{\varkappa} C(\iota,k) + \tilde{N}(\iota,1) \right] \qquad (7.4)$$

$$C(\iota,k) = \begin{matrix} +1 \\ -1 \end{matrix} \left\{ \text{if } \delta'(\iota,k) - \delta'(\iota,k-1) \right\} \begin{matrix} > +\pi \\ < -\pi \end{matrix}$$

By low-pass filtering the high-frequently speckle noise will be smoothed and the low-frequency interference fringes will be amplified. The filter operation can be described [7.3] as the weighted sum H of the gray-value GW of each pixel. For the kernel-width of n×m image points the sum H holds

$$H = A \sum_{\iota=-n/2}^{+n/2} \sum_{\varkappa=-m/2}^{+m/2} a(\iota,\varkappa) \, GW(\iota,\varkappa) \qquad (7.5)$$

where A will be determined according eq. (7.6)

$$A^{-1} = \sum_{\iota=-n/2}^{+n/2} \sum_{\varkappa=-m/2}^{+m/2} a(\iota,\varkappa) \qquad (7.6)$$

The GAUSSIAN low-pass filter can be considered as a linear filter with point symmetry. Applying the GAUSSIAN function to a finite square grating an approach only of the ideal GAUSSIAN filter will be obtained. The GAUSSIAN normal distribution over the grid structure of the CCD-chip will be approximated by a binomial distribution and the coefficients a(ι,κ) in eq.s. (7.5) and (7.6) then can be calculated according to eq. (7.7)

$$a(\iota,\varkappa) = \frac{n!}{(n/2-\varkappa!)(n/2+\varkappa!)} \cdot \frac{m!}{(m/2-\iota!)(m/2+\iota!)} \qquad (7.7)$$

The solution of eq. (7.1) yields the displacement vector $d_s(\iota,\kappa)$ in the direction of the sensitivity vector, i.e. in the direction of the angle bisector of the angle between the direction of the impinging light ray and the direction of observation. To get the spatial displacement vector the phase difference and furthermore the displacement vector d_s must be determined in at least three different directions either of observation or of illumination. Then the spatial displacement vectors d_m are to determine in the nodal points m of the grid for FEM evaluation on the surface of the

Although the measured distribution of the light intensity has been subjected to an image filtering process already eliminating the effects of noise and of outliers of the recorded values the calculated components according to eq. (7.1), related to a global co-ordinate system, may still show slightly undulatory course, which is caused by the phase shifting process. Therefore an approximation algorithm based on the least-square method is applied for smoothing.

Eq. (7.1) yields the displacement yields the displacement vector $d_s(\iota,\kappa)$ in the direction of the sensitivity vector, i.e. the angle bisector of the angle between the direction of the impinging light ray and the direction of observation. To get the spatial displacement vector $d(\iota,\kappa)$ the phase difference must be taken in at least three different directions.

Although the measured distribution of the light intensity has been subjected to an image filtering process already eliminating the effects of noise and of outliers of the registered values the calculated components of d, related to a global co-ordinate system, yet may show a slightly undulatory course caused by the phase-shifting. Therefore, an approximation procedure based on the least-squares method is applied for smoothing, yielding finally the displacements d_m in nodal points of a FE-grid on the observed object surface [7.4] for further evaluation.

7.2. Evaluation of the experimental data by means of FEM.

According to the FE-displacement method the equilibrium condition generally holds [7.5]

$$[K]\cdot(v) = (f_q) \tag{7.8}$$

with the displacement vector (v) and the stiffness matrix [K] of the structure or the section respectively,

$$[K] = \sum_j [L_j^{(e)}]^T \cdot [K^{(e)}] \cdot [L_j^{(e)}] \tag{7.9}$$

where $[L_j^{(e)}]$ denotes the incident matrix and $[K^{(e)}]$ the stiffness-matrix of the element e. With the matrix of elasticity [E], the interpolation-matrix [G] (= matrix of form-functions) and [D] the differential-matrix the stiffness-matrix of the elements holds

$$[K^{(e)}] = \int_V [D_G]^T \cdot [E] \cdot [D_G] \cdot dV \qquad (7.10)$$

With reference to the element e the vector of the surface loads q is given by

$$(f_q^{(e)}) = \int_0 [G]^T \cdot (q) \cdot d0 \qquad (7.11)$$

and furthermore the vector of the surface load for the entire structure

$$(f_q) = \sum_j [L_j^{(e)}] \cdot (f_{qj}^{(e)}) \qquad (7.12)$$

Having determined the displacement vector (d_m) in the nodal points m of the FE-grid on the object surface by the ESPI-method (Note: The method can be applied only -in principle- to load-free surfaces), the displacement vector (v) will be separated into a sub-vector (d_m) and a sub-vector (v_i).

With reference to Fig.7.1, assuming e.g. one element over the thickness h of the object to be analysed, a set of linear equations will be obtained

$$\begin{bmatrix} [K_{mm}] & [K_{mi}] \\ \hline [K_{im}] & [K_{ii}] \end{bmatrix} \times \begin{Bmatrix} (d_m) \\ \hline (v_i) \end{Bmatrix} = \begin{Bmatrix} (0) \\ \hline (f_{qi}) \end{Bmatrix} \qquad (7.13)$$

the solution of which yields the as yet unknown values of the displacement vectors in the nodal points i, either

$$(v_i) = -[K_{mi}]^{-1} \cdot [K_{mm}] \cdot (d_m) \qquad (7.14a)$$

or alternatively

$$(v_i) = [K_{ii}]^{-1} \cdot \{(f_{qi}) - [K_{im}] \cdot (d_m)\} \qquad (7.14b)$$

if and only if i = m.
These equations enable the solution of the inverse problem, i.e. the calculation of the actual values of either the external load or the material response, e.g. the YOUNG's modulus (see chap.8.)

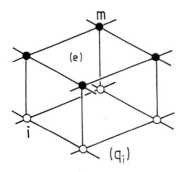

- • m nodal points in which the displacements (d_m) are measured;
- ○ i nodal points the displacements (v_i) of which are to determine

Fig.7.1: One element over thickness h.

i. control of loading:

$$(f_{qi}) = \{[K_{im}] - [K_{ii}] \cdot [K_{mi}]^{-1} \cdot [K_{mm}] \cdot (d_m)\} = [\tilde{K}_{im}] \cdot (d_m) \tag{7.15}$$

or if (qi) is known or also measured
ii. control of material response (YOUNG's modulus):

$$E \cdot [\tilde{K}^*_{im}] \cdot (d_m) = (f_{qi}) \rightarrow E \tag{7.16}$$

(for each i)

Taking into account two elements over the object thickness (Fig.7.2) the respective system of linear equations holds

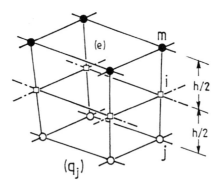

Fig.7.2: Two elements over thickness h.

$$\left[\begin{array}{c|c|c} [K_{mm}] & [K_{mi}] & [K_{mj}] \\ \hline [K_{im}] & [K_{ii}] & [K_{ij}] \\ \hline [K_{jm}] & [K_{ji}] & [K_{jj}] \end{array}\right] \times \left\{\begin{array}{c} (d_m) \\ \hline (v_i) \\ \hline (v_j) \end{array}\right\} = \left\{\begin{array}{c} (0) \\ \hline (0) \\ \hline (f_{qj}) \end{array}\right\} \quad (7.17)$$

The solution of these equations yields the displacement vectors (v_i) and (v_j) in the nodal points i and j, if i = j = m.

$$(v_j) = -[\tilde{K}_{ij}]^{-1} \cdot [\tilde{K}_{im}] \cdot (d_m)$$

$$(v_i) = -[K_{mi}]^{-1} \{[K_{mm}] - [K_{mj}] \cdot [\tilde{K}_{ij}]^{-1} \cdot [\tilde{K}_{im}]\} (d_m) \quad (7.18)$$

$$(v_j) = [\tilde{K}_{ji}]^{-1} \{(f_{qj}) - [\tilde{K}_{jm}] \cdot (d_m)\}$$

with the denotations

$$[\tilde{K}_{im}] = [K_{im}] - [K_{ii}] \cdot [K_{mi}]^{-1} \cdot [K_{mm}]$$

$$[\tilde{K}_{ij}] = [K_{ij}] - [K_{ii}] \cdot [K_{mi}]^{-1} \cdot [K_{mj}]$$

$$[\tilde{K}_{jm}] = [K_{jm}] - [K_{ij}] \cdot [K_{mi}]^{-1} \cdot [K_{mm}] \quad (7.19)$$

$$[\tilde{K}_{ji}] = [K_{ji}] - [K_{jj}] \cdot [K_{mi}]^{-1} \cdot [K_{mj}]$$

Having calculated the nodal displacements the components of the stress-tensor can be determined according to the respective FE-relations.

$$[\sigma] = [E] \cdot [D_G] \cdot (v) \quad (7.20)$$

As already mentioned before the actual internal and external parameters like the real loading conditions and the real material response can be proved.

7.3. Example of application.

As an example to explain the described hybrid method a cut-out of a cylindrical shell (pipe under internal pressure), (Fig. 7.3) will be analysed. For the calculations the FE-program ANSYS [7.6] has been used and especially the element-type "solid 45" which is described by the displacements of the eight corner nodes. To obtain the internal stress distribution over the thickness of the shell two elements over the thickness have been chosen.

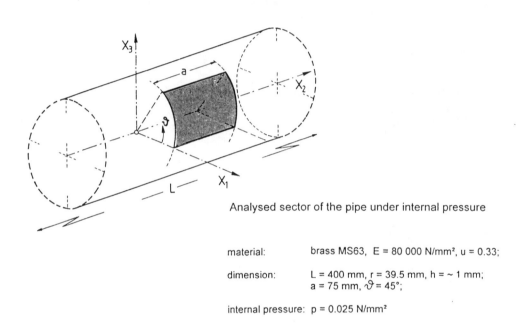

Analysed sector of the pipe under internal pressure

material: brass MS63, $E = 80\,000$ N/mm², $u = 0.33$;

dimension: $L = 400$ mm, $r = 39.5$ mm, $h = \sim 1$ mm; $a = 75$ mm, $\vartheta = 45°$;

internal pressure: $p = 0.025$ N/mm²

Fig.7.3: Position of the analysed sector of the pipe.

The distribution σ_9 on the outer surface (Fig.7.4) differs remarkably along a cross-section from the results of a theoretical analysis. However, it has been proved by separate holographic measurement of the whole pipe, that the difference is caused by the slightly variable thickness of the model, thus confirming the sensitivity of the described hybrid method.

The advantages of combining speckle-interferometry with FE-analysis are obvious

i) the displacement vectors (d_m) obtained by means of ESPI can be compared with those values (v_m) in the same nodal points m obtained by an overall FE-analysis of the whole structure;

ii) permanent or periodic measurement yields actual information on the structural response during utilisation, enabling the solution of inverse problems, i.e. the calculation of internal and external parameters on the basis of measured data;

iii) introducing the measured values (d_m) in the respective FE-formulation information on the actual strain- and stress-state under real loading and boundary conditions can be forwarded.

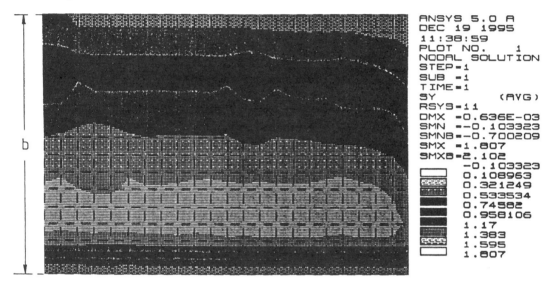

Fig.7.4: Stress-component σ_θ on the surface of the shell sector.

8. REMARKS GENERALLY TO CONSIDER IN PERFORMING EXPERIMENTAL MECHANICS.

8.1. Multidisciplinary character of experimental mechanics.

Modern optical methods in experimental mechanics demand dealing with various scientific and technical disciplines (Fig.8.1).

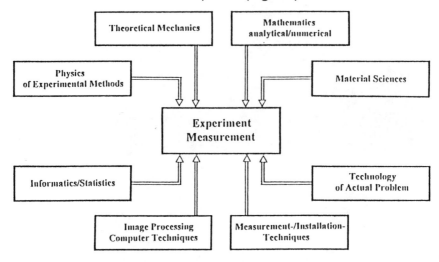

Fig.8.1.: Disciplines involved in experimental mechanics.

These methods are based on several different physical theories and laws. To apply the optical methods in measurement techniques it is necessary to understand and to master the relevant theories, as J.T. PINDERA has stated: *"There is no experiment without a theory behind it!"*

To transfer the observed and recorded optical phenomena,- mainly interference fringes, distribution of light intensities, amplitudes, phases and their changes,- into information on mechanical quantities deepened knowledge in theory and practice of image processing is unalienable. However the thus obtained "measured mechanical quantities", in principle deformations, displacements, relative displacements (strains) as well as vibration modes, ~frequencies, ~amplitudes, do not completely meet the finally wanted information on the response of the objects and structures to be analysed, as e.g. information on stresses, internal forces and internal structural parameters. Therefore the "measured" and "processed" data must run through a further phase of evaluation.

8.2. Necessity of advanced theories of mechanical problems.

Because of the high resolving power and the sensitivity of the modern optical methods proper mathematical algorithms transformed into discrete numerical procedures are required, in order to describe the mechanical problem under consideration on a higher level of theory in technical mechanics as this has been done very often as yet. This necessity will be demonstrated by two examples.

Considering a beam under pure bending with rectangular cross-section (Fig.8.2), the strains are generally calculated according to eq. (8.1).

$$\varepsilon_{11} = \frac{12M}{Ebh^3} x_2 ; \quad \varepsilon_{22} = \varepsilon_{33} = 0 \tag{8.1}$$

This may be an approach of sufficient accuracy in practice, neglecting the effect of lateral contraction on the observed optical phenomena. However because of the sensitivity of the optical methods the measured quantities are remarkably influenced by the lateral contraction. Fig.8.2b shows the actual deformation of the cross-section in the plane (x_2, x_3).

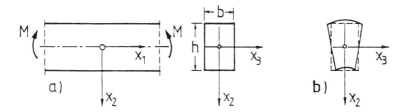

Fig.8.2.: Beam under pure bending, rectangular cross-section.

Based on a more precise theory the actual strains are given by eq. (8.2).

$$\varepsilon_{11} = \frac{12M}{Ebh^3} x_2 ; \quad \varepsilon_{22} = \varepsilon_{33} = -\mu \frac{12M}{Ebh^3} x_2 \tag{8.2}$$

The lateral displacement of the lateral surface in x_3-direction holds

$$u_3 = -\mu \frac{b}{2} \frac{12M}{Ebh^3} x_2 \tag{8.3}$$

In photoelasticity [8.1] the length of the optical light path will be changed as well as its course through the model because of the refraction at the in reality inclined lateral surfaces. Similar effects on the data measured by means of grid~ and interferometric methods must be taken into account.

A further argument to evaluate the data after image-processing on the basis of advanced theories and mathematical modelling respectively is given in chap.2. Even in the case of generally assumed small deflections of thin plates in bending in reality membrane stresses are superimposed always on the bending stresses according to the classical KIRCHHOFF-LOVE-plate theory, thus influencing the optical measurement.

8.3. Inverse problems in final evaluation.

In analysing structural problems usually effects such as deformations, displacements, strains, stresses, internal forces caused by functional, loading and perhaps environmental conditions are determined taking into account the basic information on the geometry and the response of materials, the latter one mainly given by material testing. If (w) denotes the effects, (u) the causes both are linked by an operational matrix [K], which includes all the basic information as mentioned above and the algorithms, which describe the mathematical/mechanical model of the problem to be analysed.

$$(w) = [K](u) \qquad (8.4)$$

Eq. (8.4) describes a "direct" problem, which always leads to well-defined solutions.

If however the causes are to re-calculate from measured effects a so-called inverse problem of the 1st kind exists because generally the operator matrix is neither regular nor a square matrix and therefore the inverse matrix $[K]^{-1}$ is not defined and an unequivocal solution does not exist [8.2].

Assuming the vector (w) to include s elements, the vector of the causes (u) however r elements, where r < s, the solution of such inverse problem is possible by means of the GAUSSIAN least square method minimising the quadratic functional

$$J = \sum_{i=1}^{s} (\sum_{j=1}^{r} K_{ij} u_j - w_i)^2 = \|[K](u) - (w)\|^2 = \min \qquad (8.5)$$

which leads to the unequivocal solution

$$(u) = ([K]^T \cdot [K])^{-1} \cdot [K]^T \cdot (w) \qquad (8.6)$$

If the matrix $[K]^T \times [K]$ exists, is regular, steady and positive definite then the inverse problem is well-posed.

As each measurement is full of systematic as well as of random errors (δw) the sensitivity of the solution according to eq. (8.6) against these errors is given to

$$(\delta u) = ([K]^T \cdot [K])^{-1} \cdot [K]^T \cdot (\delta w) \tag{8.7}$$

If it is the objective of the measurements to get information on the internal parameters, which are included in the elements of the operational matrix [K] and which generally are inaccessible to direct measurements, an inverse problem of the 2nd kind is on hand. To solve such problems unequivocally additional information is necessary. This can be basic physical facts like e.g. restraints, logical coherence, conditions of symmetry etc. and/or "á-priory experiences". For explanation of this statement chap.7, eqn.s (7.8) and (7.13), might be considered. The vector (f_{qi}) in eq. (7.8) denotes the vector of causes (u), vector (v) the vector of effects (w) and the question will be raised, whether it is possible to determine the YOUNG's modulus E as one of the internal parameters in the operator matrix [K]. As additional information can be considered: i = m, change of POISSON's ratio is neglegible small, the geometrical data and the external load conditions, i.e. vector (f_{qi}), are known and E should be equal in all nodal points of the FE-grid. Then the operator matrix can be written as

$$[K] = E [K^*] \tag{8.8}$$

Considering these conditions the second set of eqn.s (7.13) yields the value of the YOUNG's modulus in all nodal points i and according to the previous suppositions in the nodal points m too.

$$E = \frac{1}{i} \sum_{\iota=1}^{i} f_{q\iota} / D_\iota \tag{8.9}$$

with

$$D_\iota = \sum_{x=1}^{m} K^*_{ix} d_x + \sum_{\iota=1}^{i} K^*_{i\iota} v_\iota \tag{8.10}$$

where the elements v_ι are the solutions of the first set of eqn.s (7.13).

The shortly described simple example should demonstrate only principally how to proceed in solving inverse problems. It must be pointed out that the danger exists, that "very beautiful" results of data evaluation might be presented, which are rather speculative interpretations of measured data, and it might be impossible to recognise and to value subsequently, which information of which quality had been taken as basis of the evaluation and whether the presented results are actually truthworthy and unequivocal.

8.4. Transfer problems related to measuring systems.

The modern achievements in experimental mechanics, in measurement techniques, techniques of data acquisition, image and data processing as well as computer-oriented controlling and on-line evaluation have led to very complex and comprehensive measuring systems. It is of substantial importance not to look at the whole configuration of this measuring systems as a "black box". To assure reliability of the finally obtained information the relations between the original input signals and the output signals must be known, i.e. the flow of energy in the system must be traced very carefully [8.3]. The real transfer-functions, the impedances, the different signal-noise ratios as well as the extension of the measuring range of the single instruments in the complex configuration and their relation to one another are to regard thoroughly. This is necessary to avoid uncontrolled changes or distortions of the original input signals (Fig. 8.3 and Fig. 8.4).

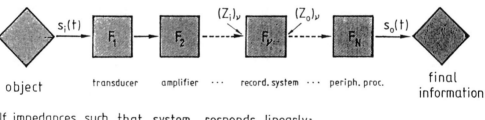

If impedances such that system responds linearly:

$$\frac{(Z_o)_{\nu-1}}{(Z_i)_\nu} \ll 1 \Rightarrow \text{system transfer function: } F_a = F_1 F_2 \cdots F_\nu \cdots F_N$$

Fig.8.3.: Operational transfer function

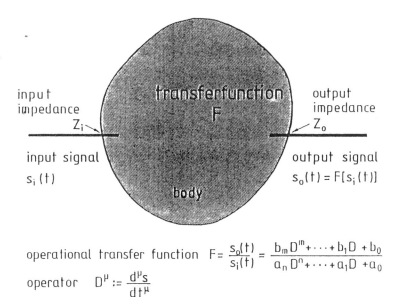

operational transfer function $F = \dfrac{s_o(t)}{s_i(t)} = \dfrac{b_m D^m + \cdots + b_1 D + b_0}{a_n D^n + \cdots + a_1 D + a_0}$

operator $D^\mu := \dfrac{d^\mu s}{dt^\mu}$

Fig.8.4.: System transfer function

It must be pointed out repeatedly, that it is of outmost importance to look at the whole analysing process as a cybernetic process. The results of any mechanical analysis are to feed back and to relate always to the real event considered, no matter whether these results are taken by mathematical/numerical, experimental/numerical or hybrid procedures.

9. REFERENCES

Chapter 2:

2.1. Laermann, K.H.: Über die Bestimmung des vollständigen Spannungszustandes in Platten mit großer Durchbiegung, VDI-Berichte Nr. 399, VDI-Verlag, Düsseldorf 1981, 45-49.
2.2. Aben,H.: Integrated Photoelasticity, McGraw-Hill, New York 1979.

Chapter 3:

3.1. Wolf,H.: Spannungsoptik, Bd.1, 2.Auflage, Springer-Verlag, Berlin/Heidelberg/New York 1976.
3.2. Kauderer,H.: Nichtlineare Mechanik, Springer-Verlag, Berlin/Göttingen/Heidelberg 1958.
3.3. Neumann, F.E.: Die Gesetze der Doppelbrechung des Lichtes in comprimierten oder ungleichförmig erwärmten unkristallinen Körpern, Abh. Königl. Akademie der Wissenschaften zu Berlin (1841).
3.4. Coker,E.G. and L.N.G. Filon: A Treatise on Photoelasticity, 2nd.ed. (Ed.H.T. Jessop), Cambridge Univ. Press 1957.
3.5. Mindlin,R.D.: A Mathematical Theory of Photo-viscoelasticity, J.Appl. Physics, Vol. 20,(1949).
3.6. Laermann, K.-H.: Ein experimentell-rechnerisches Verfahren zur Analyse zweidimensionaler Spannungszustände bei nichtlinear-elastischem Stoffverhalten, Berichte ikm, 5, Weimar 1990,37-41.
3.7. Engeln-Müllges,G. and F. Reutter: Formelsammlung zur Numerischen Mathematik mit Standard FORTRAN 77-Programmen, 6. Auflage, Wissenschaftsverlag, Mannheim/Wien/Zürich 1988.
3.8. Dowell M. and P.Jarratt: The "Pegasus"-Method for Computing the Root of an Equation, BIT 11 (1971).

Chapter 4:

4.1. Pindera,J.T.: Remarks on Properties of Photo-viscoelastic Model Materials, Exp. Mechanics, 6 (1966).
4.2. Theocaris,P.S.: A Review of the Rheo-optical Properties of Linear High Polymers, Exp. Mechanics, 5 (1065).

4.3. Dill, E.H.: Photoviscoelasticity, in: Proc.4th Symp. On Naval Structural Mechanics, Mechanics and Chemistry of Solid Propellants, Pergamon Press, 1965.
4.4. Dill, E.H.: Photoviscoelasticity, in: The Photoelastic Effect and its Application (Ed. J.Kestens), Springer-Verlag, Berlin/Göttingen/ Heidelberg 1975.
4.5. Coleman, B.D. and E.H. Dill: Photoviscoelasticity: Theory and Practice, in: The Photoelastic Effect and its Application (see ref.4.4.)
4.6. Dill, E.H. and C.Fowlkes: Photoviscoelastic Experiments, The Trend in Engineering, 7 (1964)
4.7. Coleman, B.D. and E.H. Dill: Theory of Induced Birefringence in Materials with Memory, J.Mech.Phys.Solids, Vol.19,(1971).
4.8. Flügge,W.: Viscoelasticity, Springer-Verlag, Berlin/ Göttingen/ Heidelberg 1975.
4.9. Engeln-Müllges,G. and F. Reutter: Formelsammlung zur Numerischen Mathematik mit Standard FORTRAN 77-Programmen, 6.Auflage, Wissenschaftsverlag, Mannheim/Wien/Zürich 1988.

Chapter 5.

5.1. Smirnow, W.I.: Lehrgang der Höheren Mathematik, Teil IV, VEB Deutscher Verlag d. Wissenschaften, Berlin 1975.
5.2. Hartmann, F.: Methode der Randelemente, Springer-Verlag, Berlin/ Heidelberg/New York/London/Paris/Tokyo 1987.
5.3. Laermann, K.-H.: On a Hybrid Method to Analyse Viscoelastic Problems,in: Advances in Continuum Mechanics (Ed. Brüller/ Mannl/ Najar),Springer-Verlag, Berlin/Heidelberg/New York 1991, 455-465.
5.4. Lee, E.H. and T.G. Rogers: Viscoelastic Stress Analysis using measured Creep or Relaxation Functions, J.Appl.Mech.,30 (1963).
(for further references see chapter 4.)

Chapter 6.

6.1. Wernicke, G. and W.Osten: Holographische Interferometrie, VEB-Fachbuchverlag, Leipzig 1982.
6.2. Laermann, K.-H.: On the Evaluation of Holographic Interferograms to Determine the Interior Strain State in 3-D Solids, Proc. CSME Mech. Eng. Forum 1990, Vol.II, Toronto 1990.

6.3. Leipholz, H.: Einführung in die Elastizitätslehre, Wissenschaft und Technik, Verlag G. Braun, Karlsruhe 1968.
6.4. Smirnow, W.I.: Lehrgang der Höheren Mathematik, Teil II, 8.Auflage, VEB Deutscher Verlag der Wissenschaften, Berlin 1968.

Chapter 7.

7.1. Jones, R. and C. Wykes: Holographic and Speckle Interferometry, Cambridge University Press,1989.
7.2. Ettemeyer, A.: Ein neues holografisches Verfahren zur Verformungs- und Dehnungsbestimmung, Diss. Universität Stuttgart (1988).
7.3. Zamperoni,P.: Methoden der digitalen Bildsignalverarbeitung,Viehweg & Sohn, Braunschweig 1989
7.4. Laermann, K.-H. und H.G. Monschau: Hybrid Analysis of Shell Structures by means of Electronic Speckle Interferometry combined with FEM, Acta Mechanica Slovaca, 1/1998, 3-8.
7.5. Bathe, K.-J.: Finite-Elemente-Methoden, Springer-Verlag, Berlin/Heidelberg/New York 1986.
7.6. ANSYS Users Manual, Vol.3 and 4, Swanson Analysis System Inc.

Chapter 8.

8.1. Laermann, K.-H.: Reflections on the Accuracy of Photoelastic Stress Analysis, ÖIAZ, 142.Jg.,Hft. 5/1997, 396-401.
8.2. Moritz, H.:General Considerations Regarding Inverse and Related Problems, in G. Anger et al. (eds.): Inverse Problems: Principles and Applications in Geophysics, Technology and Medicine, Akademie-Verlag, Berlin 1993, 11-23.
8.3. Pindera, J.T.: New Physical Trends in Experimental Mechanics,in CISM Courses and Lectures No.264, Springer-Verlag Wien-New York, 1981,203-327.

CHAPTER II

RECENT DEVELOPMENTS IN 3-D-PHOTOELASTICITY AND GRATING STRAIN MEASUREMENT

A. Lagarde
University of Poitiers, Poitiers, France

ABSTRACT

In photelasticity a new implementation with CCD camera permits to obtain quickly separated isoclinic and isochromatic patterns and to plot isostatics for one plane model optically sliced.

For small a large strains, using two orthogonal gratings marked upon a plane measure base, we determine, in its plane, the rotation of the rigid solid and the (algebric values of) principal extensions. For that, an optical device works by optical diffractions without contact and at a distance. Methods to improve accuracy are noticed. A set·up permits to extend the method for locally cylindrical surface. The holographic record permits the extension of the measurement to the whole of a plane surface. For dynamic event, the grating interrogation by beam laser with angular coding gives not only strains for local measure base but also rigid motions.

1. THREE-DIMENSIONAL PHOTOELASTICITY

1.1 Present practice and new possible way

Photoelasticity still gives subjects to searchers. It is the case for the study of the residual stresses in glass specially by means of integrated photoelasticity. So it is the case with the use of the isodynes in vue of taking into account the three dimensional local effects.
These fields are particular ones. We deal with the study of elasto-static problems about pieces having complex geometry. In this way, with the help of the well known stress frozen technic, photoelasticity is still often used in the test and research laboratories of motor vehicle and aeronautic industries.
This situation proves the efficiency of the photoelastic study ; this efficiency is due to the fact that the model being worked out with the loading elements, then the real boundary conditions are taken into account. In the last ten years, photoelasticity and numerical simulation have been used sometines in parallel.
The process consists in first adjusting the boundary conditions in order to get the obtained numerical values identify to the experimental values, the mechanical parameters being those corresponding to the frozen temperature. Then, the stress distribution inside the prototype is obtained with its own mechanical parameters. So, we become free from difficulty due to the quasi incompressibility of the model material at the frozen temperature.
Let us notice a recent progress : the original, necessary to make the mould, can be realised from C.A.O. data by means of the stereolithography. Now, it is even possible to directly realise the model with frozenable resin.
The exploitation of the frozen model is executed by mechanical slicing into plane sheets with a thickness from 2 to 4 mm. Each sheet polished roughly and analysed in linear and circular polariscope like in plane photoelasticity. Let us notice that slicing and analysing of one serie of parallel sheets require one qualify personal working during one week. The time of work is reduce with the use of immersion tank because it is not necessary to polish roughly the slices. The aim is to determine the parameters of interest for Ingeneers : the difference $\sigma'-\sigma''$ of the secondary principal stresses and the angle α giving their directions versus $\bar{x}$ reference axis. These quantities obtained for three series of mutually orthogonal planes make possible, using well known relations, to integrate the equilibrium equations with known boundary conditions. Then, the most general study, scarcely done, requires the making and the slicing of three models. Pratically, most often, we restrict ourself in determining only the above quantities in principal planes in order to optimize the shape of the model.

Over fifteen years ago, we have developed and used two optical slicing methods of the model : one point wise [1.1 to 1.5], the other whole field [1.4 to 1.7]. In a three-dimensional medium, these two methods give a non destructive way of investigation, based on the scattering light phenomenon. This phenomenon is intensified by introducing fine particles of silice in the epoxy resine.

The point wise method of optical slicing offers the possibility to use the light scattering phenomenon as polarizer or as analyser. This method with a linear detection of parameters is automatized in order to facilitate the use ; it permits a precise determination, in every point of the sheet, of the three optical parameters leading in general to the determination of α and $\sigma'-\sigma''$ (see § 1.2.2.). Let us notice that a number of methods have been developed to determine the three optical parameters, more particulary for the thick medium located between the sheet and the model boundary ; it is analysed in the book of Srinath [1.8]. Let us mention the iteractive process and the ones using polarizer and analyser rotations and compensator [1.9].

For example, we have determined the stress tensor along a line in a prismatic bar, under torsion, the cross section of which being an equilateral triangle. The values of the stresses are normalised by the maximum value of the shear of the cross section. So, in the base of a turbine blade, we could determine the values of the equivalent stress in the Von Mises sense at various points for a tensile load [1.3 to 1.5].

The whole field optical slicing method is based on limitation of a sheet by two parallel beams emitted from a laser.

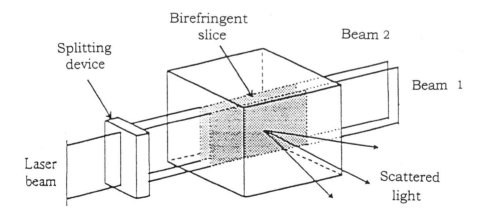

Fig. 1.1 - Isolation of a slice with two plane laser beams

Then, the model is analysed slice by slice. The scattered radiations interfere (Rayleigh's law) on the image plane of the middle of the sheet. They take into account the different polarizations that depend of optical caracteristics of the sheet. Two methods have been developped (see § 3.4.). The aim of this part is to demonstrate that it is possible to obtain the isoclinic and isochromatic fringes and to plot the isostatic patterns. It is the new possible way.

Before presenting the method principles, we shall give the actual conception of the modelisation of the light waves propagation in three dimensional medium having low anisotropy.

1.2 Propagation of the light wave through photoelastic medium

To define the light propagating direction in a photoelastic medium a basic hypothesis consist in assuming the medium to be isotropic (indeed current photoelastic materials are slightly anisotropic). It follows that for a ray light propagating along direction, the wave planes (x, y) are orthogonal to .

It can be shown that the secondary principle directions of the indice tensor and that of the stress tensor coincide and that we have the following relationship.

$$n' - n_0 = c_1 \sigma' + c_2 (\sigma'' + \sigma_z)$$
$$n'' - n_0 = c_1 \sigma'' + c_2 (\sigma' + \sigma_z) \quad (1.1)$$

where n' and n'' denote the secondary principal indices in the wave-plane (x,y) and σ', σ'' are corresponding secondary principal stresses ; c_1, c_2 are constants for a photoelastic material.

1.2.1. The classical scheme

In three-dimensional photoelasticity it is usually assumed that the directions of secondary principal stresses and their values are constant through the thickness dz of a slice having its parallel face normal to $\bar{z}$. This assumption allows to consider this slice as a birefringent plate characterized with the two following parameters:

· secondary principal angle α · angular birefringence $\phi = \dfrac{2\pi\delta}{\lambda}$,

$$\delta = dz (n' - n'') = C (\sigma' - \sigma'') dz \quad C = c_1 - c_2 \quad (1.2)$$

C being a photoelastic constant.

1.2.2 Aben schematisation

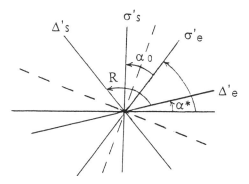

Fig. 1.2 · Orientation of the characteristic Δ' directions and the σ' directions at the entrance and the emergence

Aben in 1966 showed that when rotation of secondary principal axes was present, there were always two pairs of perpendicular conjugate «characteristic directions» (Fig.1.2). He distinguished the primary characteristic directions at the entrance of light (Δ'_e, Δ''_e), and the secondary characteristic directions (Δ'_s, Δ''_s) for the light emerging from the medium. The light linearly polarized at the entrance along one of the primary directions emerges as linearly polarized along the conjugate secondary direction. We will denote by R the angle determinated by two such directions, $R = (\Delta'_e, \Delta'_s)$ and by α^* the angle (x, Δ'_e).

The characteristic directions are generally different from the secondary principal directions of the stress tensor (or those of index tensor). At the entrance we have (σ'_e, σ''_e) and at emergence we have (σ'_s, σ''_s) from the medium. We denote $\alpha_0 = (\sigma'_e, \sigma'_s)$.

1.2.3 Hypothesis for a thin slice

For the case where dx/dz and σ'–σ'' are constant through a thickness, important conclusions follow [1.10 – 1.11]:
· The bisecting lines for the angles formed by two associated «characteristic directions» coincide with those for the angles formed by the associated secondary principal directions at the entrance and at the exit, (Fig. 1.).

Remark : The bisectors mentioned correspond to the secondary principal directions (mechanical or optical) at middle thickness ; so their directions are defined by the angles ± R/2 from the characteristic directions :
· The phase difference ϕ^* characteristic to the medium traversed by the light wave along two characteristic orthogonal directions is generally different from the angular birefringence which would result in the absence of rotation R. The values R, α^*, ϕ^* and α, α_0, ϕ are linked [1.3 – 1.7].

About the torsion strain, in a bar of equilateral triangular section, let us recall that the Aben schematisation gives good concordance between calculated and experimental results for slices inclined at an angle of $\pi/4$ from the axis of the model with the thickness of the slices being only 2 mm for classical size of model [1.3].

1.2.4 Discrete analysis into thin slices

This technique approaches the thick medium, in the direction $\vec{z}$ of ligth propagation, by n plane thin slices perpendicular to $\vec{z}$ [1.11 – 1.12]. This approach gives more realistic images in comparison with experimental images : in this case, we have variation of the stresses difference along the thickness ; the Aben's hypothesis does not respect this condition. One poster is devoted to this discrete analysis and shows this interest, in particular simplicity and good connexion with finite element method.

About the behaviour of slightly anisotropic medium, with a large thickness, crossed by light, the Poincaré's theorem permits us the representation by birefringent plate followed by a rotatory power (or by the inverse position, with the axis rotated of the value of the rotatory power)

1.3 Whole-field analysis with a plane polariscope

Here, the analysis with a light-field polariscope is presented as it corresponds to the whole-field method of optical slicing. One can conduct an analogous study for a dark polariscope.

Let us examine a slice (which should be obtained by freezing and slicing) in a plane light-field (rectilinear) polariscope. In each point the slice is represented by a birefringent plate and a rotatory power. Let I_0 designate uniform light-field illumination and x the polarizing axis of the polarizer. Then the light intensity is :

$$I = I_0[\cos^2 R - \sin 2\alpha * \sin 2(\alpha^* + R) \sin^2 \phi^*/2] \qquad (1.3.)$$

The extremum values for intensity distribution correspond to :

$$\alpha^* = R/2 + k\pi/4, k = 0,1,2,... \qquad (1.4.)$$

In order to specify the condition of analysis of fringe patterns we plotted the variations of Imax and Imin. Versus ϕ for different values of α_0 obtained following the relationships given in sect. 3.2.3. As example, curves were plotted for $\alpha_0 = \pi/9$ in Fig.1.3.

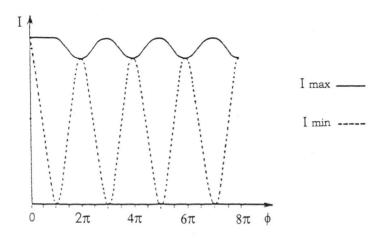

Fig. 1.3. - Variation of Imax and Imin as a function of ϕ ($\alpha_0 < \pi/9$)

The foregoing analysis indicates that for small values of α_0 ($\alpha_0 < \pi/9$) the maximum intensity Imax shows a reduced modulation. Thus, it characterizes an isoclinic zone which permits one to locate the secondary principal stress directions (or those of the indices) in the median plane of a slice. This zone corresponds to $\alpha^* = R/2 \bmod \pi/2$ (remark sect. 1.2.3.). The orientation of the polarizer then coincides with one of the secondary principal directions in the median plane. This interesting result is analogous to the one established by Hickson [1.14] for a dark-field polariscope.

It should be emphasized, that in order to avoid errors during the numerical integration procedure, the discritization points should lie on the median plane of a slices.

If α_0 increases, the Imax modulation increases and it becomes very pronounced for $\alpha_0 = \pi/3$. In this case the isoclinic zone disappears although it should be noticed that the isoclinics are discernible till α_0 value of $\pi/6$.

The term Imin which is strongly modulated for α_0 close to $\pi/6$, characterizes the isochromatic pattern. The extremum values occur for $\phi = k\pi$ ($k = 1,2...$) and it follows that localization of fringes is practically independent of the rotation of secondary principal axes.

We can now conclude by noting result : investigation of a slice within the plane (rectilinear) polariscope allows one to determinate the secondary principal stress directions in the median plane (without resorting to rotatory power measurements) and the angular birefringence ϕ for the multiple π · values when the rotation of the secondary principal axes is less than $\pi/6$.

We should point out that the condition on the orientation of the secondary principal axis is not very limiting since one is able to choose the slice-thickness for the non-destructive optical slicing method.

1.4 Whole-field optical slicing method

We have developped two methods using the limitation of the sheet by two parallel plane beams emitted from a laser. The scattered radiations interfere (Rayleigh's law) in the image plane of the middle of the sheet. The informations concerning isoclinic and isochromatic patterns of the sheet are obtained from the square of the correlation factor γ of the two speckle fields.

We show that the illumination of the speckle field is the following basement relation

$$I(x,y) = I_1(x,y) + I_2(x,y) + 2\sqrt{I_1 I_2}\, \gamma \cos(\psi_2 - \psi_1 - \eta) \qquad (1.5)$$

with
$$\gamma^2 = \cos^2 R - \sin 2\alpha^* \sin 2(\alpha^* + R) \sin \phi^*/2$$
ψ_1, ψ_2 random variables, η function of α^*, ϕ^*, R.

The speckle fields interfere in amplitude for $\gamma = 1$; they add in energy for $\gamma = 0$ (it is said that they interfere in energy).

1.4.1 Method based on the contrast measurement of one recording intensity field

This method, that uses only one recording with holographic film, is purely optical of one recording interesting field.

We note < > the average spatial.

The static study of the speckle field gives, from basement relation

$$\sigma^2 = \langle [I - \langle I \rangle]^2 \rangle = \langle I_1 \rangle^2 + \langle I_2 \rangle^2 + 2\gamma^2 \langle I_1 \rangle \langle I_2 \rangle \tag{1.6}$$

As the contrast $\rho_i = \dfrac{\sigma_i}{\langle I_i \rangle}$ is unit for the two speckle fields, we have for the variance

$$\sigma^2 = \sigma_1^2 + \sigma_2^2 + 2\gamma^2 \langle I_1 \rangle \langle I_2 \rangle \tag{1.7}$$

We suppose $\langle I_1 \rangle = \langle I_2 \rangle$ then the square contrast ρ^2 of the recording speckle is $\rho^2 = \dfrac{1+\gamma^2}{2}$. Therefore the maximum of contrast of the fringes is 1/3.

To increase the contrast of the fringes, we use a polychromatic laser (laser with variable wave length) [1.15]. In the regions of the image with $\gamma = 1$, the grain of the speckle are channeled. One pass band filtering gives theoritically a unit contrast. In practice, the noise of the film tempers somewhat these results. An example is given Fig. 1.4a for torsion strain in a bar of square section.

This method was successfully used in the context of linear fracture mechanics to determine the caracteristic parameters K_1 and σ_{on} for a semi-elliptical surface crack loaded in opening made in a bar in tension [1.16].

1.4.2 Method based on the variance measurement of the combination of three recording intensity fields [1.17 – 1.18]

The idea to use a CCD camera instead holographic film was motived by the consideration of the density of isochromatic patterns. It is possible to use more big size of speckle grain. Then CCD is available.

To take into account, one background intensity due to the fluorescence phenomenon of the material and a part of the scattered light not polarized, we add I_{1B}, I_{2B} to the values I_{1S} and I_{2S} corresponding to Rayleigh laws

$$I_1 = I_{1B} + I_{1S} \quad I_2 = I_{2B} + I_{2S} \tag{1.8}$$

and we suppose

$$k = \frac{<I_{1S}>}{<I_1>} = \frac{<I_{2S}>}{<I_2>} \qquad (1.9)$$

In these conditions, the basement relation becomes

$$I = I_1 + I_2 + 2\sqrt{I_{1S}}\sqrt{I_{2S}}\,\gamma\cos(\psi_1 + \psi_2 + \eta) \qquad (1.10)$$

and the variance σ^2 of $I - I_1 - I_2$ is

$$\sigma^2 = 2k^2\gamma^2 <I_1><I_2> \qquad (1.11)$$

The recording of the three fields I, I_1, I_2 and the determination of the variance give a value proportional to γ^2. An example is given Fig. 1.4 b always for torsion strain in a bar of square section.

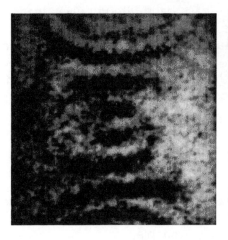 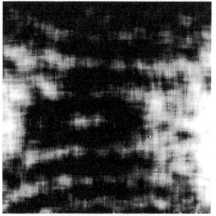

(a)
Previous method with
holographic film and optical filtering

(b)
New method with C.C.D.
camera and image processing

Fig. 1.4 · For torsion strain in a bar of square section and a plane slice inclined at an angle of $\pi/4$ from the axis, tickness 8 mm : comparison of the two methods.

1.5 Separation of isoclinic and isochromatic patterns of the slice. Isostatic plotting

We recall that the properties of polarization of the scattered ligth (Rayleigh's law) permits to realize, with two plane parallel laser beams, the optical slicing giving the analysis of the slice in ligth-field polariscope.

In the plane of the slice, we change the orientation of the beams and we record several images of the field (for example sixteen for a variation of $\pi/2$).

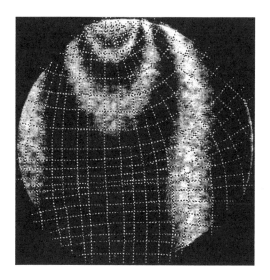

Fig. 1.5 - Isostatics and isochromatics patterns for one 4 mm thicknesses slice

Then, we have suffisant informations caracterising a periodic phenomenon, and it is possible to calculate in each pixel the Fourier transform of the correlation factor. The filtering in Fourier plane of the zero order gives the isochromatics and the same operation for the 1 order gives the isoclinic patterns [1.21].

When the isoclinic patterns are determined, the isostatics can be plotted to obtain the visualization of the orientations of the secondary principal stress directions. Fig. 1.5 shows the results for optical slice in model loaded by concentred force.

This process is important for the optimization of the design of the shape in mechanical construction.

1.6 Conclusion and perspectives

Our works, for optical slicing method in three-dimensional photoelasticity with the CCD camera and the means of image processing, show, in statics, that it is easy and rapid to determine isochromatics and isostatics of a model slice. In next future, it should be good to be able to determine the stress tensor.

1.7 References

1.1. Brillaud J., Lagarde A., "Ellipsometry in scatttered light and its application to the determination of optical characteristics of a thin slice in tridimensional photoelasticity". Symposium IUTAM "Optical Methods in Mechanics of Solids", Poitiers, septembre 1979, (Sijthoff Noordoff).

1.2. Brillaud J., Lagarde A., "Méthode ponctuelle de photoélasticimétrie tridimensionnelle". R.F.M. n° 84, 1982.

1.3. Brillaud J., "Mesures des paramètres caractéristiques en milieu photoélastique tridimensionnel. Réalisation d'un photoélasticimétrie automatique. Applications". Thèse de Doctorat d'Etat, Poitiers, 1984.

1.4. Lagarde A., "Non-obstructive Three Dimensional Photoelasticity-Finite strains Applications Photoelasticity". M. Nisida, K Kawata Springer Verlag 1986. "Photo-elasticity Proceeding" of the Inter. Symposium on Photoelasticity, Tokyo, 1986.

1.5. Lagarde A., Editor "Static and Dynamic Photoelasticity and Caustics Recent Developments", Springer Verlag, 1987.

1.6. Desailly R, Lagarde A., "Sur une méthode de photoélasticimétrie tridimensionnelle à champ complet". Journal de Mécanique Appliquée". Vol. 4, n° 1, 1980.

1.7. Desailly R., "Méthode non-destructive de découpage optique en photoélasticimétrie tridimensionnelle · Application à la mécanique de la rupture". Thèse d'Etat n°336, Poitiers, 1982.

1.8. Srinath L.S., Torta Mc Graw-Hill, "Scattered Light Photoelasticity". Publishing Company.

1.9. Srinath L.S., Keshavan S.Y., "A simple method to determine the complete photoelastic parameters using scattered light", Mechanics Research Communications, 5(2), 1978, pp. 85-90.

1.10. Aben H., "Optical phenomena in photoelastic models by the rotation of principal axes". Expr. Mechanics, Vol. 6, n° 1, 1966.

1.11. Aben H., Josepson J., "Strange interference blots in the interferometry in inhomogeneous birefringent objects", Applied Optics, vol. 36, n°28, pp. 7172-7179, Oct. 1997.

1.12. Zénina A., Dupré J.C., Lagarde A., " Découpage optique d'un milieu photoélastique épais pour l'étude des contraintes dans un milieu tridimensionnel", 13ème Congrès Français de Mécanique, Poitiers, France, Vol. 4, pp. 447-450, Sept. 1997.

1.13. Zenina A., Dupré J.C., Lagarde A., "Optical approches of a photoelastic medium for theoretical and experimental study of the stresses in a three-dimensional specimen". Symposium IUTAM "Advanced Optical Methods and Applications in Solids Mechanics" Poitiers August 31 – September 4 1998 Kluver academic publishers.

1.14. Hickson V.M., "Errors in stress determination at the free boundaries of "Frozen stress" photoelastic model". J. Appl. Phys., Vol. 3, n°6, p. 176-181, 1952.

1.15. Desailly R, Froehly C., "Whole field method in three dimensional photoelasticity : improvement in contrast fringes". Symposium IUTAM "Optical Methods in Mechanics of Solids", Poitiers, september 1979. Sijthoff Noordhoff.

1.16. Desailly R., Lagarde A., "Surface Crack Analysis by an Optical Slicing Method of Three Dimensional Photoelasticity" Experimental Stress Analysis Haïfa Israël, août 1982.

1.17. Dupré J.C., Plouzennec N., Lagarde A., "Nouvelle méthode de découpage optique à champ complet en photoélasticimétrie tridimensionnelle utilisant des moyens numériques d'acquisition et d'analyse des champs de granularité en lumière diffusée", C.R. Acad. Sci, Paris, t, 323, Série II b, pp. 239-245, 1996.

1.18. Dupré J.C, Lagarde A., "Photoelastic analysis of a three-dimensional specimen by optical slicing and digital image processing", Experimental Mechanics, Vol. 37, n°4, pp. 393-397, Dec. 1997.

1.19. Plouzennec N., Dupré J.C. and Lagarde A., "Visualisation of photoelastic fringes within three dimensional specimens using an optical slicing method". Symposium IUTAM "Advanced Optical Methods and Applications in Solids Mechanics" Poitiers August 31 – September 4 1998 Kluver academic publishers.

1.20. Zenina A., Dupré J.C., Lagarde A., "Plotting of isochromatic and isostatics patterns of slice optically isolated in a three dimensional photoelastic model". Symposium IUTAM "Advanced Optical Methods and Applications in Solids Mechanics" Poitiers August 31 – September 4 1998 Kluver academic publishers.

1.21. Morimoto Y., Morimoto Y. Jr and Hayashi T., "Separation of isochromatics and isoclinics using Fourier transform. Experimental Technique 17-5, pp. 13-16, 1994.

2. GRATING STRAIN MEASUREMENT

2.1 Introduction

The idea to realise grids at the surface of a specimen in order to accede to its strains is not new. This idea has been introduced by Rayleigh [2.1] in 1874 suggesting the use of moiré phenomenon which has been developed by Dantu [2.2] in 1957. This technique gives the field of displacements which has been also obtained using orthogonal grids.
The coherent radiation of the laser has permitted the development of the interferometric moiré [2.3] which offers a higher accuracy. Always, in view to increase the accuracy, let us notice the numerus works dealing with the interpolation between fringes and the multiplication of the fringes [2.4] and other technics [2.5]. The strains are obtained by the help of a derivation process.
The direct measurement of strains has been realised from the local observation of circle marked on the surface. Orthogonal grids have also been employed. With a smaller pitch the information has been extracted from the grating by optical diffraction either locally [2.6] or along a line using filtering [2.7]. The use of a optoelectronic device [2.8] has allowed an automatic calculation of strain distribution. With the recent development of CCD cameras and image processings, these methods are being revived [2.9 to 2.11].
Our purpose is, on an area which can be of very small dimension, the determination of algebric values of the principal extensions for large and small strains with a good accuracy and on a whole-field [2.12].

2.2 Local strain measurement

2.2.1. Recall : description of the move of a continuous medium

We suppose a plane strain state.
It is supposed that each material particle can be flowed during its displacement.
The configuration occupied by the material being in natural state, at t instant, is the reference configuration C_0, said <u>Lagrange's configuration</u> (Fig. 2.1).
The particle, being at M when $t=0$, comes to m at t instant. Then, the set of all particles constitutes the present configuration C(t) said <u>Euler's configuration</u>.

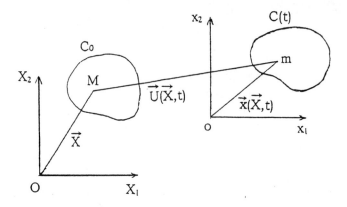

$\overrightarrow{OX_1}, \overrightarrow{OX_2}$, physical space reference.
ox_1, ox_2 intermediary reference

Fig. 2.1 · Référence spaces and configurations

The family of all the trajectories represents the mouvement of the continuous medium.

$$\vec{x} = \vec{x}(\vec{X}, t) \tag{2.1}$$

In our study, we shall consider the natural and loaded states.
Introduce the displacement vector

$$\vec{x} = \overrightarrow{oO} + \vec{X} + \vec{U}(\vec{X}, t) \tag{2.2}$$

with $\bar{\bar{F}} = \overline{\overline{\text{grad}}}\, \vec{x}$ defined by $F_{ij} = \dfrac{\partial x_i}{\partial X_j}$ we have

$$d\vec{x} = \bar{\bar{F}}\, d\vec{x} \tag{2.3}$$

It is the <u>tangent homogeneous transformation</u>.

2.2. Principle of the method

For example, let us consider a plane piece with constant thickness, which is loaded in his plan. We use tangent homogeneous transformation to study deformations of an orthogonal grating of pitch p which is engraved or marked on the plane surface of the piece. The square with $p\vec{X}_1, p\vec{X}_2$ sides, becomes one parallelogram with $p\vec{\bar{F}}\vec{X}_1, p\vec{\bar{F}}\vec{X}_2$ sides. The point M with X_1, X_2 coordonates, becomes one point m with x_1, x_2 coordonates (Fig. 2.2).

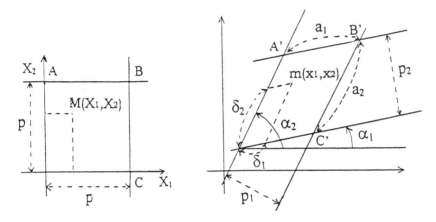

Fig. 2.2 – Deformation of a square

Because of the linearity of the transformation

$$\frac{\delta_1}{a_1} = \frac{X_1}{p} \qquad \frac{\delta_2}{a_2} = \frac{X_2}{p} \qquad (2.4)$$

Seing the figure, we have

$$x_1 = \frac{a_1}{p}\cos\alpha_1 X_1 + \frac{a_2}{p}\cos\alpha_2 Y_2$$

$$y_2 = \frac{a_1}{p}\sin\alpha_1 X_1 + \frac{a_2}{p}\sin\alpha_2 Y_2 \qquad (2.5)$$

and, more,

$$\sin(\alpha_2 - \alpha_1) = \frac{p_2}{a_2} = \frac{p_1}{a_1} \qquad (2.6)$$

Taking this relations into acount, the measurement of pitches p_1, p_2 and of orientations α_1, α_2 of the gratings make possible to determine the analytic transformation defining the strain (change from square to parallelogram) that is to say $\overline{\overline{F}}$

$$F = \begin{bmatrix} \frac{a_1}{p} \cos \alpha_1 & \frac{a_2}{p} \cos \alpha_2 \\ \frac{a_1}{p} \sin \alpha_1 & \frac{a_2}{p} \cos \alpha_2 \end{bmatrix} \qquad (2.7)$$

The measurement method consists on detemining the four parameters p_1, p_2 α_1, α_2 characterizing the gratings geometry. It is to notice that in the change from M to m, the translation is not taking into acount. Recall that the moiré method makes possible for us to determine the displacements field.
Knowing the analytic transformation which defines the strain, we are able to determine the well known tensors in mechanics. Let us indicate the Cauchy-Green's right tensor $\overline{\overline{C}} = {}^t\overline{\overline{F}}\,\overline{\overline{F}}$ (we note ${}^t\overline{\overline{F}}$ the $\overline{\overline{F}}$ transposed).

$$C = \begin{bmatrix} \left(\frac{a_1}{p}\right)^2 & \frac{a_1 a_2}{p^2} \cos(\alpha_2 - \alpha_1) \\ \frac{a_1 a_2}{p^2} \cos(\alpha_2 - \alpha_1) & \left(\frac{a_2}{p}\right)^2 \end{bmatrix} \qquad (2.8)$$

We should give also the Green-Lagrange tensor $\overline{\overline{E}} = \frac{1}{2}(\overline{\overline{C}} - \overline{\overline{1}})$.

Now we remind the polar decomposition $\overline{\overline{F}} = \overline{\overline{R}}\,\overline{\overline{U}} = \overline{\overline{U}}\,\overline{\overline{R}}$, orthogonal tensor $\overline{\overline{U}}$ pure strain tensor in Lagrange representation $\overline{\overline{U}}^2 = \overline{\overline{C}}$, $\overline{\overline{V}}$ pure strain tensor in eulerian representation $\overline{\overline{V}}^2 = \overline{\overline{F}}\,{}^t\overline{\overline{F}}$ we can express the rigid-body rotation of the solid R :

$$\tan R = \frac{a_2 \cos\alpha_2 - a_1 \sin\alpha_1}{a_2 \sin\alpha_2 + a_1 \cos\alpha_1} \tag{2.9}$$

We assume the strain to be uniform on the measurement base. Thus we obtain the orientation and the value of the principal extensions and the rigid body rotation from a knowledge of four parameters (two pitches p_1 and p_2 and two angles α_1 and α_2). These values are experimentally obtained either by optical Fourier Transform or by numerical Fourier Transform.

The first procedure gives the strain with an accuracy of 10^{-3} (see § 2.4.). For better performances we use the phase shifting technique (see § 3.1.1.). The second procedure requires an adapted interpolation process (see § 3.1.2.).

2.2.3 Grating realisation

We use [2.13] the interferences of two beams laser for the exposure of a photosensitive coating used directly for diffraction. The repeate of these interferences allows direct engraving at the surface of the material (epoxy, steel, ...) [2.14] with the help of a yag laser. We use also the Post replication technic to obtain phase grating. All technics give grating which are disturbed when large strains appear. We develop a technic to make small viscoelastic pavements. Other ways to realise grating with lines or points consist to use print, transfer, inking pad, mold, point or issued from the specimen structure clothes, sails. The choice depend of the size of the measurement base and the nature of the problem.

2.2.4 Measurement by optical diffraction

The diffraction phenomena of a parallel beam of coherent light passing in normal incidence through a plane grating is well known.

We have represented in Fig. 2.3 the diffraction image by a grating of parallel crossing lines. We notice that the direction formed by the diffraction points is perpendicular to the orientation of the family of corresponding lines. It is now easy to describe α_1, α_2 as functions of δ_1, δ_2.

$$\alpha_1 = \delta_1 + \frac{\pi}{2}, \quad \alpha_2 = \delta_2 - \frac{\pi}{2} \tag{2.10}$$

The figure 2.3 gives an idea of the decreasing of the spots intensities from zero order using blots with decreasing areas. In fact, on the screen, all the spots have the same diameter.

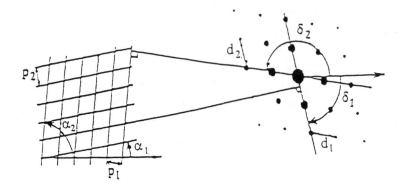

Fig. 2.3 – Diffraction image by a grating

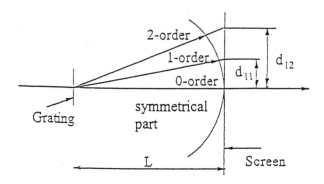

Fig. 2.4 – Diffraction order

Let us consider the grating of pitch p_1. On the screen, situated at distance L from the grating (Fig. 2.4), the blots of principal diffraction are located on right lines perpendicular to the line of this grating and containing 0 order center. For small diffraction angles, the blots are regularly spaced out, the m order blot being defined by $\beta_m = m\beta_1$ with

$$\beta_1 = \frac{d_1}{L} \qquad (2.11)$$

For small strains, we have

$$p_1 = \frac{\lambda L}{d_1} \quad \text{and likewise} \quad p_2 = \frac{\lambda L}{d_2} \tag{2.12}$$

For large strains or if we take into acount the hight order diffraction blots, we take

$$p_1 = \frac{m\lambda L}{\text{Arc tg}\frac{d_{1m}}{L}} \quad \text{and} \quad p_2 = \frac{m\lambda L}{\text{Arc tg}\frac{d_{2m}}{L}} \tag{2.13}$$

The diffraction image is recorded by a CCD camera, and the centroïd (x, y) of the spots computed from an intensity analysis.

The grating analysis using the optical diffraction allows measurements at distance and is very convenient for strain determination in hostile environment. For example, we give the evolution of longitudinal and transversal strains determined by this way on a specimen in epoxy resin and subjected to an uniaxial test at the frozen temperature. The measure base was 0,5 x 0,5 mm² and the line density 300 by millimeter.

From this test we can evaluate the strain accuracy to 10^{-3}.

Let us notice that the method has been adapted in dynamics [2.15] (see § 5) and also, in statics, to studies of cylindrical specimens [2.16] (see § 6).

2.2.5 Measurement by spectral analysis

The crossed grating is recorded by CCD camera and an algorithm of bidimensional DFT is used. The location of the 5 peaks of the spectrum (central order and 4 peaks order ± 1) gives searched parameters as for the optical analysis. See Fig. 2.5 following page.

2.3 Improvement of accuracy

2.3.1 The tools of the accuracy

2.3.1.1. The phase shifting method [2.17]

The problem is to measure the difference of phasis $\phi(x)$ between two coherent beams, expressed by a grating of interference fringes of intensity

$$I(\bar{x}) = a(\bar{x}) + b(\bar{x}) \cos \phi(\bar{x}) \tag{2.14}$$

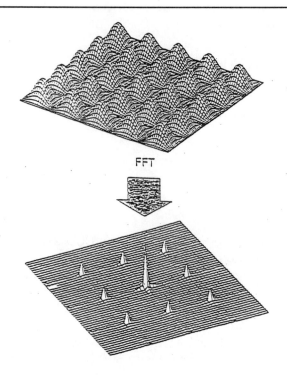

Fig. 2.5 – A sinusoidal cross grating and its two-dimensional FFT

It is possible with the use CCD camera to perform the detection in all the field for the acquisition, the processing and the analysis of image.
Let us consider the signal

$$I(x,y) = a(x,y) + b(x,y) \cos \phi(x,y) + B(x,y) \qquad (2.15)$$

a, b and ϕ are unknown and B (x, y) is a white noise.

The phase shifting method consist to introduce successively three known ϕ_i and measure the corresponding intensity :
We assume that variations of ϕ do not change the noise

$$I_i = a + b \cos(\phi + \phi_i) + B \qquad i = 1, 2, 3 \qquad (2.16)$$

So, we have a system of three equations with three unknowns which gives

$$\tan\phi = \frac{(I_3 - I_2)\cos\phi_1 + (I_1 - I_3)\cos\phi_2 + (I_2 - I_1)\cos\phi_3}{(I_3 - I_2)\sin\phi_1 + (I_1 - I_3)\sin\phi_2 + (I_2 - I_1)\sin\phi_3} \qquad (2.17)$$

$$b = \frac{I_1 - I_2}{\cos(\phi + \phi_1) - \cos(\phi + \phi_2)} \qquad (2.18)$$

$$a = I_1 - b\cos(\phi + \phi_1) - B \qquad (2.19)$$

We can remark that the expression of $\tan\phi$ is independant of the noise B (x, y). Of course a ($\bar{x}$) stay affected by the noise. But in practise, one takes an interest only on ϕ. For example, let us consider the unidimensional signals corresponding to $\phi_1 = 0, \phi_2 = 2\pi/3$ and $\phi_3 = -2\pi/3$ (Fig. 2.6).

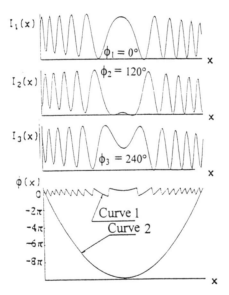

Fig. 2.6 – Signal for different phases and unwraping

The curve 1 Fig. 2.6 present the calculated arctangent included $-\pi/2$ and $\pi/2$. The curve 2 shows the graph obtained by unwrapping ϕ.

Let us consider the usual case it is possible to introduce the phase shifted ϕ_i constant in the whole field.

If in one part of the field we have periodic fringes it is possible to determine ϕ_i with FFT.

If in one part of the field we have periodic fringes it is possible to determine ϕ_i with FFT.

In fact, let be the signal

$$I_1 = a + b \cos(\frac{2\pi x}{p} + \psi_1) \tag{2.20}$$

We have

$$TF(I_1) = A_{01} \delta(N) + A_{11} \left[\delta(N - \frac{1}{p}) e^{j\psi_1} + \delta(N + \frac{1}{p}) e^{-j\psi_1} \right] \tag{2.21}$$

N being the spatial frequency.

In the neigbourhood of order 1 peak

$$\begin{aligned} Re_1 &= A_{11} \cos \psi_1 \\ Jm_1 &= A_{11} \sin \psi_1 \\ \psi_1 &= \text{Arc tg} \frac{Jm_1}{Re_1} \end{aligned} \tag{2.22}$$

Then it is no more necessary to adjust, with precision, the experimental device for introducing ϕ_i.

Notice that the inaccuracy on ϕ_i determination introduce a same order inaccuracy on the research phase ϕ_i.

In the case of a periodic signal the determination of ϕ gives the period p of the signal.

The phase measurement algorithms using three or four steps has been developped by Creath 1988 [2.18] and Wyant 1982 [2.19]. One particular four step technique has used by Carré 1966[2.20]. The advantage of this last one is that the phase shifted does not need to be calibrated. The various phase-measurement algorithms including comparison are discribed by Cloud [2.16].

3.1.2. The spectral interpolation method

The crossed grating is recorded by CCD camera. The location of the peaks of the spectrum gives searched parameters as for the optical analysis. We will

To illustrate the effect of discrete Fourier transform (DFT) we have computed the discrete Fourier transform of the cosine function illustrated in Fig. 2.7. Note that the thirty-two samples define exactly four periods of the periodic waveform. In Fig. 2.7, we also plotted the magnitude of the discrete Fourier Transform of these samples as computed by a fast Fourier transform (FFT) algorithm. The results are zero except at the desired frequency located in position 4 and 28. The frequency f_0 is given by the location of the peak divided by the number of samples.

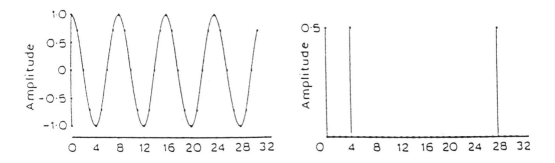

Fig. 2.7 – Fourier transform of a cosinus signal truncation interval equal to a multiple of the period

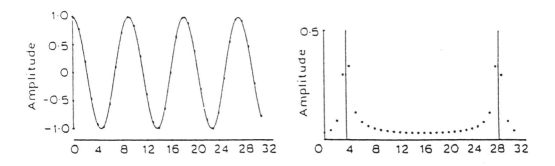

Fig. 2.8 – Fourier transform of a cosinus signal truncation interval not equal to a multiple of the period

For example :

$$f_0 = 4/32 = 1/8$$

If the truncation interval is chosen to a multiple of the period which is the practical case, the side-lobe characteristics of the (sin f)/f frequency function results in a considerable difference in discrete and continuous Fourier transform results. To illustrate this effect, consider the cosine waveform in Fig. 2.8. Note that the thirty-two points are not a multiple of the period and as a result a sharp discontinuity has been introduced. In Fig. 2.8 we also show the magnitude of the DFT of the samples. There exist non-zero frequency components at all frequencies of the discrete transform. As stated previously, the additional frequencies components are termed leakage and are a result of the side-lobe characteristics of the (sin f)/f function. In this case, the extraction of the signal frequency is more difficult and requires a spectral interpolation.

Thus we use the method of spectral analysis [2.21 – 2.22] for the general case of periodic and pseudo-periodic components of various signals with or without noise. The unknown frequency, f, can be related to two successive integer multiples of Δf (where Δf represents the sampling period of the spectrum) as follows :

$$\begin{cases} K \Delta f \leq f \leq (K+1) \Delta f \\ (K-1) \Delta f \leq f \leq K \Delta f \end{cases} \quad (2.23)$$

In practise K corresponds to the position n of the peak in the spectrum represented by the values H_n, even if the leakage effect is present (Fig. 2.9). These two relations can be expressed as follows :

$$f = (K + \beta) \Delta f \quad \text{with} \quad -0,5 < \beta < 0,5 \quad (2.24)$$

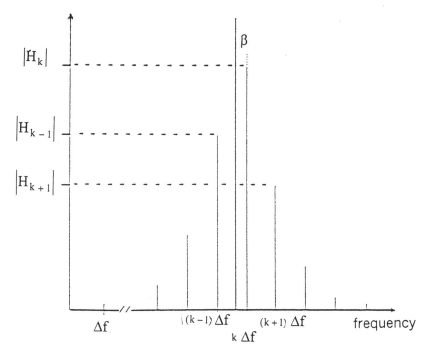

Fig. 2.9 – Leakage effect on the k-order peak

The interpolation process consists of computing the parameter β. We can show that this coefficient can be expressed, from the complex values H_n of the DFT in the neighbourhood of the peak $n = k$ by the relationship :

$$\beta = R_e \left(\frac{H_{k-1} - H_{k+1}}{2H_k - H_{k+1} - H_{k-1}} \right) \quad (2.25)$$

This technique does not use amplitude, but real and complex parts of the spectrum. Furthermore considering the substraction in the expression of β, we note the β is insensitive to noise which changes H_{k-1}, H_k, H_{k+1} by nearly the same quantity.

We applied this numerical procedure to the study of strain on a sheet of cloth in tension. The weft is then used as a crossed line grating. The specimen is illuminated by transmission and the image is recorded by a CCD camera on 256x256 pixels which corresponds to 12 lines in the Y-direction normal to the loading. Experimental results using the Lagrangien description.

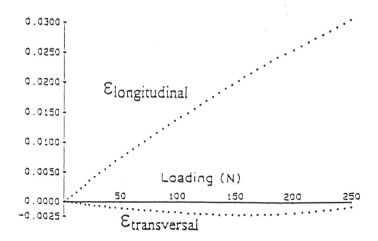

Fig. 2.10 – Strain distribution on a sheet of cloth

As show in Fig. 2.10, a perfect regularity of strains is seen and we note an accuracy of 10^{-4} is reached for ε_y in spite of a coarse.

2.3.2 The device

Using a numerical Fourier transform of the grating image or the optical transform by laser beam diffraction, an out of plane translation between the reference state and the deformed state of the specimen leads to an error in the strain determination. We present a measurement method at distance insensitive to the translations and with better accuracy by the use of quasi-heterodyne detection [17-18].

We consider in normal incidence the diffraction of a laser beam by a parallel equidistant lines grating. The optical device (Fig. 2.11) is made by a cylindrical mirror having the same axis as the incident beam. After reflecting, the two diffracted beams of the order ± 1 interfere on the level of the axis of the cylinder. The interference field is composed of parallel fringes and its analysis allows to characterize the geometry of the grating bounded to the piece. The interference field has a depth of several millimeters, function of the transverse dimension of the diffracted beams where the geometry of the fringes is identical. There is so no problem of focusing and an out of plane displacement of the piece and therefore of the interference zone does not lead to an error measurement. To be free from the translation in the plane of the grating, we use a dimension of grating superior to the diameter of the incident beam.

The interference field can be analysed using the numerical spectral interpolation. We have shown that the use of the quasi-heterodyne technique gives better results than the previous one. We applied this by moduling the phase of one of the diffracted beam using a Bravais compensator. For an unidimensional signal this technique consists of

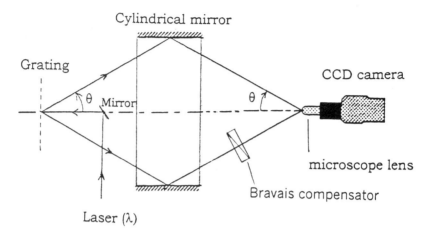

Fig. 2.11 – Optical device

the use of three dephasing ψ_k (calculated by Fourier transform) for the determination of the phase ϕ (x) and so of the value of the period of the signal. The same procedure is applied for a two-dimensional signal. In order to show the performances of the developped optical device, we give results [2.24] from an uniaxial traction test on a plexiglas specimen submitted to step of strain of 2×10^{-5}.

2.4 Holo-grating analysis

2.4.1 Holo-grating recording

We have developped a whole field strain measurement method [2.25] which uses the recording of the grating by holography (Fig. 2.12). The image of the crossed grating illuminated in normal incidence is formed in an optical device that realises successively two Fourier transforms. We obtain so in the second Fourier plane the interferences of the diffracted beams which create a new grating. A filter located in the first Fourier plane allows to eliminate the zero order and the orders superior to ± 1. Few different orientations of the reference beam are generated to record so many object states on the same holographic plate.

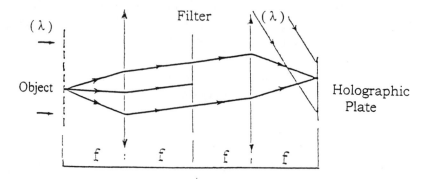

Fig. 2.12 - Holographic recording device

The figure 2.12 shows the optical device for the recording of the image of the grating. This device allows to generate 4 different reference beams.

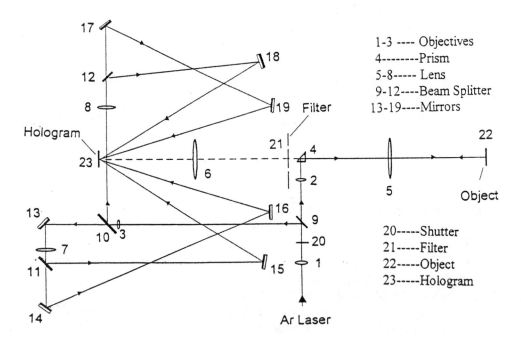

Fig. 2.13 – Set up for the holographic recording device

2.4.2 Holo-grating reconstruction

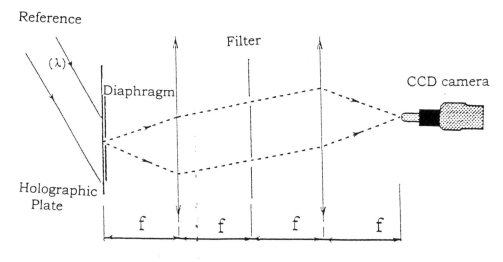

Fig. 2.14 – Set-up for holographic reconstruction

At the reconstruction, the illumination of the holographic plate using the reference beam gives the order ± 1 in each point of the plate, image of an object point. The analysis of the reconstructed grating using the device with a phase shifting procedure gives the determination point by point of the strain components in the whole-field. It is also possible for the strain determination to employ the set-up for the recording procedure. This method has been used for a ductile fracture investigation.

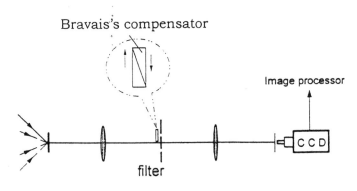

Fig. 2.15 – Set-up to introduce phase shifting

2.4.3 Application to the ductile fracture [2.25 – 2.26]

For a power hardening elastoplastic material, the assumption that the stress and strain components can be written as separable forms in polar coordinates r and θ induces the possibility to characterise the amplitude of the singularity by the path-independant J integral. With reference to these coordinates centered at the crack tip, the asymptotic tip strain fields (referred as the HRR field) are :

$$\varepsilon_{ij} = \alpha\varepsilon_0 \left[\frac{J}{\alpha\sigma_0\varepsilon_0 L_n r} \right]^{n/n+1} \tilde{\varepsilon}_{ij}(\theta) \tag{2.26}$$

where σ_0 is the yield stress, ε_0 the yield strain, α a material constant and n represents the strain hardening exponent. The dimensionless constant I_n and the θ variations of $\tilde{\sigma}_{ij}(\theta), \tilde{\varepsilon}_{ij}(\theta)$ depend on n, on the fracture mode and on the state of plane stress or plane strain conditions. It is not the development of asymptotic expansions that agree with the H.R.R. field for $r \to 0$.

The J integral is defined by :

$$J = \int_\Gamma (Wdy - \vec{T}\frac{\partial \vec{u}}{\partial x}ds) \tag{2.27}$$

where W is the strain energy density, $\vec{T}$ is the stress vector defined according to the out-ward normal along the integration outline Γ surrounding the crack tip, $\vec{u}$ is the displacement vector and ds an element of small length along Γ. The J integral is so interpreted as a measure of the intensity of the crack-tip field for a ductile material like the stress intensity factor K for a brittle one.
The evaluation of the J-integral requires the strain components, the stress component at each data point along the contour. In our study, the component ε_{ij} and $\frac{\partial \vec{u}}{\partial x}$ can be quantified directly by the holography grating method, and the stress components are calculated using the inverted form of the J_2 deformation theory.
J-measurement which was derived for rectangular contours surrounding the crack-tip, is divided into line integrals along the vertical and horizontal segments shown in Fig. 2.16 following page.

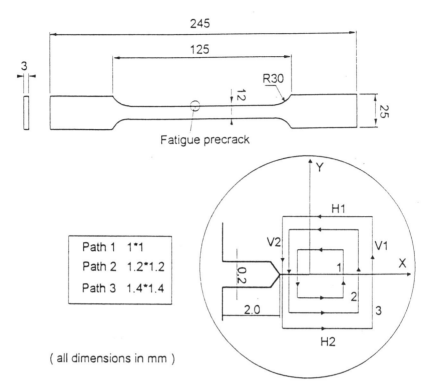

Fig. 2.16 – Integration path around the crack tip

For an Aluminium 2017 specimen loaded in mode I, on which we have realised a phase greating, we find for a stable crack growth up to 1 mm corresponding to a load equal to 4000 N that the near field J integrals are path independant.

Path number	1	2	3
J(Kpa.m)	3.52	3.54	3.54

Using the J values, we can compare the strains expressed by the H.R.R. field with those experimently determined. The comparison of these strain distributions [Fig. 2.17] in the crack direction shows that to the one parameter characterization of ductile fracture by the J term seems to be not sufficient to correctly described the strains singularity. Analogous results have been obtained from displacement measurements using moiré Interferometry [2.28].

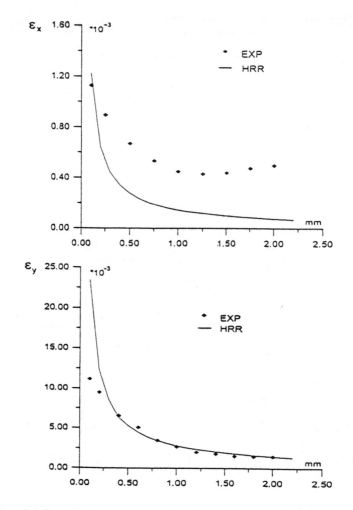

Fig. 2.17 – Strain components along the crack direction

2.5 Local measurement of strains in dynamic [2.15, 2.19, 2.30]

The considered surface element is supposed to be plane, on which is marked a crossed grating formed with two orthogonal sets of parallel lines.

We have a photographic film, we mind to record, at t moment, the spots of 0 and ± 1 order given by beams diffracted by the grating lighted by a parallel beam of coherent light. In order to separate on the film the recordings made at different moments, the grating will be lighted upon different oblique incidences (angular coding).

5.1 Diffraction in oblique incidence (Fig. 2.18)

Let us T, an orthonormed reference system $\bar{x}, \bar{y}, \bar{z}$ and let be

X_I, Y_I, Z_I the direction cosines of the incident beam $\bar{I}$.
X_D, Y_D, Z_D the direction cosines of the diffracted beam D.

Let us note T_R, an orthonormed reference system $\bar{x}_R, \bar{y}_R, \bar{z}_R$ where $\overline{Oy}_R$ and $\overline{Oz}_R$ are linked to the measurement base.

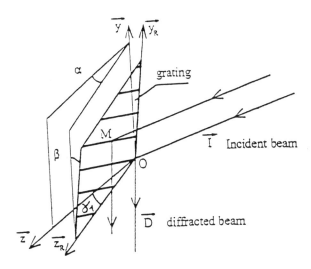

(for clarity of the figure, only one grating is represented)
Fig. 2.18 · Position of one grating in space

Let us note γ_1 and γ_2 the angles between $\bar{z}_R$ and the lines of the two gratings of pitches p_1 and p_2 respectively. Let us note k et k' the diffraction order of the two gratings.

α, β the rotations of T_R respectively around y and z, relatively to T.

We directly proved that the two diffraction relations under oblique incidence

$$\left.\begin{aligned}(X_D-X_I)\cos\alpha\sin\beta+(Y_D-Y_I)\cos\beta+(Z_D-Z_I)\sin\alpha\sin\beta &= \frac{k\lambda}{P_1}\cos\gamma_1+\frac{k'\gamma}{P_2}\cos\gamma_2 \\ (X_D-X_I)\sin\alpha+(Z_D-Z_I)\sin\alpha &= -\frac{k\lambda}{P_1}\sin\gamma_1-\frac{k'\lambda}{P_2}\sin\gamma_2\end{aligned}\right\} \quad (2.28)$$

with the conditions for the unitary vectors

$$X_I^2 + Y_I^2 \; Z_I^2 = 1$$
$$X_D^2 + Y_D^2 \; Z_D^2 = 1 \qquad (2.29)$$

Let us note T_F the orthonormal reference system (o', x, y_F, z_F) linked to the plane of the recording film (y_F, z_F) with $Oo' = L_x \vec{X} + L_y \vec{Y} + L_z \vec{Z}$ and α_p, β_p the rotations of T_R respectively around y_F and z_F relatively to T_F. We notice that, knowing Oo' and the position of one diffracted spot M_j, in T_F, we can deduce the components of corresponding unitary vector D.

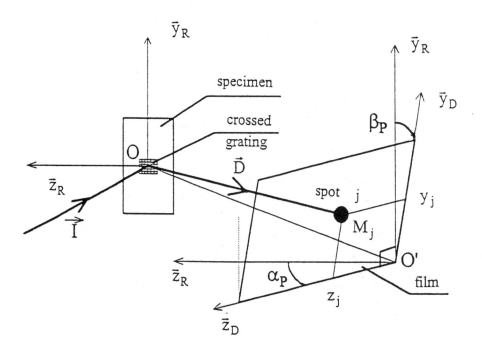

Fig. 2.19 – Definition of geometrical quantities

2.5.2 Principle of the method

At sampled moments of the dynamic event, the measurement area is lighted by means of a cylindrical beam whose axis is centred on that area and is on the surface of a cone with circular base.

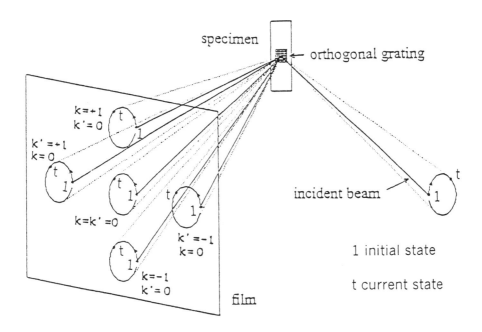

Fig. 2.20 – Schema of angular coding

By a numerical simulation for a uniaxial loading of a plane and isotropic specimen ($\alpha=\beta=0$), without translation of the measurement base, we obtain, for 20 states, the positions of the diffraction spots on the screen, in two situations : one without loading, other with traction increasing and then decreasing.

The unknown quantities of the system are
- 4 parameters characterizing the measurement base, orientations and pitches of the two gratings taking into acount pure strains and the rotation of the rigid solid in the plane of this base.
- 2 angular parameters of the base orientation.
- 3 parameters defining the position of the base.
- 5 distances from the measurement base to the position of the spots on the screen.

In all, there are 14 unknown quantities.

At each moment, the experimental data are the orientation of the incident beam and the position of the 5 corresponding spots on the screen. As previously, we use the five brightest spots, 0 order (k = k' = 0), ± 1 order (k = ± 1, k' = 0) and (k = 0, k' = ± 1).

It can be proved that, at each sampled moment, the unknown quantities are the solutions of one non linear system with 15 equations characterizing the 5 considered beams.

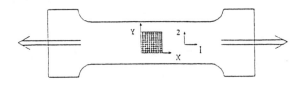

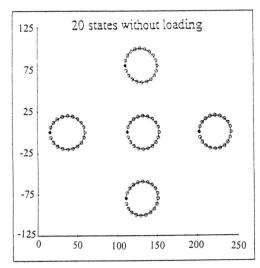

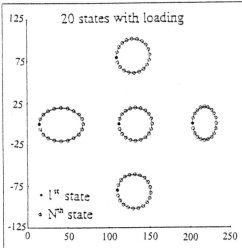

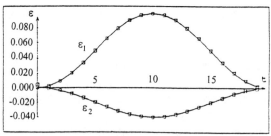

Fig. 2.21 – Numerical simulation

2.5.3 Implementation

The optical device requires several acousto-optic modulators (shutter of beam and deflectors) making possible, at each moment of photographic recording, to associate a particular orientation of the incident laser beam. The acousto-optic deflectors give beams deviating from optical axis. These elements make possible the recording of 23 sequential informations during the dynamic event at the maximum frequency near 1 MHz.

An acousto-optic shutter authorizes exposure time equal 30 ns. One optical element, made with 24 couples of two plane adjustable mirrors, gives circular (diameter 1 cm) cylindrical beams converging on the measurement base (diameter 1 mm) and engraved crossed gratings of 200 lines by millimeter.

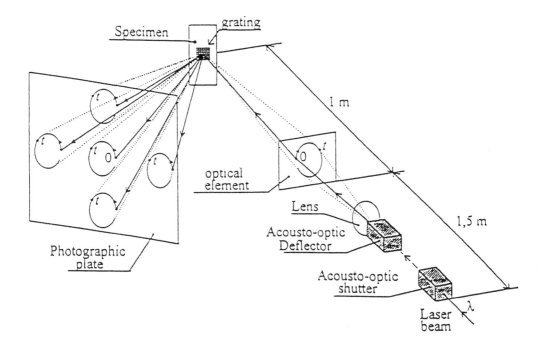

Fig. 2.22 – Schema of optical recording device

After this first realisation we built a small recording device about (30 x 50 x 20 cm³) which allows the use of this technique on most experimental or industrial site.

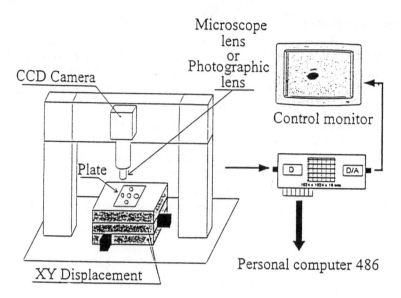

Fig. 2.23 Acquisition and analysis device

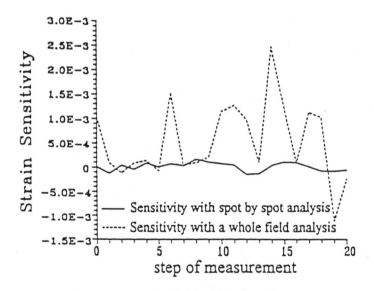

Fig. 2.24 – Strain sensitivity comparison

The accuracy of the strain measurements being directly linked to the appreciation of the position of the diffracted spots on the film, we have developed an original system for recording analysis. First, we carry out the whole acquisition of the plane film by CCD camera fitted with photographic focus in order to locate the diffraction spots. This process is followed by spot-by-spot visualization, in order to determine, with accuracy, the position of each of them by means of the CCD camera fitted with microscopic focus, by acquisition of information all along the micrometric displacement of the film (Fig. 2.23). This process, entirely automatic, multiplies by 10 the strain sensibility of the method (Fig. 2.24) relatively to a whole analysis of the film. We obtained a strain sensibility of approximely 2.10^{-4}.

2.5.4 Experimental tests

For all these experimental investigations the diameter of the measurement base is about 1 mm.

First we have glued on the specimen a strain gauge on one side and made a grating of 200 lines per millimeter on the other, then we have applied this measurement method to static loading using a uniaxial tension test [2.25]. We have compared the strains optically determined with those obtained from classical extensometry and formed a very good agreement between these two sensors.

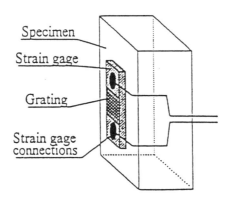

Fig. 2.25 – Sensors disposition for dynamic tests

In order to show the performance of the strain measurement method, we have performed simple dynamic experiments. We have created an impact using the potential energy of a mass which hits the moving grip of the specimen loaded in tension first and in compression in the second test (Fig. 2.26 – 2.27). For these dynamic test we have made a grating of 200 lines per millimeter on the strain gauge for better agreement on the localisation of the two sensors.

For the recording we use an Argon laser (λ=514.5nm) and a photographic plate (9 x 12 cm^2) with a 400 ASA sensitivity. Because of the duration of the impact, the frame rate (i.e. the frequency of the acousto-optic deflector) is chosen equal to 1.5 kHz for the first test and 35 kHz for the second. The time exposure is respectively 100 µs (15 % of the duration time between two successive spots) and 2.8 µs (10 % of the duration time between two successive spots). We can see (Fig. 2.26) and (Fig. 2.27) the good agreement between the strains measured with this optical method and those given by a strain gauge.

The frame rate and the exposure time can be adapted to the loading speed. The frame rate can be equal to 1 MHz and the exposure time can be very short even with conventional equipement. For example, a study of the energetical efficiency of our experimental device shows that the use of a 300 mW laser power allows the recording on a plate of 400 ASA sensitivity with a minimum exposure time equal to 0.4 µs. In these conditions the developped recording device can be applied for the strain measurements on dynamic loading of a minimal duration equal to 50 µs.

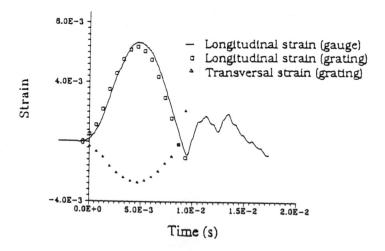

Fig. 2.26 – Dynamic uniaxial tension test

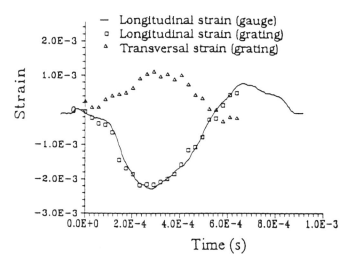

Fig. 2.27 – Dynamic uniaxial compression test

2.5.5 Hopkinson bar investigation

On (Fig. 2.28) we present a compression test performing with a Hopkinson bar loading. We have adjusted the framing rate to 1 MHz according to the duration of the loading (20 µs). In this test, the exposure time is 10 % (0,1 µs) of the duration time between two successive spots.

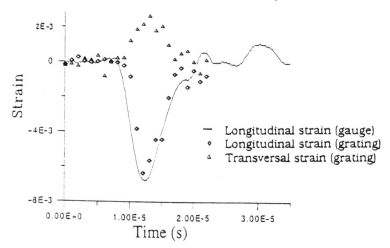

Fig. 2.28 – High speed compression test with CCD camera

One laser of 50 mW is convenient for recording with CCD camera.

2.6 Local strains measurement on cylindrical specimen [2.16, 2.31]

The cylindrical specimen is submitted to uniaxial load. On reflecting surface, without grating, an incident laser beam returns wit a divergent shape by the reflection laws. When a grating is present, the diffraction phenomenon is superimposed (Fig. 2.29). As a consequence the light spots are elongated, the less the radius is small, the more the spots are stretched. For small diffraction angles, the spots can join together (Fig. 2.31).

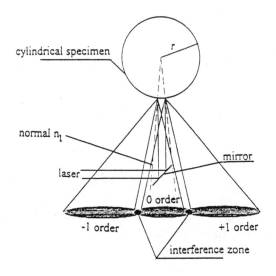

Fig. 2.29 – Diffraction images on cylindrical surfaces

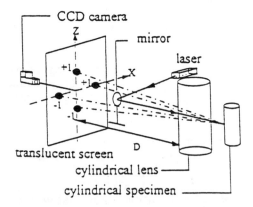

Fig. 2.30 – Experimental set-up for reflexion specimens

For small diffraction angles, we saw that the spots are joined together. In practise, we use a cylindrical lens of radius equal to 15 mm placed at 5 mm before the cylindrical specimen of duralumin of radius equal to 5 mm. This distance is sufficient for the diffracted orders −1, 0 and + 1 to pass through the lens (Fig. 2.30). As the grating is not located on the focal point of the lens, the diffraction spots are circular in a plane situated at a distance D = 120 mm from the lens. The spots are then collected on a translucent screen placed in this plane. Of course, due to the presence of the mirror the zero order is intercepted and it is not recorded by the CCD camera. The spots cover an area of about 5 or 6 pixels on the CCD grating. The strains are then calculated between the first order of diffraction.

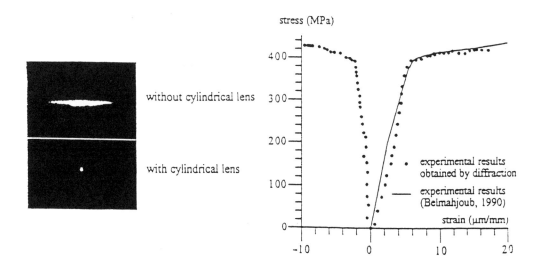

Fig. 2.31 – Diffraction spots In reflexion

Fig. 2.32 – Experimental results obtained on duralumin

The experimental results are presented in Fig. 2.32 in good concordance with given results in litterature [2.32].

2.7 Conclusion and perspectives

The optical method based on the grating interrogation on the local plane surface, can make measurement at large distance without contact and without difficulties linked to the use of electrical alimentation of the electrical strain-gages. This is convenient for the control and the measurement in hostil environment. The accuracy is smaller 10^{-5} against 10^{-6} for the strain gage but the size of the measurement base can be more reduced than that necessary for rosette strain gage. The recourse to holography allows to extend the measure to quasi-plane surface.

The measurement of the large strains is only limited by the quality of the markage.

This optical method can allow to extend the metrology in extensometry to domains badly adapted to use the electrical strain gage (bonded joints, crack, sails, soft materials).

For one dynamic event, the local grating interrogation gives not only the strains but also the rigids motions of the measurement base. Present technical means should permit the extension to large plane surface.

2.8 References

2.1   Rayleigh, "Sur la production et la théorie de réseaux de diffraction" Phil. Mag. 47 (81) 193-(1874).

2.2   Dantu M.P., "Utilisation des réseaux pour l'étude des déformations" Annales de l'I.T.B.T.P., n° 12 (reprend une conférence du 5 mars 1957) 1958.

2.3   Post D., "High Sensitivity Displacement Measurement by Moiré Interferometry" Pro. 7th. Int. Conf. Exp. Stress Analysis Haïfa Israël 1982.

2.4   Sciammarella C.A., "Technique of fringe interpolation in Moiré patterns" Exp. Mech., pp. 19 A-29 A 7 (11) Novembre 1967.

2.5   Fu-Pen Chiang, "Techniques of optical spatial filtering applied to the processing of Moiré Fringe Patterns" Exp. Mech. pp. 525-526, November 1969.

2.6 Bell J.F., 1959, "Diffraction grating strain gage" S.E.S.A. Proceeding XVIII (2) 51-64.

2.7 Boone P.M., "A method for directly determining surface strain - Fields using diffraction grating" Exp. Mech. 11, Noll (1971).

2.8 Sevenhuijsen P.J., 1978 "The development of a laser grating method for the measurement of strain distribution in plane, opaque surfaces, V.D.I. Berichte, N°313, pp. 143-147.

2.9 Brémand F., Lagarde A., "Méthode optique de mesure des déformations utilisant le phénomène de diffraction". C.R. Acad. Sciences. Paris, 303, série II, 1986, p. 515-520.

2.10 Brémand F., Lagarde A., "Optical Method of Strain Measurement on a Small Size Area Application" IUTAM Symposium on Yielding, Damage and Failure of Anisotropic Solids, Villars de Lans, Août 1987.

2.11 Brémand F., "Photoélasticimétrie en grandes déformations. Méthodes de mesure de petites et grandes déformations". Thèse de Doctorat de l'Université de Poitiers, 1998.

2.12 Brémand F., Dupré J.C., Lagarde A., "Non-contact and non disturbing local strain measurement methods I Principle", Eur. J. Mech. A/Solids 11 n°3, pp. 349-366, 1992.

2.13 Cottron M., Brémand F., Lagarde A., "Non-contact and non-disturbing local strain measurement methods II Applications", Eur. J. Mech. A/Solids 11 n°3, pp. 367-379.

2.14 Post D. "Developments in moiré interferometry" Optical Engineering p. 458-467, vol. 21, 1982.

2.15 Valle V., Cottron M., Lagarde A., "A new optical method for dynamic strain measurement". 10th International Conference on Experimental Mechanics Lisbon 1994.

2.16 Meva'a L., Brémand F., Lagarde A., "Optical methods of strain measurement of cylindrical specimen submitted to uniaxial load". 10th International Conference on Experimental Mechanics Lisbon 1994.

2.17 Cloud G.L., "Optical methods of engineering analysis" Cambridge University Press, pp. 447-491.

2.18 Dupré J.C., Cottron M., Lagarde A., "Phase shifting technique for local measurement of small strains by grid method". Proc. 10th International Conference of Experimental Mechanical Lisbon, 1994.

2.19 Creath K, (1988) "Phase measurement interferometry methods". In progress in Optics XXVI, Ed. E. Wolf, pp. 349-93. Amsterdam : Elsevier Science Publishers.

2.20 Wyant J.C., (1982) "Interferometric optical metrology, basic systems and principles". Laser Focus, May 1982 : 65-71.

2.21 Carré P. (1996) "Installation et utilisation du comparateur photoélectrique et interférentiel du Bureau International des Poids et Mesures ; Metrologia, 2, 1 : 13-23.

2.22 Rajaona D.R., Sulmont P., "A method of spectral analysis applied to periodic and pseudo-periodic signals", Jour. Comput. Phys., Vol. 61, n°1, pp. 186-193, October 1985.

2.23 Dupré J.C., "Traitement et analyse d'images pour la mesure de grandeurs cinématiques, déplacements et déformations à partir de la granularité laser et de réseaux croisés, et pour l'étude de couplage thermomécaniques". Thèse de Doctorat de l'Université de Poitiers, 1992.

2.24 Dupré J.C., Lagarde A., "Optical method of the measurement at distance of local strains", XVIIIth International Congress of Theoretical and Applied Mechanics, Haïfa, Israël, August 22-28, 1992.

2.25 Dupré J.C., Cottron M., Lagarde A., "Méthode indépendante des petites translations de l'objet, pour la mesure locale et à distance avec quasi-hétérodynage de l'état de déformations", C.R. Acad. Sci. Paris, Tome 315, Série II, n°4, pp. 393-398, 1992.

2.26 Wang S.B., Cottron M., Lagarde A, "An holographic grid technic for a whole strain measurement and application to elastoplastic structure". 10th International Conference on Experimental Mechanics Lisbon 1994.

2.27 Wang S.B., Méthode optique de mesure de déformations à champ complet associant holographie et analyse numérique de réseaux. Application à la mécanique de la rupture". Thèse de Doctorat de l'Université de Poitiers, 1994.

2.28 Dadkhah M.S., Kobayashi A.S., "Further studies on the H.R.R. field of a moving crack, an experimental analysis". Int. Journal of Plasticity, vol. 6, pp. 635-650., 1990.

2.29 Valle V., "Développement et mise en œuvre d'une méthode optique de mesure locales des déformations en dynamique" Thèse de Doctorat de l'Université de Poitiers, 1994.

2.30 Valle V, Cottron M., Lagarde A., "High speed local strain determination from grating diffraction" Symposium IUTAM "Advanced Optical Methods and Applications in Solids Mechanics", Poitiers August 31 – September 4, 1998, Kluver academic publishers.

2.31 Meva'a L., "Méthodes optiques de mesure à distance des déformations, sur éprouvettes planes à température de figeage et sur éprouvettes cylindriques pour des essais de traction". Thèse de Doctorat de l'Université de Poitiers, 1994.

2.32 Belmahjoub F., "Comportement thermomécanique de matériaux métalliques sous divers trajets de chargement uniaxe", Thèse de Doctorat, Université de Montpellier II, 1990.

CHAPTER III

AUTOMATED IN-PLANE MOIRÉ TECHNIQUES AND GRATING INTERFEROMETRY

M. Kujawinska
Warsaw University of Technology, Warsaw, Poland

ABSTRACT

Several problems in experimental solid mechanics and material engineering require determination of in-plane displacement / strain fields. It is especially true if we consider flat samples under simple loading arrangement. The effective experimental tools working under these assumptions are in-plane grid / moiré technique and high sensitivity grating (moiré) interferometry. Below the principles of both techniques and modern solutions of grating / grid technology and design of moiré and interferometric systems are presented. As the fringe patterns obtained at the output of the systems require automatic analysis, the overview of the phase methods of fringe pattern analysis especially suited for various opto-mechanical configuration of the systems are described. Also the interaction of the results with FEM is presented, while referring to various concepts of hybrid experimental-numerical analysis. The result advances in measurement technology expand significantly the applications of the in-plane moiré and grating interferometry techniques. The numerous examples refer to the most challenging applications including local material constant determination, micromeasurements, residual stress analysis and monitoring of various engineering structures.

1. THE BASIC RULES IN FULL-FIELD GRID TECHNIQUES

1.1. Basic grid moiré techniques

Methods for the measurement of deformations of structures are of interest because local and global deformation can reveal the severity of loads on the structure. The most easy way to see the deformation is to apply markings on the surface of structures to track the displacements of markings when the structure is loaded. This approach is also the basic for whole range of grid techniques [1-3]. From the point of view of applications to technical problems, there are four basic grid techniques:

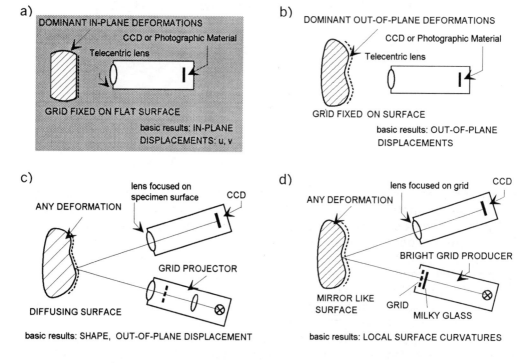

Fig. 1.1, Optical schemes for four basic grid methods: a) fixed grid method for in-plane displacements, b) fixed grid method for out-of-plane displacements, c) projected grid method for out-of-plane displacements and d) reflected grid method for local surface curvatures determination.

1. The *intrinsic moiré* or *fixed grid* technique in which the in-plane displacement of a flat surface to which a grid is attached is determined (Fig. 1.1a).

2. The intrinsic moiré technique may also be used on curved surfaces to measure out-of-plane deformations and displacements of surface points (Fig. 1.1b). This technique can only be used when in-plane deformation at the surface of the object remains negligible in relations to the out-of-plane deformations.
3. The *projected and shadow moiré* technique in which a grid is projected on a light diffusing surface and the shape (surface contours) of the object is determined (Fig. 1.1c).
4. The *reflected moiré* requires a mirror like surface. A grating and this reflection give moiré pattern which determine the loci of equal projected slopes (local surface curvatures) (Fig. 1.1d).

The grid based phenomena are analysed by using a common framework in which the optical system plays the main role. Moiré techniques differ from the pure grid technique in that the optical system produces the low frequency fringes to highlight displacements [4,5].

In this chapter we focus on the fixed grid method and related techniques based on moiré and interferometric phenomena for in plane displacement determination.

1.2. Fixed grid method

The fixed grid method is one of the oldest ways to demonstrate and measure deformations [6,7]. A grid is a regular pattern of either one set of parallel lines spaced at one pitch (linear grid) or two perpendicular sets of parallel lines (cross grid). A grid is applied to the surface of structural part or a test specimen. For measurement of deformation it is necessary that differences between successive images of these patterns are related only and directly to deformation.

Application of a grid technique involves four basic steps:
- choosing of the best applicable grid and preparation of the specimen surface and the test setup;
- recording of the grid images, at different loads;
- reading of the grid images and processing of the data;
- interpreting and presenting the results.

In a pure grid technique, characteristics like pitches, locations and directions of lines are measured directly at each separate grid or grid image per load situation. By tracking the corner points of elementary squares (Fig. 1.2) of a grid a true Lagrangian strain distribution can be measured:

$$\varepsilon_{PQ} = \frac{P'Q' - PQ}{PQ} \quad \text{or} \quad \varepsilon_{PQ} = \frac{\frac{q'}{\cos(\alpha - \beta)} - q}{q} \tag{1.1a}$$

$$\varepsilon_{PR} = \frac{P'R' - PR}{PR} \quad \text{or} \quad \varepsilon_{PR} = \frac{\frac{r'}{\cos(\alpha - \beta)} - r}{r} \tag{1.1b}$$

$$\gamma_{QPR} = \angle Q'P'R' - \angle QPR \quad \text{or} \quad \gamma_{QPR} = \alpha - \beta \tag{1.1c}$$

These are the basic relations for the fixed grid technique, which uses as the input data the direct images of the grid (as shown in Fig. 1.3). In theory it is possible to work with grids with infinitely thin lines and infinitely small points but in practise lines are stripes and points are spots. For in-plane displacement measurement it is convenient to work with pattern of spots (areas between crossing lines). The locations of spots are measured relative to each other and for this, a specific reproducible point of a spot has to be defined. Experiences and investigations have proved that a usable specific and reproducible point of a spot is the light intensity centroid, which coordinates are calculated from the neighbourhood pixels intensities values [8]. For this purposes nowdays sophisticated image processing techniques are employed including correlation approach and subpixel procedures [9]. However the sensitivity and resolution in respect to the field of measurement and the resolution of detector (CCD camera) of this direct grid method is restricted and often the measurement process is difficult to automate.

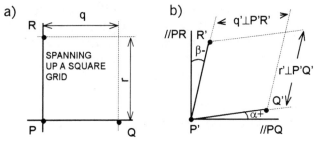

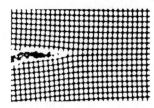

Fig. 1.2, Elements for the description of in-plane deformations to be expressed in strains: a) unloaded and b) loaded

Fig. 1.3, Example of a grid image showing the deformation distribution near a crack tip

1.3 Principles of grid (grating) theory

In order to simplify the analysis process, enhance the quantities measured and increase the sensitivity of the method two basic approaches are applied: geometrical moiré and classical interferometry.
Both approaches rely on the information coded into grid attached to the specimen measured, which is treated as a reference system on the analyzed

surface. The grid (grating) is a carrier of displacement information and the spatial frequency of the carrier ($f_0=1/p$, p the period of a grating) has been modulated by the applied displacement [1,10,11,12]. The mathematical description of both regular (reference) and deformed grating are given by their transmission function described by Fourier expansion (the treatment is restricted for simplicity to one dimension; the two-dimensional treatment follows analogously). For regular grating the transmission function, T_u, is given by:

$$T_u(x) = a_0 + \sum_{n=1}^{\infty} a_n \cos n\left(\frac{2\pi}{p}x + \alpha\right) = \sum_{n=-\infty}^{+\infty} a_n \exp\left[in\left(\frac{2\pi}{p}x + \alpha\right)\right] \tag{1.2}$$

where p is the grating period and the a_n values are determined by the grating profile, α is phase shift.
For the grating deformed by a displacement $u(x)$ the transmission function, T_d, equals:

$$T_d(x) = T_u[x - u(x)] = a_0 + \sum_{n=1}^{\infty} a_n \cos n\left(\frac{2\pi}{p}[x - u(x)] + \alpha\right) =$$

$$= \sum_{n=-\infty}^{+\infty} a_n \exp\left\{in\left[\frac{2\pi}{p}(x - u(x)) + \alpha\right]\right\} \tag{1.3}$$

In the case of a sinusoidal profile of the grating, equation (1.3) takes form

$$T_d(x) = a_0 + a_1 \cos\{2\pi f_0[x - u(x)]\} \tag{1.4}$$

in which the modulation of the carrier frequency f_0 by the $u(x)$ displacement is clearly seen.
Basing on the equations (1.3) and (1.4) the principal relationships for the various decoding techniques may be derived:
- the formation of moiré in incoherent light
- interferometric techniques.

1.3.1 Formation of moiré in incoherent light
The illumination of reference and deformed gratings in superimposition in transmission gives the resultant intensity [10]:

$$I_e = I_i T_u T_d$$

where I_i and I_e are the intensities of the illuminating and emergent beams respectively.
This gives:

$$I_e(x) = I_i \left[a_0^2 + a_0 \left(\sum_{n=1}^{\infty} a_n \cos n \left(\frac{2\pi x}{p} + \alpha \right) + \sum_{m=1}^{\infty} a_m \cos m \left\{ \frac{2\pi}{p}[x - u(x)] + \alpha \right\} \right) \right. +$$

$$+ \sum_{n=1}^{\infty} \sum_{m=1}^{\infty} \frac{a_n a_m}{2} \cdot \left(\cos \left\{ \frac{2\pi}{p}[(n+m)x - mu(x)] + n\alpha + m\alpha \right\} + \right.$$

$$\left. + \cos \left\{ \frac{2\pi}{p}[(n-m)x + mu(x)] + n\alpha - m\alpha \right\} \right) \quad (1.5)$$

which is of general form
$$I_e = I_0 + I_1 + I_2$$
where:
I_0 · a constant term
I_1 · a series of cosine terms whose argument contains $2\pi kx/p$, k=integer > 0
I_2 · a series of cosine terms whose argument consists solely of $2\pi mu(x)/p$ (m = 1, 2, 3, ...), arising when n = m

The I_1 terms are of much higher spatial frequencies than the others. Using suitable means, usually optical, the high frequencies are 'filtered out' or 'smoothed out', leaving the moiré intensity [13,14]

$$I_m(x) = I_0 + b_1 \cos\left[2\pi \frac{u(x)}{p} + \beta\right] + b_2 \cos\left[2\pi \frac{2u(x)}{p} + 2\beta\right] +$$

$$+ b_3 \cos\left[2\pi \frac{3u(x)}{p} + 3\beta\right] + \ldots = I_0 + \sum_{k=1}^{K} b_k \cos k\left[2\pi \frac{u(x)}{p} + \beta\right] \quad (1.6)$$

where the b and β values result from the manipulations of the trigonometric terms in equation (1.6) and describe the profile and phase shift of moiré fringes. Considering only the first two terms in equation (1.6) the intensity of moiré fringe becomes:

$$I_m = I_0 + b_1 \cos\left[\frac{2\pi u(x)}{p} + \beta\right] \quad (1.7)$$

and as opposed to the discrete law [11] this equation represents a continuous relationship between intensity

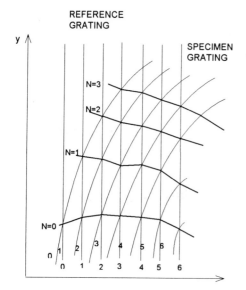

Fig. 1.4, Detail of moiré fringe formation between reference and specimen (deformed gratings)

and displacement, that forms the basis for the analysis of the moiré fringes and more specificly the phase of moiré fringes.

A two-dimensional case of moiré fringes formation is shown in Fig. 1.4. The moiré fringes can be considered as isothetics of displacement in the direction perpendicular to the specimen grating and the displacement of the specimen is given by:

$$u(x,y) = N_x p \qquad (1.8)$$

where N_x is the current fringe number [3,5,13].

The moiré fringes given by equations (1.6) and (1.7) may be modified by introducing carrier frequency (Fig. 1.5), by superposition of gratings with slightly different pitches, p and p', (pitch mismatch) or set at an angle θ (rotational mismatch). The carrier moiré fringes orientation ϕ, and spacing S are the functions of interacting gratings (Fig. 1.5)

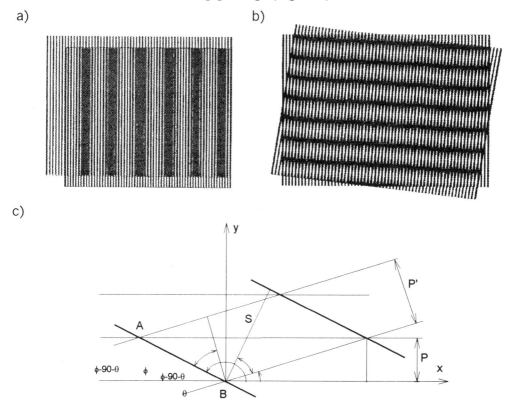

Fig. 1.5, Examples of carrier moiré patterns due to a) pitch mismatch, b) rotational mismatch, c) the geometrical configuration describing their formation

$$\phi = \arctan\frac{p\sin\theta}{p\cos\theta - p'} \qquad (1.9a)$$

$$S = \frac{pp'}{\sqrt{p^2\sin^2\theta + (p\cos\theta - p')^2}} \qquad (1.9b)$$

where θ · the angle between gratings.

1.3.2. Interferometric techniques

The conventional moiré method for studying in-plane displacements of the objects under load [4,5] displays sensitivity limitations caused by technological difficulties with copying amplitude type binary diffraction gratings and a rapid decrease of the moiré fringe contrast due to a finite gap between two superimposed structures. The upper frequency of the gratings to be copied is assumed to be 50 lines/mm what corresponds to the method basic sensitivity of 20 μm per fringe order. This sensitivity value is highly insufficient to conduct various investigations within the elastic deformation region.

In general, when higher sensitivity is needed it is necessary to switch to coherent illumination of the gratings and interferometric methods [15-18]. The information on the departure of grating lines from straightness is coded in the departure of wave fronts of grating diffraction orders from the plane wave front. Let as assume a very simple configuration. The deformed specimen grating (Eq. 1.3) with encoded displacement u(x) is illuminated with coherent plane wavefront, Σ^A, at the angle tuned to the first diffraction angle of specimen grating, SG (Fig. 1.6a) i.e.

$$E^A \cong \exp\left(-i2\pi\frac{1}{p}x\right) \qquad (1.10)$$

(for simplicity of considerations the wavefront amplitude is omitted).
The wavefront is diffracted on SG and multiply of diffracted wavefronts occur as shown below:

$$E^{A'} = E^A \cdot T_d = \exp\left(-i2\pi\frac{1}{p}x\right) \cdot \sum_{n=-\infty}^{+\infty} a_n \exp\left\{in\left[\frac{2\pi}{p}(x - u(x)) + \alpha\right]\right\} \qquad (1.11)$$

For further consideration we take the wavefront which propagate along the optical axis i.e. the wavefront originating from the first diffraction order (n=+1)

$$E^{A'}_{+1} = \exp\left(\frac{-i2\pi}{p}x\right)a_1\exp\left\{i\left[\frac{2\pi}{p}(x - u(x)) + \alpha\right]\right\} = a_1\exp\left\{i\left[\frac{2\pi}{p}(u(x)) + \alpha\right]\right\} \qquad (1.12)$$

It is clearly seen that this wavefront carries information about the u(x) in-plane displacement. It can be visualized if the wavefront $\Sigma^{A'}_{+1}$ is recombined with the other mutually coherent plane reference wavefront.
Let the reference wavefront Σ^R propagate along the optical axis (Fig. 1.6b)

$$E^R \cong \exp(\cdot i\alpha_R) \qquad (1.13)$$

The intensity distribution of the interferogram obtained becomes

$$\left|E^{A'}_{+1} + E^R\right|^2 \cong 2\left\{1 + \cos\left[\frac{2\pi}{d}u(x) + \alpha'\right]\right\} \qquad (1.14)$$

where $\alpha' = \alpha \cdot \alpha_R$ is an arbitrary phase shift of the interferogram.

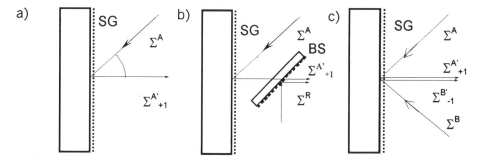

Fig. 1.6, The scheme for grid based interferometric techniques: a) introducing phase information from grating into wavefront Σ^A, b) recombining with plane reference wavefront Σ^R, c) recombining with the conjugate wavefront Σ^B_{-1}. BS – beam splitter, SG specimen grating.

If the reference wavefront recombines with the higher diffraction order the sensitivity of u(x) decoding increases directly proportional to the number of diffraction order applied [1,18,19]. The other possibility is to compare the wavefront Σ^A_{+1}, with its longitudinally reversed counterpart (frequently called as the conjugate wave front). In the latter case the sensitivity is doubled. High diffraction orders of the same number and opposite sign can be brought into interference by using symmetrical double beam illumination. This approach was used in the first studies on sensitivity increase of the moiré method by Post [20]. An excellent improvement of the approach to using higher order mutually conjugate diffraction orders is moiré (grating) interferometer [17]. Additionally in this configuration the influence of non-flat sample or out-of-plane displacement can be to some extend eliminated. The detailed description of moiré interferometry method is given in Chapter 5.

2. GRID (GRATING) TECHNOLOGY

A basic practical problem to use any of the fixed grid based techniques is the application of a grid (grating) onto the surface of the structural component. There are several ways to apply a grid [17,18,21]: scratching as in a hardness tester; etching as in microelectronics, photographic engraving, inking through a hole pattern or along a ruler; sputtering through a hole pattern, "burning" by means of laser or an electron beam in a scanning electron microscope; cementing the pixture layer of a strippable film; or cementing the pixture layer of photographic paper; exposing photoresist by a field formed by interfering coherent waves etc. Important questions include the following: What kind of grid application equipment is at hand? What is the desired pitch? What is the deformation the grid has to sustain? What is the procedure for the reading of the grid images?

Within a variety of technologies for applying grids, two groups of methods may be considered (however sometimes these groups may be connected e.g. by bleaching or etching procedures):
- techniques for producing amplitude type gratings (mainly black/white lines or dots),
- techniques for producing phase type gratings (mainly relief type)

Below some of these techniques are described in reference to the measurement techniques.

2.1 Amplitude grids

For application of geometric moiré, grids of relatively large pitch has to be attached to the specimen by a variety of processes. To do so the master grating has to be avaliable [21]. Pitches of amplitude master grating used for geometrical moiré vary from 1 mm to 10 μm (i.e. 1 to 100 lines per mm).

The coarse gratings, 1 to 5 lines/mm, can be obtained from graphics arts suppliers, in sheets of 25 by 50 mm. Commercial artists use both cross gratings and one-way gratings, of various line densities, for shading. The cross gratings are often rectangular dot arrays. Finer gratings (5 to 10 lines/mm) are available from high quality lithographers or photoengravers and theirs suppliers. A few suppliers have gratings with frequencies greater than 40 lines/mm. Optical grating companies supply gratings beyond 40 lines/mm. Suppliers to the stress analysis industry provide 20 to 40 lines/mm rulings and cross gratings, on glass, up to 20 by 25 cm, as well as metallic dot arrays of 20 lines/mm. Most commercial gratings have a line width of approximately 50% of the pitch, which are optimum to achieve maximum fringe contrast.

Photography can be used to duplicate gratings and in special cases to vary line width, by varying exposure and developing times. An amplitude grating can be also made by exposing a photographic plate to two laser beams from the same source. The angle between the beams determines the pitch, however this method is usually applied for high frequency amplitude gratings, not applicable in geometrical moiré.

Having the master grating, it has to be attached or replicated at the specimen surface [22,23]. A technique which is common for producing specimen grid is "stripping film", in which the black/white dots (cross pattern) are photographically reproduced from "master" grids in which the halide emulsion was carried on a duplex backing. The developed material is made to adhere to the specimen surface with glue, with the antihalo layer outward. This is then peeled off, leaving the grid on the specimen. A fairly smooth and clean surface is generally desirable for good results. A similar concept to the "stripping film" lies behind commercially available "nickel dot" grids separable from a steel backing sheet. Black epoxy is generally used as the transfer medium, leaving reflective dots at the black background. For high temperature applications a white ceramic cement is preferred: the nickel turns dark as it oxidizes, giving a reverse contrast.

Soft or elastic materials require the pattern application without any supportive layer. Such an application can be realized by means of stencilling through a fine nickel mesh with an array of holes of up to 80 holes/mm. The mesh is temporally fixed and properly oriented to the surface by using a nondrying adhesive. After drying, the surface is sprayed with a white pigment e.g. titanium dioxide, which is recommended for both ambient and high-temperature applications. For very large, percentage strains, or for large fields of view, quite mundane process with large dots is possible e.g. printing with an etched roller or an use of an adhered simple woven textile or wallpaper with striper [23]. For the pitch values in the range 2.5 to 6 dots/mm ink-printed, lithocopies of a negative called a "film tint", on the sheets of paper are commercially available. These printed paper patterns can be fixed at the objects (usually large engineering structures: houses, bridges) using standard paper hanging paste. Table 2.1 summarize the avaliable methods to produce an amplitude specimen grid depending on the material of the object under measurement [18,23,24].

Table 2.1 Specimen grating patterns associated with different material

Metals	Masonary	Plastic, composites	Others
Non-ferrous 1-4　Steel 1-4	Concrete 1-3, 6　Brick, 1,2, 6　Timber, 1, 2　Clay 4	PCV, 4　Carbon fibre, 4　Rubber, 4, 6	Textiles, 5　Graphite, 4　Skin 3　Card 3
1, Printed paper patterns; 2, "Stripping film"; 3, Direct printing; 4, Stencilled patterns; 5, Untreated materials; 6, Other: textile, random patterns with tuned imaging optics.			

2.2. Phase gratings

There are two types of phase gratings, PG:
1. PG basing on periodical variation of refractive index. These gratings usually works in transmission and typically are obtained by bleaching of amplitude-type gratings. Transferring an amplitude grating into a phase one increases its diffraction efficiency [21],
2. PG basing on periodical variation of the relief. These gratings usually are applied for reflection arrangements. Here we will mainly focus on the relief gratings used for various moiré interferometry setups [17].

The specimen relief grating is produced by moulding process or direct exposure of the element covered with photosensitive emulsion. A typical moulding process is shown in Fig. 2.1. A pool of liquid adhesive is poured on the specimen and squeezed into a thin film by pressing against the mould. Epoxy adhesives are suitable. After polymerization, the mold (e.g. photographic plate) is pried off, leaving a reflective diffraction grating bonded to the surface of the specimen. The weakest interface in the system occurs between the gelatin of the photographic plate and the evaporated aluminium or gold, which accounts for the transfer of reflective film to the specimen. The result is a reflective, high-frequency, phase-type diffraction grating formed on the specimen. Its thickness is about 0.025 mm and the most popular frequency is 1200 l/mm. The master grid, from which the moulds themselves are produced, generally originates for patterns produced by two-beam interference, which are recorded on holographic plates or in a photoresist. Sometimes the master grid is used to produce a submaster in silicone rubber and this submaster is used to replicate the phase grating in epoxy on the specimen surface [25], as explained above, but not using Al or Au layer in between. In thise moulding process the surface is not particularly important - apart from cleanliness - since the curing materials, being 100 per cent solids, fill the surface imperfections.

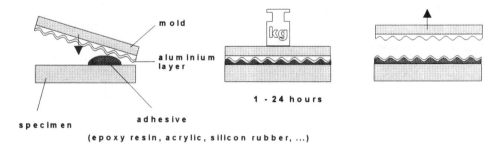

Fig. 2.1, Steps in producing the specimen grating by casting or replication process

This is a great advantage, however in numerous applications the considerable thickness of the epoxy layer cannot be accepted. In such cases the zero-thickness grating are produced on the specimen [26]. It mainly refers to micromeasurements in which the additional layer may change the performance of the element under test [27]. Also the experiments performed in high temperature require special grating technology [28]. The basis for production of zero-thickness grating is covering the specimen with thin (~1 μm) photoresist layer and then exposing it with two interfering laser beams (Fig. 2.2)). If simple, reflectance grating is required (e.g. for material or fracture mechanics studies), the photoresist is developed and the relief is covered (by evaporation) with a thin aluminium or gold layer (Fig. 2.2b). If the grating totally integrated with silicon element is needed (MEMS, MOEMS), the development is followed by etching procedure (Fig. 2.2a) and the grating is produced directly in the material of the specimen. The most sophisticated technology is required for high-temperature resistant gratings (Fig. 2.2c).

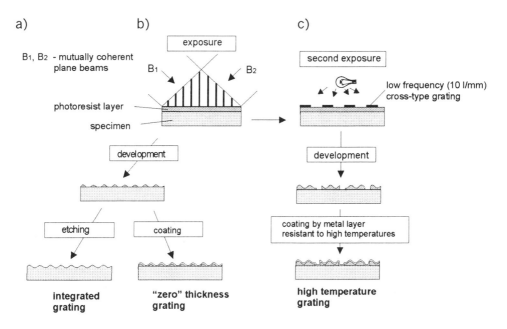

Fig. 2.2, The stages of producing zero-thickness specimen gratings: a) grating integrated with the specimen, b) "conventional" zero-thickness grating and c) high-temperature resistant grating

After first exposure the second exposure through low frequency cross-type grating (e.g. 10 l/mm) is performed and than the photoresist is developed. It leaves the specimen with the islands of high frequency grating, which is coated by metal layer resistant to high temperatures. At elevated temperatures (up to 1200 C) the photoresist degrades completly, but the metal film retains sufficient integrity to give good quality interferometric fringes.

All zero-tickness gratings require good surface polish. The highest frequency required for optical moiré interferometry with immersion is up 3000 lines/mm. However the works connected with electron beam moiré interferometry aim now in achieving 10 000 lines/mm specimen grating. It may be reached by using electron beam to write the specimen array [29,30].

3. PRINCIPLES OF AUTOMATIC ANALYSIS OF RESULTS

3.1. Introduction

It was shown in Section 1 that displacement information is encoded in the phase modulation function of the carrier gratings or the moiré patterns or interferograms formed by these gratings upon plane wave illumination. Irrespective of their source, this information is given as the intensity distribution of a general form:

$$I(x,y) = a_0(x,y) + \sum_{n=1}^{\infty} a_n(x,y)\cos n(\phi(x,y)+\alpha) \qquad (3.1)$$

where a_0 is the background, a_n is local contrast of n-th Fourier component of a fringe pattern and ϕ is the unknown phase.

This intensity must be processed, phase has to be retrieved and derivatives computed. The unknown phase should be extracted from Eq (3.1), although it is screened by several other functions. Also the intensity depends periodically on the phase which causes two additional problems for evaluation:
- due to periodicity, the phase is determined mod 2π (the "discrete" moiré law [11] $\cos(\phi)=\cos(\phi+2N\pi)$)
- due to the even character of the cosine function i.e. $\cos\phi=\cos(-\phi)$ the sign of ϕ cannot be extracted from a single measurement of I(x,y) without "a priori" knowledge.

Because of the above features of a fringe pattern, FP, the proper acquisition and data processing system is not usually sufficient to retrieve uniquely the phase from a single fringe pattern. Many of the early attempts to automated fringe pattern analysis used image processing techniques such as skeletonizing (reduction of the fringes to lines by erosion) and fringe tracking. Although these intensity - based (or passive methods) are still sometimes used [31], methods based on extraction of the underlying phase distribution are more common (these methods are known as phase or active FP analysis). They have significant advantages over the intensity-based methods; data are obtained over the full field, not just at the fringe maxima and minima; the sign of deformation is given; and immunity from noise is usually better. Below only phase methods will be considered; intensity - based are, however covered in details in references [32-34].

The relationship between phase $\phi(x,y)$ and the displacement field at the surface of the specimen is determined by the optical configuration used to acquire data. In geometric moiré, in which a specimen grating bonded to the specimen surface is superimposed on to a stationary reference grating, the in-plane displacement field is related to the phase changes of the resulting moiré fringe pattern through the equation:

$$u(x,y) = \frac{p\phi(x,y)}{2\pi} \tag{3.2}$$

where grating lines are assumed to be oriented parallel to the y axis.

In the interferometric techniques, in which two conjugate beams are recombined through diffraction at the specimen surface, the relationship between the displacement component u and ϕ is:

$$u(x,y) = \frac{\lambda\phi(x,y)}{4\pi \sin\theta} \tag{3.3}$$

and for usual moiré (grating) interferometry arrangement the values of illumination angle θ is related to the pitch of the specimen grating:

$$p = \frac{\lambda}{\sin\theta} \tag{3.4}$$

The relations (3.2) and (3.3) give the basis for the proper phase scaling after fringe pattern analysis.

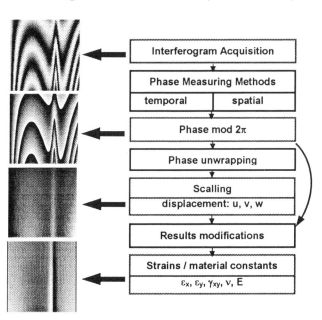

Fig. 3.1, The main stages of automatic fringe pattern analysis

3.2. Fringe pattern analysis

The phase methods of automatic fringe pattern analysis rely on active modifications of its phase to provide more information to solve the regularization and sign ambiguity problems. The main stages of the FP analysis (Fig. 3.1) involve a fringe

pattern acquisition, its analysis by temporal (static events) or spatial (dynamic events) phase measuring methods and numerical derivation of strains or material constants distribution. The result of the principal algorithms is usually phase mod (2π), so it has to be unwrapped in order to obtain continuos phase function. Then the phase is scaled, often smoothed (filtered) and numerically differentiated to get strain maps. Sometimes the strains are calculated directly from the scaled values of phase mod (2π).
Below we focus on the principles and practical aspects of two methods:
- temporal phase shifting [34] supported by its spatial (SPS) and spatial carrier (SCPS) versions [35],
- Fourier transform, FT, method, called also spatial heterodyning method [35,36], which are most often used in the automatic moiré and grating interferometry systems.

3.2.1 Phase shifting methods

The principle of phase shifting method (PSM) resets on setting a file of phase shift values $\delta=\delta_i$ i=1, 2, ...N and fitting a cosine function to the corresponding set of intensity values I_i at a single point. The basic fringe pattern which may be evaluated by PSM is given by:

$$I_i(x,y)=a_0(x,y)+a_1(x,y)\cos[\phi(x,y)+\delta_i] \qquad (3.5)$$

where sequential intensities I_i may be obtained at discrete time intervals (temporal PSM or simultaneously in time but, are separated in space (spatial PSM)). TPSM and SPSM are mathematically equivalent, although TPSM was developed first [37] and is used in the majority of commercial interferometers. If $\delta_i=i2\pi/N$ then the phase is calculated according to formulae:

$$\phi(x,y)=\arctan\frac{\sum_{i=1}^{N}I_i(x,y)\sin\delta_i}{\sum_{i=1}^{N}I_i(x,y)\cos\delta_i} \qquad (3.6a)$$

or in more general term

$$\phi(x,y)=\arctan\frac{\sum_{j=0}^{M-1}\beta_j I_i(x,y)}{\sum_{j=0}^{M-1}\alpha_j I_i(x,y)} \qquad (3.6b)$$

where α_j and β_j are the real sampling coefficients of the sinusoidal signal (Eq. 3.5) and M is the total number of frames acquired. The set (α_j, β_j) and value $N=2\pi/\delta_i$ (number of frames per cycle) completly specify the performance of a phase shifting algorithm.

For optimal performance of the method in the given experimental conditions (profile of fringe pattern, temporal stability, type of an event), a proper selection of the phase shifting algorithm should be performed [34,38]. The aim is to minimize the systematic and random errors in ϕ while keeping the number of frames M, to a minimum. Errors of order fringe/100 are obtained for smooth-wavefront interferograms (low noise, sinusoidal fringes), however for mechanical specimen and moiré fringes the usually achievable accuracy is between fringe/40 and fringe/20. The main error sources are as follows:

- systematic: effects of nonsinusoidal fringe pattern (effect of higher harmonics, q). If the signal contains significant spectral energy up to a frequency $v=qv_0$ the number of phase shifted frames per 2π required to avoid such errors is $N>1+q$ [38,39],
- systematic errors: effect of phase shifter miscalibration. If the phase shifts differ from the chosen value δ_i the retrieved phase suffers from the error at twice fringe frequency. Several approaches have beem developed to reduce the influence of this type of error [34,38,39]. One class of algorithms is especially efficient since it requires only one extra frame $M=N+1$. If significant harmonic content is simultaneously present the number of samples should be increased to $M=2N-2$ [38]
- random errors: in the presence of random intensity errors the phase has an average standard deviation σ_ϕ given by [40]

$$\sigma_\phi^2 = \frac{2\sigma_1^2}{\eta^2 M a_1^2} \tag{3.7}$$

where σ_1 is the standard deviation in intensity and η is a numerical factor ranged from 0.8 to 1.

Table 3.1 shows some of the more important phase shifting algorithms, taken from the extensive collection listed in [40]. Algorithms 1 and 2 require the minimum number of samples (N=3 and 4). Algorithms 3 and 5 are the corresponding N+1 algorithms i.e. are insensitive to linear phase shift miscalibration, while algorithms 4 and 6 are insensitive to both higher harmonics and linear phase shift error.

A useful alternative to the temporal and spatial PSM is a spatial carrier PSM [42,43], which involves recording just a sinple image with carrier fringes. The equation (3.5) is replaced by the following spatial variation:

$$I(x,y)=a_0(x,y) + a_1(x,y) \cos[\phi(x,y) + 2\pi f_0 x \text{ (or } 2\pi f_0 y)] \tag{3.8}$$

where a proper relation between the sampling frequency of the detector (f_d =1/K, K – number of pixels) and the spatial carrier of fringe pattern, f_0, has to be fulfilled in order to provide the chosen phase shift δ_i between sampling points. Also the assumption of slowly varying phase $\phi(x,y)$ background $a_0(x,y)$ and contrast $a_1(x,y)$ functions should be kept. The most often used algorithm in engineering application of SCPSM is 5-point algorithm with $\pi/2$ phase shift between pixels with modified coefficients due to the necessity to correct for the linear phase term introduced by carrier fringes [43]. The phase at discrete sampling points (i,j) is calculated according to the equation:

$$\phi(i,j) = \arctan \frac{-I(i-2,j)+4I(i-1,j)-4I(i+1,j)+I(i+2,j)}{-I(i-2,j)+2I(i-1,j)-6I_3(i,j)+2I(i+1,j)+I(i+2,j)} \quad (3.9)$$

where i, j are actual pixel numbers, i=1, ..., K, j=1, ... L, with K, L the sampling frequencies of the detector in the x and y directions.

Table 3.1 Common phase shifting algorithms. The coefficients are given in the form $(b_0, b_1,..., b_{M-1})/(a_0, a_1, ..., a_{M-1})$ [41]

Number	N	M	Algorithm sampling coefficients
1	3	3	$\frac{\sqrt{3}(0,-1,1)}{2,-1,-1}$
2	4	4	$\frac{0,-1,0,1}{1,0,-1,0}$
3	4	5	$\frac{0,-2,0,2,0}{1,0,-2,0,1}$
4	4	6	$\frac{0,-2,-2,2,2,0,}{1,1,-2,-2,1,1}$
5	6	7	$\frac{\sqrt{3}(-1,3,3,0,-3,-3,1)}{3(-1,-1,1,2,1,-1,-1)}$
6	6	10	$\frac{\sqrt{3}(-1,-3,-3,1,6,6,1,-3,-3,1)}{1,-1,-7,-11,-6,6,11,7,1,-1}$

Fig.3.2 presents exemplary fringe patterns used in TPSM and SCPSM together with the calculated phase mod (2π).
In strain analysis it is sometimes necessary to measure two or even three displacement maps (u, v, w) simultaneously. This can be done by spatial frequency multiplexing. Each fringe pattern is set up with uniquely oriented spatial carrier fringes e.g.

$I(x,y) = a_0(x,y) + a_1(x,y)\cos[\phi_x(x,y) + 2\pi f_{0x}x] + b_1(x,y)\cos[(\phi_y(x,y) + 2\pi f_{0y}y)]$ (3.10)

The component phases can be separated either in the spatial domain by two-directional SCPS [44], as explained in Fig.3.3 or in the frequency domain [45].

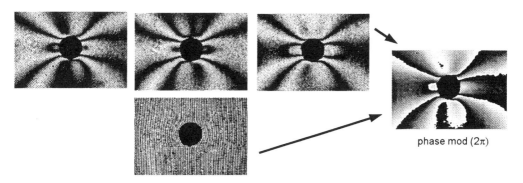

phase mod (2π)

Fig.3.2, The exemplary fringe patterns required for TPS and SCPS methods of analysis and calculated phase mod (2π)

3.2.2. Fourier transform method

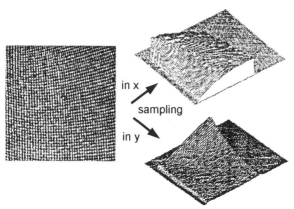

Fig.3.3, An exemplary analysis of cross-fringe interferogram by two-directional SCPS method

Since the spatial carrier frequency f_0 (Eq. 3.8) is introduced into fringe pattern, its analysis may be performed in the frequency domain [35,36]. The Fourier transform of an exemplary interferogram is shown in Fig.3.4. The intensity function is Fourier transformed and the information peak is separated in the frequency domain with a filter with a sharp frequency cut-off. The separated spectrum is inversly Fourier transformed (usually after its translating by f_0 on the frequency axis) and the phase function is calculated from the relation

$$\phi(x,y) = \arctan\frac{\text{Im}[c(x,y)]}{\text{Re}[c(x,y)]}$$ (3.11)

a) 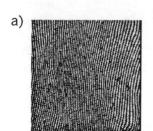 b)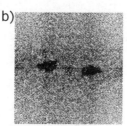

Fig.3.4, a) An exemplary fringe pattern with spatial carrier frequency and b) its Fourier spectrum

where $c(x,y) = \frac{1}{2} a_1(x,y) \exp[i\phi(x,y)]$.

The Fourier transform method is convenient way of FP analysis however a few disadvantages occur when compared with phase shifting methods:
- filtering in frequency domain is a global operation and due to non-carefull processing some information about displacement function may be easily lost, especially in the case when higher order components of information (nonsinusoidal fringes) occurre in the spectrum,
- the sharp frequency-domain filter means that significant leakage occurs between pixels in the spatial domain, causing significant errors near discontinuities such as cracks and specimen boundaries [35].

On the other hand the proper filtering in frequency domain enables to obtain smooth phase function without additional image or/and phase processing. Also filtering and transforming the sequential components of a multiplexed interferogram allows simultaneous recording and latter analysis of complex information (e.g. u and v displacements) [45].

3.3 Phase unwrapping

The absolute phase of a point $P(x,y)$ located within the analysed domain is defined as [46]:

$$\phi(x,y) = N(x,y)2\pi = \left|N_R + \tilde{N} + \hat{N}\right|2\pi \qquad (3.12)$$

where N – refers to the actual fringe number, N_R is the known (integer) fringe order of a reference point P_R, $\tilde{N}$ is the integer part and $\hat{N}$ fractional part of fringe order in $P(x,y)$.

If spatial or temporal phase measurement methods are used, the fractional fringe order $\hat{N}$ can be calculated from the phase mod (2π) delivered by these methods i.e.:

$$\hat{N} = \frac{1}{2\pi}\phi\bmod(2\pi) \qquad (3.13)$$

However the fringe counting problem remains since the integer part of $\hat{N}$ has to be provided. It is realized by phase unwrapping (or phase demodulation) techniques [47,48] which may be defined as the process of resolution of the 2π- discontinuities by adding a step function consisting only of 2π ·

steps. This process can be performed under the condition that the neighbouring phase samples satisfy the relation: $\cdot \pi \le \Delta\phi(i,j) < \pi$ with $\Delta\phi(i,j) = \phi(i,j) \cdot \phi(i-1,j)$. Unfortunately unique phase unwrapping is often hindered by the existence of singular points or defects in the phase mod (2π) map which are generated by noise and/or by under sampling and/or discontinuous height steps. In order to avoid some of the problems connected with these singularities several strategies have been proposed [47] including:
- blocking up inconsistent unwrapping paths by cut-line method [48] or by masking bad pixels [49],
- building up a reliable unwrapping path by minimum/maximum spanning tree method [50]
- phase unwrapping without path integration (these are computative extensive methods, e.g. [51,52], usually not applied in our applications)
- phase unwrapping using higher dimensions by additional exploring a time axis (temporal unwrapping [53])

The phase unwrapping strategies listed above and the related algorithms have various capability to cope with the singular points. If the object to be anlysed is not simply connected, i.e. isolated objects or shaded regions occur, then the mod (2π) unwrapping procedure fails because of violated neighbourhood conditions and unknown phase jumps greater than π. However usually phase fringes obtained from moiré techniques and grating interferometry are successfully unwrapped by medium sophisticated phase unwrapping methods e.g. minimum /maximum spanning tree method.

The value N_R (Eq. 3.12) required for the absolute phase determination usually is not important in experimental mechanics, as during differentiation performed for strain calculation it turns to zero.

3.4 Calculation of strain fields

The result of analysis described in the previous sections is one of the two components of displacement (u, v). In practise, there is often interest in in-plane strains which require differentiation of displacement data. The in-plane strains are given by the equations:

$$\varepsilon_x = \frac{\partial u}{\partial x} = \frac{u_{i+p} - u_{i-p}}{2p\Delta x'} \qquad (3.13a)$$

$$\varepsilon_y = \frac{\partial v}{\partial y} = \frac{u_{i+p} - u_{i-p}}{2p\Delta y'} \qquad (3.13b)$$

$$\gamma_{xy} = \frac{\partial u}{\partial y} + \frac{\partial v}{\partial x} = \frac{u_{i+p} - u_{i-p}}{2p\Delta y'} + \frac{u_{i+p} - u_{i-p}}{2p\Delta x'} \qquad (3.13c)$$

where $2p\Delta x'$, $2p\Delta y'$ – distance between the points at the sample

$\Delta x' = (1/\beta)\Delta x$ and $\Delta y' = (1/\beta)\Delta y$

$\Delta x, \Delta y$ – distance between pixels at CCD matrix plane, 2p – number of pixels over which differentiation is performed.

Numerical differentiation of noisy data is very difficult. Finite difference formulae (Eq. 3.13) are normally insufficiently robust and it is necessary to filter the data by, for example, fitting low-order polynominals over a small region of the field or by performing filtering in spatial or frequency domain [41]. The simplest method is a linear fit of the form

$$(u(m,n) \text{ or } v(m,n)) = \alpha m + \beta n + \gamma \qquad (3.14)$$

over a square subregion of side (2p + 1) pixels [54]. If the central pixel of the subregion is defined to be the origin (m=n=0) and none of the pixels are excluded, then the estimators α and β become:

$$\hat{\alpha} = \frac{\sum mu(m,n)}{\sum m^2}, \qquad \hat{\beta} = \frac{\sum nu(m,n)}{\sum n^2} \qquad (3.15)$$

If all the u are independent random variables with standard deviations σ_u, then it is easy to show that for p>>1 the standard deviation of the estimated strain values is given by:

$$\sigma_{\varepsilon_x} = \frac{\sigma_\alpha}{L} = \frac{\sqrt{3}}{2} \frac{\sigma_u}{p^2 \Delta x'} \qquad (3.16)$$

The quantity $2p\Delta x'$ can be regarded as the effective gauge length, L, of the strain sensor and defines the spatial resolution (smallest resolvable feature) of the strain field. Equation (3.16) shows that as progressively more and more detail within the strain map is sought, so the resulting strain become less and less reliable. For an order-of-magnitude example, the following conditions are given: a moiré interferometer with a field of view of 50 mm and camera resolution of 512x512 ($\Delta x'$=100 μm), a subregion of 15x15 pixels (p=7) and a typical displacement measurement error σ_u=10 nm to obtain σ_ε=2 μstrain.

4. THE MOIRÉ FRINGE METHODS

As it was stated in Chapter 1 the possible way to detect the modulation is to use moiré, an optical means of enlarging the changes that the carrier grating has experienced. Although in most of the cases the carrier itself cannot be detected the changes experienced by the carrier because of the spatial modulation can be made visible by producing moiré fringes. This is

achieved by superposing to the modulated grating, SG, a second grating called the reference or master grating, R, (see Eqs.(1.5) – (1.7))
Different optical arrangements can be used to record moiré patterns as shown in Fig.4.1.

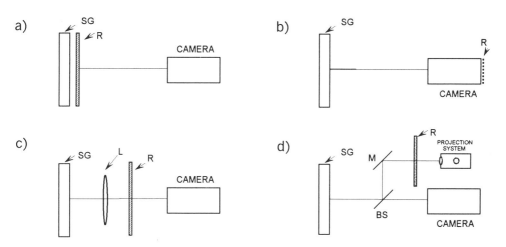

Fig.4.1, Different optical arrangements to generate moiré patterns: a) physical superimposition and b), c), d) configurations of projection superimposition of gratings; R – reference grating, SG – specimen grating

4.1. Physical superimposition of gratings

This is the most obvious method, in which an undeformed master grating is laid directly on a deformed specimen grid producing moiré, which is then recorded by a camera (Fig.4.1a). This has the benefit of having no need to resolve the individual lines, so that the observer may, stand well back, and so be able to view a reasonably large area [2,3,5,18]. Problems with physical superimposition in this way are that out-of-plane movements will produce apparent in-plane deformation and that diffraction effects place a limitation on how fine a pitch may be used, particularly when working in reflection, when the light has to pass from the specimen back through the analyser. Another factor, generic to all system, is that it is necessary either to be fastidious in producing the specimen grid such that it is of good regularity (i.e. very sparse fringes when undeformed) or to record those 'zero-field' fringes and perfome a subsequent operation of substrating the initial spurious displacement fields from those observed under load. This is not a great problem with modern computerized systems, but too many 'zero-field' fringes can render rather difficult the real-time appreciation of what is happening. A very practical optical arrangement for recording of high quality moiré fringes is shown in Fig.4.2a [18] together with exemplary u and v – displacement moiré fringes obtained during bending of polyurethan beam (Fig.4.2b,c). In such arrangements an oil film is often added between two gratings to improve the contrast of moiré fringes.

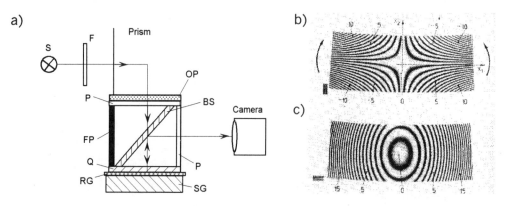

Fig.4.2, Geometric moiré a) the practical arrangement for recording of moiré fringes and exemplary fringe patterns representing: b) u(x,y) and c) v(x,y) displacements of purely bended polyurethan beam [18], S – light source, F – filter, SG, RG – specimen and reference gratings, P – polarizer, OP – opal glass, Q · λ/4 wave plate

4.2. Projected superimposition

With relatively coarse gratings, it is feasible to project reference grating on to the specimen using an imaging system (Fig.4.1d) or, in reserve, project an image of the specimen on to reference (Fig.4.1b,c). These have the feature that by defocusing the lens, or using a small aperture, what is projected contains effectively no harmonic of the grating higher than the first, so that the simpler algorithms of fringe pattern analysis may be employed. The difficulty is that non-flatness of the specimen, or distortions due to lens aberrations, will cause errors that are not obvious. It is of some importance that any instability in the connection between the two gratings will result in movement of the fringes. This is more than a mere inconvenience; any relative rotation will produce spurious shear strain. On the other hand this effect may be used for implementing temporal phase shifting method [34].

4.3. Double exposures and moiré photography

To circumvent the problems of lens aberration, double exposure is a possibility. In this method, a first exposure is made of the grid undeformed and a second of the deformed state. The developed film, with suitable filtering to separate x and y fields, exhibits the moiré (computerized analysis may achieve the same filtering effect). The problem here is that the moiré fringes cannot be manipulated to facilitate the fringe pattern analysis by temporal or spatial methods. The better solution is to record two negatives with the grating before and after loading of the specimen. Due to manipulation of the negatives the required features of a moiré fringe patterns may be designed. In the case of higher frequency grating negative (f>100 lines/mm), a "sandwich" optical analyser [35] (Fig.4.3) is used. The reference, N0, and deformed, N1, negatives are in close contact, separated by the film base layer or controlled gap. Here temporal PSM is applied by introducing sequential phase shifts through tilting the negative pair about an axis normal to the diffraction direction. The presence of the base layer introduces a parallax between images and corresponding phase shift (displacement) of moiré fringe pattern occurs according to the relation:

$$d_x = t/n \tan\theta \qquad (4.1)$$

where t, n are the thickness and refractive index of the base layer, θ is the rotation angle of the negative sandwich.

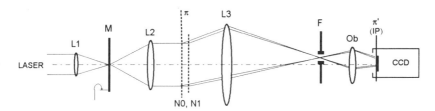

Fig. 4.3, The sandwich optical analyzer: L1 – L3, Ob. – lenses, M – rotating diffuser, N0, N1 - the sandwich of two negatives, IP – image plane

Fig. 4.4, The v – displacement fringe pattern obtained due to loading of Victorial Canal bridge [24]

This approach to moiré fringe analysis is especially useful in so called high resolution moiré photography [23,24]. In this optical arrangement a modified objective is used to record high frequency gratings. By means of limiting the aperture of a camera lens to slits positioned so as to tune the lens to the grid frequency (usually 300 lines/mm), much improved depth of focus and contrast of the gratings are obtained, permitting a larger field of view relative to the resolvable pitch. Again, excellent stability in the positioning between the exposures is essential, and out-of-plane effects caused by the loading will affect the accuracy. The studies of several big civil, engineering structures as bridges and building were done with moiré photography. Fig.4.4 shows the result of applying high resolution moiré photography on Victorial Canal bridge [24].

5. GRATING (MOIRÉ) INTERFEROMETRY

5.1. Introduction

The grating (moiré) interferometry method (both names are used in literature) [17,55-57] represents an substantial improvement of the geometric moiré. It employs high frequency reflection type phase grating fixed to the plane surface of the object under load. It has excellent characteristics like real-time whole field mapping, submicron sensitivity, high interference contrast, wide strain range, and easy alignment and operation. Grating interferometer (GI) is used in several laboratories for complete strain analysis in the elastic and plastic region including the residual strain analysis. The method is applied for material behaviour studies - it suits ideally to investigation of composite materials in both the macro- and micro-mechanics domains. Other exemplary applications include fracture mechanics, studies of joints of various types and verification of engineering designs by supplementing the method of finite element analysis.

The aim of this chapter is to present the principles of grating interferometry and various optomechanical GI systems coupled with automated interferogram analysis which have the capabilities to become the experimental tool in hybrid techniques for analysis and modelling modern problems in experimental mechanics and engineering.

5.2 Principle of grating interferometer

The principle of high sensitivity grating interferometry (GI) with conjugate wavefronts is shown schematically in Fig. 5.1. Two mutually coherent illuminating beams Σ_A and Σ_B impinge on the reflection type specimen grating SG at the angles tuned to the first and minus first diffraction order angle of SG. The +1 diffraction order of Σ_A and the −1 order of Σ_B propagate coaxially along the grating normal. Their wave fronts $\Sigma_{A'}$ and $\Sigma_{B'}$ are no longer plane.

The amplitudes in the detector plane D conjugate to SG can be described as

$$E^A_{+1}(x,y) \cong \exp\left\{ i \left[\frac{2\pi}{p} u(x,y) + \frac{2\pi}{\lambda} w(x,y) \right] \right\} \qquad (5.1)$$

$$E^A_{-1}(x,y) \cong \exp\left\{-i\left[\frac{2\pi}{p}u(x,y) - \frac{2\pi}{\lambda}w(x,y)\right]\right\} \quad (5.2)$$

where p is the spatial period of the specimen grating whose lines are perpendicular to the x axis (lying in the figure plane), u(x,y) is the in-plane displacement function corresponding to the departure of the grating lines from straightness, w(x,y) is the out-of-plane displacement function corresponding to the deformation of the specimen surface under load. For simplicity, the amplitude of the diffraction orders has been normalized to unity.

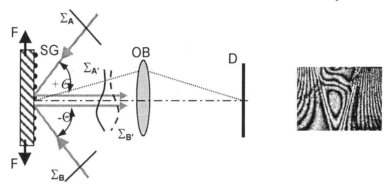

Fig. 5.1, Schematic representation of a double illuminating beam grating interferometry for in-plane displacement studies. SG – specimen grating; OB – imaging optics; D – detector plane; Σ_A, Σ_B – wavefronts of +1 and −1 diffraction orders.

It can be shown that wave front deformations caused by in-plane displacements are equal in both diffraction orders but longitudinally reserved. On the other hand, the wavefront deformations due to out-of-plane displacements have the same value and sign in both interfering beams. Therefore the influence of out-of-plane displacements is eliminated by the interference. Thus assuming that out-of-plane displacements give only small variations in slope of the wave fronts (the departure from this condition was theoretically in investigated in [58]), the intensity distribution of the interferogram becomes

$$\left|E^A_{+1} + E^B_{-1}\right|^2 \cong 2\left[1 + \cos\left(\frac{4\pi}{p}\right)u(x,y)\right] \quad (5.3)$$

The fringes observed in the interferometer represent a contour map of in-plane displacements with half a period sensitivity. For example, when using the specimen grating of spatial frequency 1200 l/mm the basic sensitivity is 0.47 μm per fringe order.

When the double beam illumination system is angularly misaligned, the carrier fringes are introduced into the interferogram. The last equation transforms into

$$I(x,y) = 2\left\{1 + \cos\left[2k\theta_x + \frac{4\pi}{p}u(x,y)\right]\right\} \quad (5.4)$$

where θ_x designates the angle between the diffracted orders and the grating normal, $k=2\pi/\lambda$.

To obtain a complete strain information a crossline specimen grating is commonly used. A crossline grating requires two pairs of illuminating beams for its readout as shown in Fig. 5.2. such configuration provides the interferograms representing u - displacement (A+B), v - displacement (C+D) or both informations simultaneously.

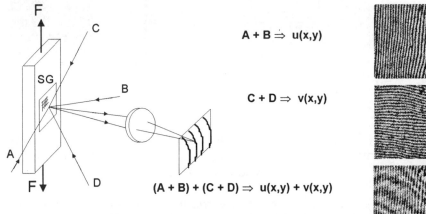

Fig. 5.2, The scheme of four – beam GI for sequential and simultaneous u(x,y) and v(x,y) interferograms capturing

5.3. Grating interferometer systems

Due to great variation of the measurement requirements, they cannot be simultaneously fulfield by a single system. This is the reason why four basic types of automated grating interferometers (AGI) are proposed and designed: the laboratory (LGI) [59-61], workshop (WGI) [55,62], sensor type (FOS) [63] and microinterferometer systems [64,65].

The laboratory system is understood as open optomechanical arrangement easy to modify for various experimental requirements, providing high accuracy results, however, demanding stable conditions (stabilized tables, constant temperature, etc.) and operated by a specialist. The laboratory system can be easily modified to provide additional information about out-of-plane d2isplacements.

The workshop portable system is a compact system which is much less sensitive to vibrations, easy to operate, however, the accuracy of the measurement may be lower and it cannot be easily modified. It is used for the analysis of influence of environmental changes (in temperature chamber), analysis of transient events and the measurement directly on a standard loading machine.

The sensor-type system is a specialized fibre optic realization of grating interferometer. Its small size, compactness and resistance to vibrations enable to use the sensor directly at machine and civil engineering structures. This system is indicated here more like a future recommendation than the actually operating sensor, although the preliminary works on its modelling and design have been done.

The microinterferometer based on waveguide head is a compact setup designed to work together with commercial optical microscope (small fields of view from a few mm^2 to hundreds of μm^2) with enhanced sensitivity due to the possibility to use immersion between waveguide glass block and specimen grating.

Although the optomechanical arrangements in all cases are different, the methodology of displacement measurement and analysis of the results is similar [27,59]. It is presented in Fig. 5.3.

As the system has to enable the analysis of interferogram captured under various conditions, including transient events and measurement in unstable environment, the special attention should be paid to the problem of fringe pattern design and its capturing in one- or multichannel optomechanical arrangements. The analysis of interferograms is performed alternatively by temporal, spatial or spatial carrier phase shifting methods.

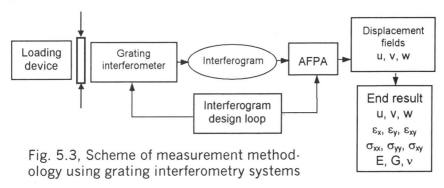

Fig. 5.3, Scheme of measurement methodology using grating interferometry systems

5.3.1. The laboratory system (LGI)

The three-mirror four-beam grating interferometer (3MGI) developed at the Virginia Polytechnic Institute and State University [60] was adopted and modified in author's research group to enable out-of-plane displacement measurement and automatic interferogram analysis [59]. Figure 5.4 shows a

schematic representation of the basic configuration of one-channel, combined 3MGI/Twyman-Green interferometer, which may be used for sequential u, v and w displacement measurement. To facilitate the automatic computer-aided analysis two approaches to implementation of temporal phase shifting technique have been proposed:
- the polarized light approach [66] (Fig. 4.4 - solid line). The specimen grating is illuminated by two beams with orthogonal polarization and the interference is facilitated by analyser (A). In the case of the u displacement field, opposite directions of polarizations are due to the different number of reflections of two illuminating beams (A1 and A2). In the case of measuring the v displacement field, the number of reflections is the same for both (A1 and A2) illuminating beams so the change of polarization in one illuminating beam should be introduced by the half-wave plate (H1). Similar situation occurs for the w displacement. In order to supply the orthogonal polarized beams in the Twyman-Green interferometer configuration used for specimen shape measurement, the additional quarter wave-plate is inserted in front of the mirror MR. This configuration enables alternative measurement of u, v and w displacements, while the proper parts of beam are cut off and the phase shift is performed by the rotating analyser according to the equation:

$$I(x,y) = a(x,y) + b(x,y) \cos[\phi(x,y) + 2\alpha] \tag{5.5}$$

where: $\phi(x,y) = (4\pi/d)[u(x,y) \text{ or } v(x,y)]$, and α - angle by which analyser was rotated.

- the tilting parallel plate approach [67] (Fig. 5.4 - dotted line). In this case, the phase shift is introduced by tilting the plate PP inserted either into the beam A2 (for v measurements) or into the beam B2 (for u measurements). The same method is used for automated measurement of the w(x,y) by inserting PP into PS-MR arm of the Twyman-Green interferometer.

The temporal phase shifting method, TPSM, requires stable conditions of measurement, while capturing at least three-phase shifted interferograms. If this requirement is not fulfilled, the GI system should be modified for spatial PSM or spatial carrier PSM.

One of the solutions for the spatial PSM is shown in Fig. 5.5 a. The principle interferogram replication at the output plane is obtained by inserting a sinusoidal transmittance diffraction grating near the focus of objective O2. The required phase shifts are introduced by three properly oriented analysers A, however the normalization of intensities of the images is needed [59].

The spatial carrier PSM requires introduction into interferogram, a controlled number of carrier fringes [61]. This is achieved by tilting either the mirror M1 (for u), mirror M2 and M3 (for v) or mirror MR (for w). The other option for introducing carrier into u and v interferograms is translating the objective CO in the direction perpendicular to the optical axis of the system. This option is also used to compensate for large linear displacements during measurement.

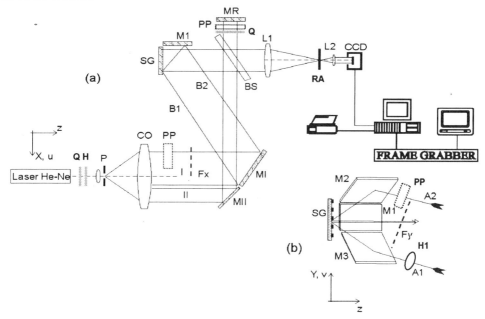

Fig. 5.4, Optomechanical configuration of the three-mirror four-beam grating interferometer modified for w(x,y) measurements. UP – pinhole system, CO – collimating objective, M1, M2, M3, MR MI and MII – mirrors, F_x and F_y – slot filters, SG – specimen gratings, PS – beam-splitter plate, Q, H, H1 · λ/4 and λ/2 phase plates, PP – parallel plate, L1, L2 – imaging optics, RA – rotating analyser

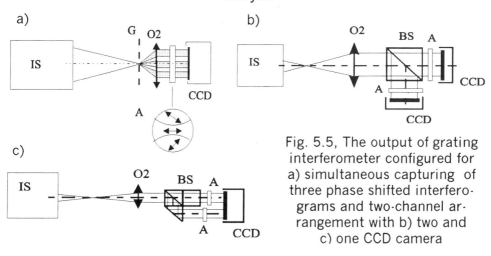

Fig. 5.5, The output of grating interferometer configured for a) simultaneous capturing of three phase shifted interferograms and two-channel arrangement with b) two and c) one CCD camera

The one-channel interferometric arrangement which allows sequential measurements of u, v and w is sufficient for static events analysis. However the system can be modified for two- or three-channel arrangements in which u, v and w interferograms are produced simultaneously and are analysed by one- or two-dimensional spatial carrier PSM [59,44]. Fig. 5.5b, c show the possible modifications of the GI output based on controlled polarization states of the interfering beams, which allows analysis of dynamic events.

5.3.2. The workshop portable system

The workshop, portable system enables analysis of transient events, displacements and strain measurements directly on a standard loading machine (Instron, Schenck) and/or unstable (often industrial) environment. The optomechanical arrangement, shown in Fig. 5.6 is based on achromatic grating interferometer (AGI) scheme [60,62] modified for the automated analysis of interferograms captured by CCD camera. Additionally a video recorder for storing sequential fringe patterns during dynamic loading may be applied. It was proved that due to applying the compensating grating (CG), the system is relatively insensitive to vibrations and can be used directly on a loading machine.

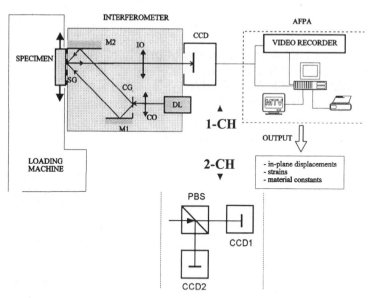

Fig. 5.6, Scheme of the portable, automatic GI system. DL – laser diode, CO – collimating objective, CG – compensating grating, M1, M2 – mirrors, IO – imaging optics, PBS – polarization beam splitter, 1-CH, 2-CH – one and two-channel detecting head

The system may work in one-channel (1-CH) and two-channel configurations (2-CH). In 1-CH option, the analysis may be performed alternatively for u and v displacements by SCPSM [61], or for two-directional SCPSM for simultaneous analysis of both displacements [44]. In the latter case, the sum-type cross-interferogram is obtained, while two orthogonally polarized pairs of beams interact. In 2-CH option, the single interferograms for u and v displacements are obtained simultaneously at two CCD cameras due to inserting the second polarization beam-splitter at the output of the system. This second option allows to enlarge the measurement range and accuracy of simultaneous u and v analysis, when compared with two-directional SCPSM.

5.3.3. Fibre optics grating interefrometer sensor (FOS)

The implementation of bulk optics version of grating interferometers is often restricted due to the relatively large size of the system. The fibre optics version of GI allows a reduction in the number of optical components to be adjusted. Due to mechanical flexibility of fibres and the small size of fibre optics components, miniaturisation of the sensor can be achieved. A simple sensor may be built for one-channel system. Fig.5.7 shows more advanced two-channel system in which two single mode laser diodes with different wavelengths are employed in conjunction with properly selected interferometric filters coupled with CCD cameras [63]. In this system the advantages of using fibre optics and laser diodes may be applied e.g. tuning the wavelength of LD may be used for introducing proper carrier frequency into interferogram, while streching fibre optics on PZT cylinder may introduce the required phase shift into fringe pattern.

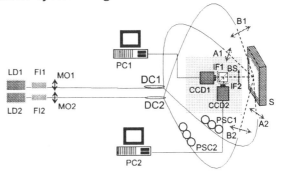

Fig. 5.7, Sensor type GI system. LD - laser diode, FI – Faraday isolator, MO – microscopic objectives, DC - directional coupler 50÷50, BS – beam splitter, PSC – controller of polarization, IF – interferometric filter

5.3.4. Waveguide grating microinterferometer (WGI)

Many fields of study require displacement measurements of tiny specimens or tiny regions of larger specimens. The small size and the need for high spatial resolution requires microscopic viewing of the specimen. Most often for this purposes the optical microscope with visible light is used, although more and more often the experiments with GI at electron beam microscopes are performed [30]. The concepts of microinterferometer are similar as in the case of the previous designs however usually they are realized by means of waveguide head [64,65]. The illuminating beams are introduced into a glass plate GP, with parallel plane surfaces and they are guided by the glass plate due to a total internal reflections at top and bottom surfaces of GP or reflections by mirrorised top and bottom surfaces of GP. A modern realization of such waveguide microinterferometer is shown in Fig. 5.8. Due to the ussage of the glass plate as the light guiding medium the immersion can be easily applied between the interferometer head and the specimen grating. The presence of immersion allows to enhance the sensitivity of displacement measurement i.e. the frequency of the virtual grating formed in a refractive medium is equal:

$$f_m = \frac{2\sin\alpha}{\lambda_m} = \frac{2n\sin\alpha}{\lambda} \tag{5.5}$$

where λ_m – the light ware in medium, i.e. the frequency increased by a factor of n (where n is the refractive index of the immersion fluid).

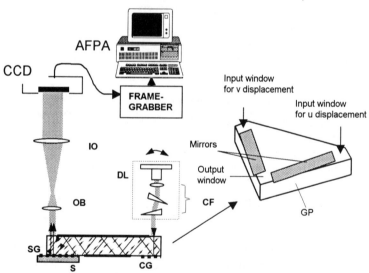

Fig.5.8, The scheme of the waveguide grating microinterferometer. DL · laser diode, CF · collimating and beam forming system, SG · specimen grating, OB and IO · imaging, optics, GP – glass plate, AFPA · analysis system

Table 5.1 gives comparison of various conventional and waveguide realisations of grating interferometers including their sensitivity to vibrations and achromatic light, which are substantial for the engineering applications of the method.

Table 5.1, Conventional versus waveguide realisation of GI systems.

Conventional	Waveguide	Sensitivity to vibration	Sensitivity to $\Delta\lambda$
	prism / GP / immersion	high	high
mirror	mirrorised / GP / immersion	high	high
Compensating grating	GP / immersion / Compensating grating	high	low
Compensating grating	GP / immersion / Compensating grating	low	low

5.4. Engineerig features of automated grating interferometers

The excellent characteristics of grating interferometry as a versatile tool for full-field in-plane displacement/strain measurement induced a variety of opto-mechanical and image processing solutions for whole range of applications. Table 5.2 summarizes the engineering features of the systems presented in this chapter.

This summary does not include the systems which works with electron beam microscopes and use the specimen gratings up to 10 000 lines/mm. These systems are under extensive development now due to the rapid growth of the metrology of MEMS.

Table 5.2, The engineering features of automated grating interferometry systems

FEATURES GI	GRATING FREQUENCY (l/mm)	MEASURING AREA (mm)	SENSITIVITY PER FRINGE (nm)	ACCURACY (PART OF FRINGE)	RANGE OF DISPLACEMENT (μm)	SENSITIVITY TO VIBRATIONS / ACROMATIC	METHOD OF ANALYSIS	MODIFICATIONS FOR w	multichannel
BASIC SYSTEM	400÷1500	up to 100x100	1250÷333	±1/40	120÷30	high / −	TPS SPS SCPS	+	+
LGI	900÷1500	2x2 ÷ 30x3	556÷385	±1/40	60÷40	high / −	TPS SPS SCPS	+	+
PGI	1200	up to 25X25	417	±1/20	40	low / +	SPS SCPS	−	+
SGI	single working frequency from the range 400÷1500	recommended 2x2	1250÷333	±1/10	120÷30	low / −	SCPS	−	+
WGI	1000÷1500 1500÷3000 (with immersion)	recommended 0.2x0.2 ÷ 1.5x1.5	50÷333 333÷166	±1/20	50÷30 30÷15	low / +	SPS SCPS	−	+

6. MATERIAL ENGINEERING AND MICRO-MECHANICS

6.1. Introduction

Current trends in development of micro-scale components, structures and systems, both mechanical and electronic, introduce unprecedented requirements on their designs. To satisfy these demands new materials and structural designs are being employed, which behaviour cannot be easily predicted by FEM methods [69]. The possible misuse of FEM while applied to microdevices lies additionally in the necessity to input material data, which often are not completly available or not adequately describe material behaviour of the micro-scale structures. This is because only a small number of grains of crystals are across the micro-scale features of these geometries, and published data are based on large scale features with hundreds, or thousands, of grains across the bulk samples used in determination of material properties. In small components the grain orientation in metals or fibre orientation in glass or carbon filled plastics have profound local effects on material behaviour including Young's modulus, Poisson ratio, thermal coefficients of expansions or stress relaxation curves. The other problem is connected with the geometry irregularities of the structures especially manufactured by etching. These irregularities although small on the absolute scale, are large portions of the dimensions of the micro-scale structures.
The conventional experimental procedures involving strain gauges, photoelasticity, mechanical probing, etc. are not generally applicable to these

measurements. An alternative to the conventional methods can be provided by full-field optical methods [27] including the most popular: speckle interferometry (ESPI) [70,71], shearography [72] and grating (moiré) interferometry [17,27] supported by fringe projection and computerized interference microscopy [73].

Below we will consider mainly application of automated grating intererferometry with configurations providing convenient tools for micromeasurement under static and / or dynamic conditions.

High quality of interferograms obtained in GI recommends this method for analysis of phenomena with high strain/stress gradients (contact area, fracture mechanics) and on the other hand for small local changes determination, where additional filtering process may distort the final results. GI employed for micromeasurements often requires application of so called "zero thickness" grating (see Section 2.2). Depending on the technical problem various GI systems may be used including:

- the three mirror system extended for out-of-plane displacement measurement by the addition of Twymann-Green interferometer,
- fibre optics version for simultaneous measurement of u and v displacements,
- the waveguide grating interferometer, which sensitivity may be highly increased while using immersion and a very high frequency grating (up to 3000 lines/mm).

Nowdays micromeasurements are strongly required in the case of three basic situation:

- local approach to experimental mechanics, i.e. analysing micro regions in bulk samples [74]. This is especially valuable when gathering experimental data for analysis of contact, fracture mechaniques and fatigue problems in the regions with ambiguous or unknown analytical solution. These interesting experimental mechanics problems are connected with conventionally sized structural elements analysed locally,
- local approach to material engineering i.e. analysing microregions in bulk samples made of various polycrystalline [75,76], composite [77] or "smart" materials [78]. The main aim of these measurements includes material structure studies, which take into account the history of its manufacturing, temperature and environmental dependencies, and determination of local material constants, which are the basis for proper FEM modelling and analysis of microelements. Here we will focus on the anisotropic behaviour of polycrystalline material caused by coarse grain structure and / or privileged orientation developed during plastic working of materials.
- analysis of microobjects. This includes two basic group of structures:
  Micro Electro Mechanical Systems, MEMS, which are micronsized mechanisms (x · xxx μm), often manufactured using VLSI techniques

adopted from the microelectronics industry and used in sensor and actuator applications in biomedicine, microrobotics and nanometrology. Many of these components are (at least in the prototype design) miniature replicas of much larger ones (gears, motors, sensors) only designed to operate on a much smaller scale. Additionally specific groups of microsized mechanical elements are electronic microconnectors which start to be extensively used for high density multichip interconnects (from ap. 400 to 12.000 contacts per device) in electronic packaging [79,80].
Electronic Components and Assemblies in respect to electronic packaging; the main problems connected with these objects include heat transfer phenomena and vibrational analysis.
In this chapter we will address at first to the problem of analysing local material constants of selected materials and latter we will describe the methodology of optimization of the design of microcomponents and electronic chips. Finally we will refer to local analysis of exemplary samples with joints and cracks.

6.2. Local approach to material engineering

Recent technologies enable production of various types of materials including polycrystalline materials, various alloys, composites, and smart structures [27,77] with enhanced mechanical and environmental properties. These properties especially considered in local regions, depend on the chemical and geometrical composition of material and the history of its manufacturing. This is the reason to introduce the special procedures for local and global material properties studies based on hybrid experimental − numerical approach [78] (Fig. 6.1).
Let us at first focus on the problems connected with polycrystalline materials. The global (published) material properties are based on the measurements of bulk samples with more or less isotropic features. In the case of microelements it is necessary to assume, in general, that the polycrystalline materials, due to crystallographic grain anisotropy, have directional properties. Additionally this anisotropy may be enhanced by different type of mechanical (e.g., cold work) and thermal (e.g., annealing) processes.
The knowledge about this anisotropy and possible non-homogeneous plastic strains is very important for proper application of the material. Here the high sensitivity automatic grating interferometry is applied to indicate several unexpected phenomena which occur in micro-scale material engineering.

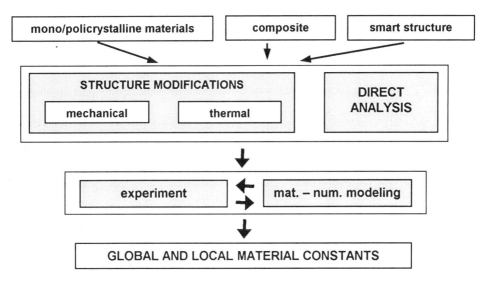

Fig. 6.1, General procedure for material properties studies

Initial experimental investigations were performed by grating interferometry on a simple model of aluminium bicrystal (Fig.6.2) with grain orientation [001]. The specimen was compressed up to the stress equal 28 MPa. The example of in-plane ε_y strain map and its crossection for load F=280N is shown in Fig.4b,c. The regular, band distribution of ε_y strain caused by initial dislocation glide in the glide system {111}<110> can be observed (Fig. 6.2b). Also γ_{xy} shear strain values indicated the rotation of component crystals ($\Delta\gamma_{xy}=10°$).

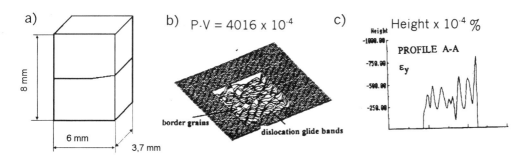

Fig. 6.2, The strain analysis of aluminium bicrystal under compresed load F=280N: a) the geometry of sample, b) 3D map of ε_y strain, c) profile A-A of ε_y map

In general, the analysis of crystal behaviour enables to observe and measure directional dislocation of structure defects such as: glide and climb dislocation, diffusion of point effects, grain boundary sliding and mechanical twin. Further analysis refers to the local tests of polycrystalline aluminium sample [75], Fig.6.3 clearly shows the displacement / strain unhomogeneity correlated with grain borders and grains or grains' complexes which were formed due to the sample pre-processing.

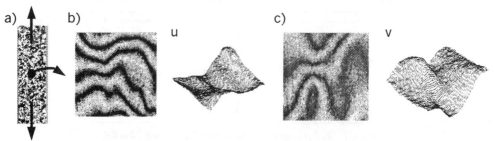

Fig. 6.3, The local analysis of a) polycrystalline aluminium sample, b) local u - interferogram and u - displacement 3D map, c) local v - inteferogram and v - displacement 3D map.

This last effect is shown in Fig. 6.4 and 6.5, which indicate the behaviour of material after hot rolling. Hot rolling causes curing, recrystalization and processes induced by plastic deformation. The intensity of the a.m. phenomena depends on the duration of the rolling process. The Al-Li alloy samples were tensile loaded up to 2% of total strain, while the measurements were performed at 1.5% strain. Fig.6.4 shows ε_x, ε_y strain maps with their statistics for Al-Li alloy 4 mm thick sheet. Due to rolling up to 4 mm thickness, the structure of material is not crystallized yet and the variations of strain distribution are low as indicated by diagrams. After further rolling the sheet of 1mm thickness was obtained (see Fig.6.5), in which partial recrystallization has occurred, as shown in Table 6.1. The increase of the percentage of Cube and Gross components in the crystalographical texture in the second sample causes its significant variations of strain distribution.

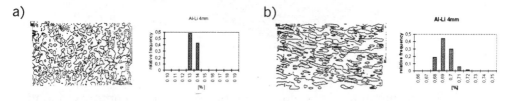

Fig.6.4, Analysis of Al-Li alloy 4 mm thick sheet under tensile load: a) ε_x and b) ε_y contour maps and their variation diagrams.

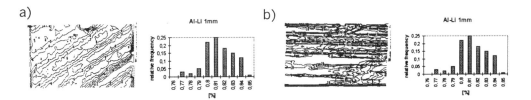

Fig.6.5, Analysis of Al-Li alloy 1 mm thick sheet under tensile load:
a) ε_x and b) ε_y contour maps and their variation diagrams

Table 6.1, The main crystallographic texture components in Al-Li alloy samples

material	main components of texture				
	S (123)<634>	Cu (112)<111>	brass (011)<211>	Cube (001)<100>	Gross (011)<100>
Al-Li 4 mm	45%	35%	17%	·	·
Al-Li 1 mm	30%	18%	8 %	33%	35

These variations of ε_x and ε_y may introduce local significant changes of material constants such as Poisson ratio and Young's modulus, which should be taken into consideration during FEM analysis and design process of microelements.

Other interesting issues in material engineering refer to composite and smart material testing.

The basic for the high performance of composite materials lies in their anisotropic and heterogeneous character. Grating interferometry helps to analyse experimentally the global and local behaviour of various composites under load [3,12]. Fig.6.6a presents an exemplary carbon fibre/epoxy laminate lay-up geometry (45°,-45°)/(-45°,45°)/(90°)$_2$/(0°)/(90°)$_4$/(0°)/ (90°)$_2$ /(45°,-45°)/(-45°,45°) [77]. One ply terminates in the middle to form a tapered panel. The specimen was subject to four-point bending and u and v displacement fields were monitored (Fig. 6.6b,c).

Differentiation of the displacement maps brought strain distribution ε_x, ε_y, γ_{xy} as shown in Fig. 6.6d. The experimental knowledge of these maps enables checking of the existing and forming new FEM models of local behaviour of composite materials with the chosen geometry.

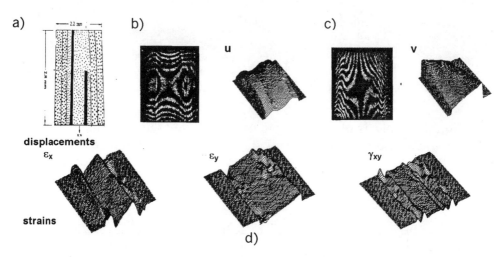

Fig. 6.6, Composite specimen (tapered panel) under four-point bending: a) laminate lay-up geometry interferograms and related displacement maps b) u(x,y), c) v(x,y), and d) strain maps ε_x, ε_y, γ_{xy}.

Another problem which may be considered is the strain determination and crack detection in a flat panel with two ply drop off. The specimen (Fig. 6.7a) with laminate lay-up geometry $|0°|_3|90°|_2|0°||90°|_4|0°||90°|_4|0°|$ $|90°|_4|0°|90°|_2|0°|$ was subjected to four point bending (here 1.35 Nm). The strain maps before and after crack formation at the same loading were calculated from experimental in-plane displacement maps (Fig. 6.7b,c). After the crack is formed, the load released by the crack transfers to neighbouring regions and redistributes the state of strain which results in strain concentration around the crack. It is also noted that the strains ε_x, ε_y and γ_{xy} everywhere are higher after crack forming. It indicates that some forms of plastic deformation have been taken place when a crack was formed at the play drop-off. Away from the crack, the difference before and after crack represents the residual strain in specimen. While designing composite materials for particular application e.g. avionic industry, an important issue is the

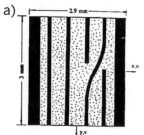

proper choice of the sequence and orientation of layers. Experiments show that e.g. experimental Poisson ratio distribution in glass/epoxy composite materials may vary significantly even in the case of identical mass fracture ($m_r=0.5$) and volume fracture ($v_t=0.389$) [67]. Exemplary results obtained for Iterglass 92145 fabric impregnated with epoxy matrix under tensile load are shown in Fig. 6.8.

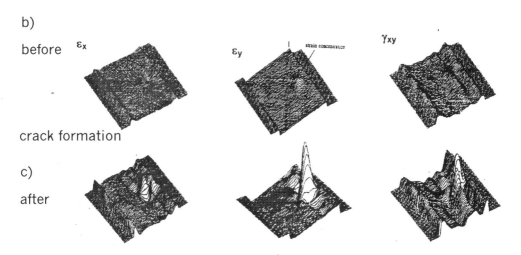

Fig. 6.7, Composite specimen with two ply drop off under four point bending a) laminate lay-up geometry and strains ε_x, ε_y, γ_{xy} b) before and c) after a crack is formed.

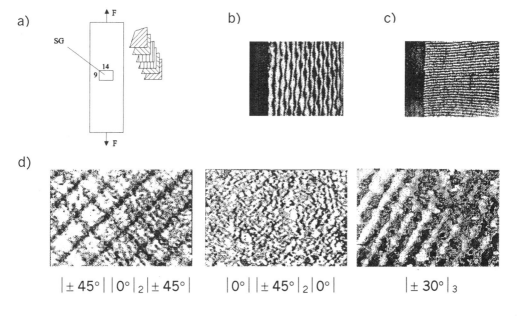

Fig. 6.8, Glass/epoxy composite analysis a) specimen geometry with exemplary sequence of layers, interferograms represented b) u(x,y) and c) v(x,y) displacements under tensile load F=1000N and d) grey level representation of Poisson ratio distributions obtained for different sequence of layers.

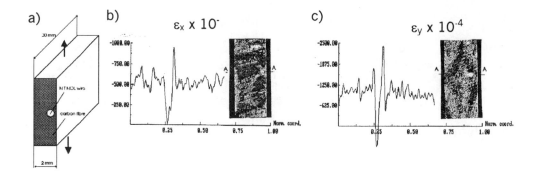

Fig. 6.9, Smart structure with NiTiNOL wire actuator under tensile load: a) the scheme of the specimen and grey level strain maps and their crossections b) ε_x c) ε_y.

Recently, a lot of effort is given to produce „smart structures" i.e. materials which manifest their own functions depending on sensed environmental changes. A large group of these materials are composites with appropriate optical fibre sensors or wire fibre actuators embedded into the structure during manufacturing. There are a lot of problems with fabrication and usage of these materials [78]. One of them is the question how does the fibre effect the composite strength characteristic. Grating interferometry gives the chance to monitor strain fields in such structures. Fig.6.9a shows the geometry of exemplary sample tested: NiTiNOL wire fibre actuator embedded in parallel to the reinforcement in carbon / epoxy composite. Significant ε_x, and ε_y strain gradients were noticed at the interface of the wire (Fig.6.9b,c), indicating the necessity to change the structure geometry. The results presented were obtained for longitudinal strain equal 0.25%. For strain equal 0.33% the specimen was broken.

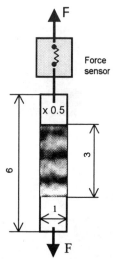

Fig. 6.10, The loading configuration and silicon beam geometry for in-plane displacement measurement by grating interferometry. The interferogram represents v-displacement map for F=0.2 N.

6.3. Microelements testing

To increase the quality and the lifetime of microcomponents an experimental determination of their features (material constants, monitoring of performance, relaxation etc.) must be carried out. Below we present three examples

of microelement studies in which grating interferometry has supported their numerical design.
- silicon micro beam studies

The aim of this experiment was to determine strain and Young's modulus of a silicon beam under tensile load [81] (Fig. 6.10). The specimen was loaded in range 0 ÷ 0.8 N (with step 0.1 N). The force was monitored by the specially designed strain gauge based sensor. The measurement field was 3x1mm. The "zero thickness' grating was produced at the specimen surface. For measurements the microinterferometer combined with glass block waveguide and designed to work with optical microscope was applied. The interferograms obtained for sequential loads were analysed by spatial carrier phase shifting method and then v − displacement maps were differentiated in order to get ε_y strain maps. The plot of average values of ε_y in the sequential stress values (calculated according to $\sigma=F/s$, where s is the field of the sample crossection) is shown in Fig. 6.11. The Young's modulus was calculated ($E=\sigma/\varepsilon_y$) for each measurement point and its average values equals 155 GPa, which is near to the literature values of about 166 GPa. The more advanced studies of silicon microelements should be focused on in-situ measurement of elements with thickness of a few tens of μm, preferably in their own environment. This elements require different approach to specimen grating i.e. producing the sinusoidal microgrid directly in the silicon materials (special etching procedures). Also the problems of proper, controlled loading and specimen fixtures are extremly difficult.

- brass micro beam studies

The aim of this experiment was to test the influence of material pre-processing on the operational applicability of beam-like microconnector [74]. A beam manufactured of 0.25 thick brass sheet was tensile loaded (Fig 6.12a) within elastic range. The interferograms representing u(x,y) and v(x,y) displacements were re-

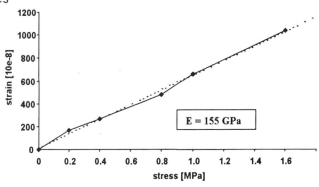

Fig. 6.11, The strain/stress relation obtained for tensile test performed on the silicon beam

corded and analysed. The resultant ε_x and ε_y maps were calculated for the several loads defined by the pusher positions. Figs. 6.12b,c show exemplary 3D ε_x and ε_y plots pusher displacement (for 59 μm) and the x-crossections for the sequential loads (13 μm, 31 μm, 59 μm) pusher displacement. Significant variations of ε_x and ε_y values with a similar character for various loads are clearly shown. They result in variations of Poisson ratio $v=(\varepsilon_{x\,k\cdot 0}\,/\,\varepsilon_{y\,k\cdot 0})$ of the range $\pm\,25\%$, which may effect significantly the strength of the element and its operational applicability.

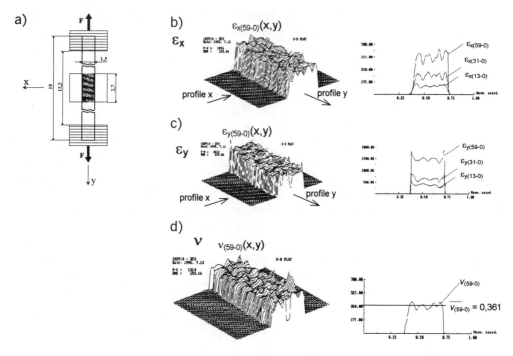

Fig. 6.12, Strain analysis of a brass microbeam under tensile load: a) specimen geometry, b) 3D plot of ε_x distribution and ε_x x-profiles for sequential loads, c) 3D plot of ε_y distribution and relevant ε_y x-profiles, d) 3D plot of Poisson ratio distribution and its x-profile

- MicroInterposer studies

These studies were performed in order to monitor performance of a newly designed type of microelectronic connector with a special emphasis paid to determination of the stress relaxation characteristic of the contact [81]. The new geometry of microconnector specially designed to be arranged in raws staggered with respect to each other has been proposed. It resembles a Greek letter Ω and is called a MicroInterposer. The contact is stamped from 114 μm thick BeCu, plated with nickel and gold and placed in individual cartridges (Fig. 6.13). The experiment had to provide long term monitoring capabilities. For in-plane displacements u, v and out-of-plane displacement with accuracy better that 20 nm. This was performed by means of 3-mirror

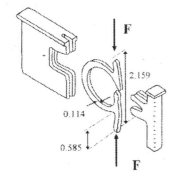

grating interferometer head supported by Twyman-Green interferometer. Exemplary 3D plots for u, v and w displacements are shown in Fig.6.14 together with the relevant FEM simulations.

Fig. 6.13, The MicroInterposer assembly: exploded view of the contact and components of the cartridge

a)   b)   c)

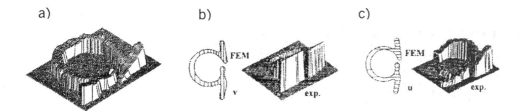

Fig. 6.14, The performance of the MicroInterposer under load:
a) out-of-plane displacement w(x,y), and in-plane displacements b) v(x,y) and c) u(x,y) with the equivalent FEM computed maps

Since the contact was designed to be used in applications where quick connects and disconnects are needed, knowledge of the stress relaxation characteristics of this contact is essential to assure its functionality. The stress relaxation rate; R_σ, can be determined using the relationship

$$R_\sigma(x,y,t) = \frac{\partial\,[\sigma(x,y,t)]}{\partial t} = f\{R_u(x,y,t), R_v(x,y,t)\} \tag{6.1}$$

where: $R_u(x,y,t) = \dfrac{\partial\,[u(x,y,t)]}{\partial t}$; $R_v(x,y,t) = \dfrac{\partial\,[v(x,y,t)]}{\partial t}$

and f denotes a functional dependence which is based on the experimental results.
Experimental monitoring of changes in v displacement in the connector's "leg" is illustrated in Fig. 6.15. The v displacement has changed by up to 225 nm over the time 36 hours with constant load applied, which indicate the possible change of connecting force during performance of the MicroInterposer.

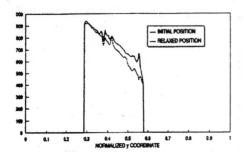

Fig. 6.15, The changes displacement values in connector's in leg after 36 hours relaxation

6.4. Electronic chips and packages studies

The reliability of electronic packages has become an increasing concern for variety of reasons including the advent of higher integrated circuit densities, power density levels and operating temperatures. The reliability depends, among other, on stresses caused by thermal and mechanical loading occuring in electronic packages and their components. The thermal stresses can be produced during fabrication steps and due to non-uniform thermal expansion of the materials comprising the package and semiconductor die. Mechanical loading can be transmitted to the package through the printed circuit board to which it is mounted. Another critical issue is the non-linear behaviour of the materials used in electronic packaging, especially solders and adhesives. Stress analyses of electronics packages have been performed using analytical, numerical, experimental and „hybrid" methods [79]. Often for

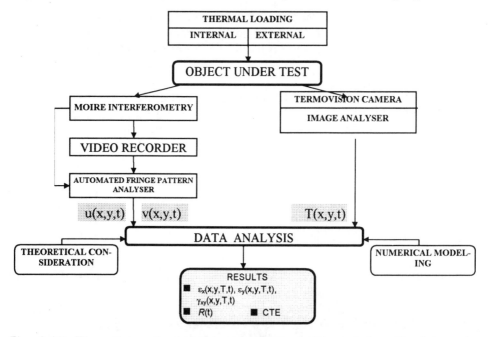

Fig. 6.16, The scheme for hybrid numerical-experimental methodology for electronic packaging testing

experimental procedure a combination of moiré interferometer and thermovision methods [82] is applied as shown in Fig. 6.16.

Several good examples showing the importance of experimental investigations were by IBM Microelectronics [79,83]. Fig. 6.17 illustrates the behaviour of an active silicon chip with an aluminium heat sink adhesively bonded to its lower surface. Fringe patterns introduced by isothermal loading represent v – displacement field of the assembly (sensitivity 417 nm/fringe) due to the temperature excursion ΔT=-60°C, from 80°C to 20°C (Fig. 6.17a) and ΔT=-120°C, from 80°C to -40°C. If all materials had constant properties, independent of temperature, the deformation in (b) would be twice that in (a). However, the bending deformation of the chip was almost an order of magnitude larger in (b). The strong nonlinearity of the structural behaviour was ascribed to the temperature dependent properties of the adhesive layer.

Below 0°C Young's modulus and the shear modulus of the adhesive were in order of magnitude greater than at room temperature. Consequently the coupling and bending were greatly increased at the low temperature.

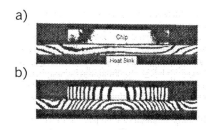

Fig. 6.17, Illustration of the effect of temperature dependent material. Moiré fringe patterns represent the vertical displacement fields of a chip/heat sink assembly, obtained by controlling (a) from 80°C to room and (b) 80°C to -40°C. In (b), the vertical displacement is ten times greater (instead of two times), proving nonlineary [79]

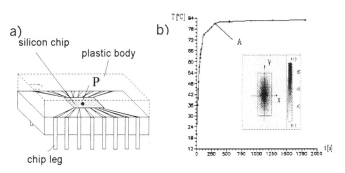

Fig.6.18, a) Scheme of electronics chip; b) temperature at the central point of the top of chip surface. A - the spatial map of temperature distribution

Interesting studies were performed on UCY74S405N chip [82,83], which is a silicon chip connected with external legs by metal wires and placed inside the plastic (epoxy resin) body (Fig. 6.18a). The main role of plastic body is mechanical retention and protection of semiconductor, protection from environment and the disipation of internal heat to the outside. Due to the internal

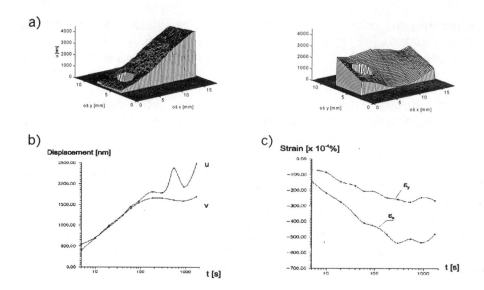

Fig.6.19, Interferograms and 3-D in-plane displacement maps (a) and temporally diagrams of displacements (b) and strains (c) measured at the point P.

heating the thermal stresses deform the chip elements and can lead to its destruction. Here, the in-plane displacement and strain at the top chip surface and the chip legs were monitored and measured for the temperature changes from room (20°C) up to operating temperature (~80°C) by 20 min.
The plot of variations of temperature in time at the central point of the chip top surface (after plugging the chip) is shown in Fig. 6.18b. The temperature was stabilised after about 300 seconds and was equal 82°C. However the spatial distribution of the temperature at the surface was not uniform and it was additionally checked by thermovision camera. During the time of 20 minutes the in-plane displacements **u** and **v** of the top surfaces were monitored. In Fig. 6.19a the interferograms obtained after this time and displacement 3-D maps calculated from these interferograms are shown. Fig. 6.19b and c show the temporal variations of the displacements and strains measured in the point P. During the first five minutes (while the temperature rised) the strains increase up to approximately -0.02% for ε_x and -0.05% for ε_y. Next, in the time when temperature is constant, both strains slightly oscillate in the range ±0.005%. It may be caused by oscillations of heat dissipation process. This effect can be treated as a fatigue load and in the longer time may decrease the life of the chip.

6.5. Mechanical / material joint testing

Recent technology brings new possibilities for unconventional joints of materials which gives extra wear and corrosion resistance or may be used as thermal or electric insulators. However joining some of the materials provides several problems, which arises from significant difference in the physical and chemical properties of joint materials. Good example of such structure is a ceramic-to-metal joint [84]. The conventional numerical and experimental methods for determination of stress in C-M joint have several limitation: the experiment usually is expensive (x-ray diffraction) and the stress is determined at only a few points with averaging over a significant length of measurement base, while the finite element method is based on a model having several simplification or requiring the knowledge of real material properties. Here grating interferometer was applied for determination of Young's modulus distribution in the joint and base materials [85]. The beam with the brazed Al_2O_3-FeNi42 joint was subjected to four – point bending (Fig. 6.20) for the series of loads (P=50÷250N).

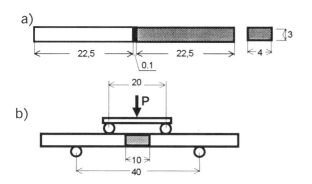

Fig. 6.20, The C-M joint: a) the sample geometry and b) loading configuration.

Fig. 6.21 shows exemplary displacement maps obtained for P=100N bending force. The nonsymetrical character of u and v plots clearly indicates the influence of different base materials at both sides of the joint and a certain transition area within the joint itself and the joint interface. The experiments were repeated for two fields of measurement: i.e. global analysis: 10x6.6 mm², local analysis: 3.2x2.2 mm². The u – displacements for both cases were numerically differentiated and ε_x - strain map was obtained (Fig. 6.22). Having ε_x distribution in the beam, the Young modulus at any point of the calculated according to the simplified relation:

$$E = \frac{M_g}{\varepsilon_x I} y \qquad (6.2)$$

where $M_g = \frac{P}{2} e$ · bending moment, $I = \frac{bh^3}{12}$ · momemt of inertia for the beam cross-section

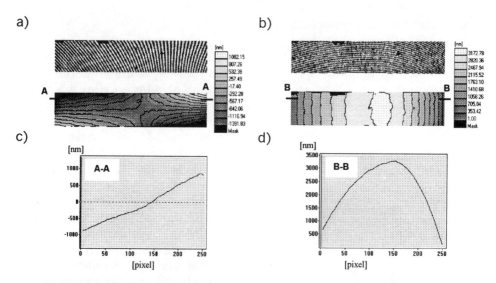

Fig. 6.21, Experimental analysis of C-M joint under P=10N bending force. The interferograms and grey level contour maps of a) u(x,y) and b) v(x,y) and their horizontal cross-sections C A-A and d) B-B

The values of the Young modulus in the base material (Fig. 6.23a) and in the joint (Fig. 6.23b) were calculated separately on the base of experimental data obtained for two different fields of view. The values for base materials show good agreement with the published data (E_c=300 GPa, E_m=120 GPa) while the values for the joint and vicinity are not avaliable in literature and were obtained for the first time [85].

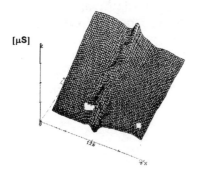

Fig. 6.22, The 3D representation of ε_x strain distribution in C-M joint its vicinity

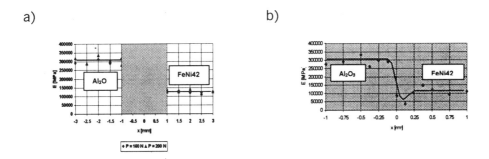

Fig. 6.23, The calculated profiles of Young modulus along horizontal cross-section of C-M joint measured with a) 0x6.6 mm² and b) 3.2x2.2 mm² fields of view.

7. HYBRID METHODS OF RESIDUAL STRESS ANALYSIS

7.1. Introduction

Manufacturing processes, the action of load and thermal stresses create residual stresses in engineering elements [86]. These stresses, acting in conjunction with those produced by live loads, can threaten the safety of an engineering structure operation by increasing the hazard of creation and propagation of fatigue cracks. Difficulty in measuring them nondestructively; the unpredictability of their magnitude, sense, and direction; their adverse ability to combine with stress corrosion, environmental and fatigue situations; the difficulty in removing them can render residual stresses to be extremely troublesome. Residual stresses should be evaluated under operating conditions to assess their effect on the intended service capability of the component. The problem of the experimental analysis of the residual state in elements has been investigated for many years in search of an inexpensive and reliable method of determination of those stresses.
In order to carry out experimental examination of residual stresses one has to answer two questions: the first one is *how to determine* residual stresses from mechanical point of view i.e. how to reach into inside phenomena of a 3D body, the second one is *how to measure* those stresses from experimental point of view, i.e. which experimental technique to apply. Nondestructive techniques that can reach and measure inside phenomena of a 3D body are for instance: rentgenography, neutronography and ultrasonography. They have capability; to an extend, to penetrate the interior of the investigated body and measure the stress state in it. However the results obtained are of different kind and values, depend on the material of the element, and often have qualitative character only.
Another way of determination of residual stresses is relaying on comparative measurements of displacement/strain values before and after stress reliev-

ing. The measurements are performed by strain gauges, photoelastic coatings, moiré or interferometric techniques. Stresses may be relieved by hole drilling (trepanation), sectioning of a body into small pieces or by annealing. Here the automated moiré (grating) interferometry, GI, method has been selected as an experimental technique leading to strain determination on the surface of a sample [87]. This technique, in conjunction with numerical analysis based on an theoretical model of stresses, enables solving the problem of determination of the original 3D residual stresses.

The paper describes the methodology and instrumentation for measurement for two types of engineering elements, namely railway rail and laser beam weldment. Although these elements are very different, but they both suffer from high values of localized residual stresses caused by manufacturing and thermal processes.

7.2. Objects of measurements

7.2.1. Railway rail

For verification analytical and experimental methods of residual stress analysis in railway rails a controlled laboratory loading process is required. It can be realized by rolling the special wheel along the top of the testing rail (see Fig. 1a) with known contact force. The part of new UIC 60 rail (made in Huta Katowice steel works) was rolled on EMS-60 testing machine [288] in central Research Institute of the Polish State Railways, Warsaw. The rolling process was performed with the load 150 kN, at a reciprocation frequency 1 Hz and the number of cycles applied equals to 5×10^5.

Usually so called Battle technique [89] is used for 3D residual stress determination in rails. It is based on the general concept of destructive stress relief and consists of two main sectioning steps for determination of the residual stresses:
- Yasojima – Machii (YM) slicing (for δ_{xx}, δ_{yy}), which is done by cutting out of a rail slab a thin transverse slice and dicing it;
- Meier sectioning (for δ_{zz}), which is done by cutting a rail into longitudinal rods (4x4x500 mm size).

Here in-plane displacement fields at YM slices were determined for comparison purposes by three methods [87]:
1. by strain gauges placed at the sections of rail slice with the stress relieved by cutting. The cutting was performed with great care (slow speed, heavy cooling) in order to prevent introducing additional stresses,
2. by moiré interferometry while the in-plane displacement was measured in the center of the same sections of rail slice as described above,
3. by moiré interferometry while the residual stress in the samples was released by the annealing process in 600° C during 3 hours.

In the case of using moiré interferometry, the YM slices were additionally prepared by attaching to the upper part of slice the cross-type diffraction grating with frequency 1200 lines/mm. For the case in which stress was released by sectioning, the alluminium master grating was transferred through epoxy. While the annealing process of stress relieving required application of a high temperature resistant grating. Herein the process developed out at National Physical Laboratory, UK, had been adopted [28].

The analytical model for prediction of residual stress in rails was developed by Orkisz [90] and it relies on the notion of a shakedown state established by the highest load to which the rail was subjected. This model was simplified by Magiera [91] for the in-plane strain and stress components in the rail slice, which can be compared with our experimental results. Calculations were performed on a geometrical model of the rail which includes 2480 elements and 2768 nodes in the rail cross-section (see Fig. 1b).

a)   b)

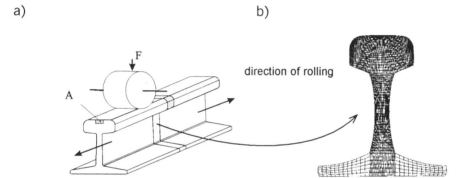

Fig. 7.1, Scheme of the rolling process (a) and the geometrical model of a rail slice (b). A –the area considered for the analysis of wheel/rail contact region.

7.2.2. Laser beam weldment

Introducing extensively laser beam welding into industry, it is necessary to provide extended knowledge about laser weldment properties e.g. strength and ductility. This knowledge is required from one hand to design properly the technological process and on the other to develop the measurement method which is able to perform on-line quality control of the weldment. The special interest in laser beam welding process originates from the density of heat source generating a vapour key-hole, which deeply penetrates into material [92] and huge residual stresses at small cross-section appear (Fig. 2). Three types of welded specimen were considered (Fig. 7.3):

1. a ferritic steel, overlap-joint welded specimen, FS, produced by a beam power of 4 kW using welded speed of 1.8 m/min
2. a structural steel welded specimen, SS, produced by a beam power of 10 kW, using welding speed of 0.8 m/min
3. TA6V alloy specimen, TA

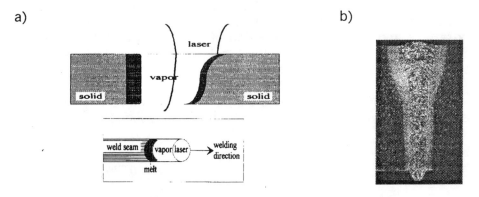

Fig. 7.2, The scheme of laser beam welding process (a) and a laser weld geometry (b)

The problem of residual stress determination was here approached by relaying on comparative measurements of in-plane displacement/strain values before and after stress releasing. The cross-type reflection, diffraction gratings (frequency 1200 l/mm) were replicated on the welded samples as shown in Fig. 7.3. Stresses were relived by sectioning of the first two samples:
- for FS by cutting through the gratings perpendicular to the weld direction (Fig 3a),
- for SS by cutting the sample along the weld (Fig. 3b).

For the TA6V alloy specimen, an incremental hole drilling method [93] in the vicinity of the was applied. The flat bottom hole ($\phi 2$ mm) with location as shown in Fig. 3c was drilled carefully with an incremental depth of 0,1 mm [93]. The analysis of fringes in regions R1 – R4 enable to determine the residual stress profile in depth.

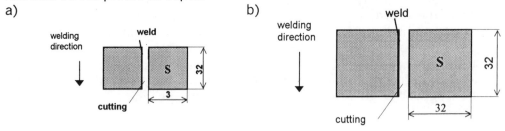

c)

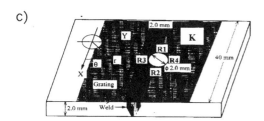

Fig. 7.3, The welded specimens under study: (a) ferritic steel, FS (b), steel, SS and (c) TA6V alloy specimen; TA [93] P, S, K – areas with grating, R1 – R4 regions taken for analysis

7.3. Experimental setup and procedure

Determination of residual stresses in rails and welded samples has to be performed in a system which allows the analysis of big elements. The one-channel moiré interferometer presented in Fig. 4 is designed to test components with the measuring area up to 100x100mm^2 [94]. In the system two mutually coherent light beams illuminate symmetrically the specimen grating. The first diffraction orders of two illuminating beams interfere and produce a fringe pattern with information about in-plane displacement in the direction perpendicular to the grating lines. Residual stresses cause the state of deformation which, when released by cutting drilling, or annealing, causes in turn the opposite deformation of the sample under test. The measurement displacement depends on this opposite deformation.

In the system presented, one mirror head and the rotary mounts of the specimen and CCD camera enable the required measurement of u and v displacement. The rotary mounts provide high accuracy (a fraction of a pixel), so that u and v displacement maps are correlated for further strain

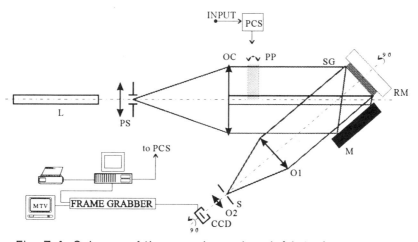

Fig. 7.4, Scheme of the one-channel moiré interferometry system: L – He-Ne laser; PS – pinhole system; OC – collimator objective; SG – specimen grating; RM, M – mirrors; PP – parallel glass plate; PSC – phase-shifting control.

calculations. Also high accuracy of repositioning of the specimen (a fraction of a fringe) is achieved, and the systematic errors connected with the specimen grating and aberration of the system can be eliminated by simple subtraction of the displacement maps before and after stress releasing.

The interferograms obtained in the system are expressed by the usual equation:

$$I_i(x,y) = a(x,y) + b(x,y)\cos\left[\frac{4\pi}{d}(u(x,y) \text{ or } v(x,y)) + \delta_i\right] \qquad (7.1)$$

where $a(x,y)$, $b(x,y)$ are background and local contrast modulation functions, respectively, d is the period of the specimen grating, δ_i is the i-th phase shift of the interferogram, $\delta_i = i2\pi/N$, $i=1, ...N$.

Interferograms are captured by CCD camera using imaging objectives O1 and O2 and video signals are digitized using frame grabber. The image is analysed by automatic fringe pattern analyser which employs temporal phase shifting method [95]. The phase shifts, are introduced automatically by rotating parallel glass plate PP. In order to minimize the errors connected with possible phase shifter misalignment and some nonlinearities of the detector, the five-intensity self calibrating algorithm, with the phase shift $\delta_i = i\pi/2$ [96] was applied.

The phase was than unwrapped using minimum spanning tree algorithm and scaled for displacement values (in nm). Latter, according to the requirements, the strains ε_x, ε_y, γ_{xy} were calculated.

The methodology of measurements was straightforward in the case of the samples with the cross-grating transferred by epoxy, as they did not require capturing of initial interferograms for calculating and subtracting the, so called, "zero displacement" fields. The high temperature gratings suffer from significant distortion (a few fringes in the field of view) and therefore the measurement process was carried out as follows:

1. Mounting and adjusting the sample on the rotary mount RM,
2. Registration of the interferograms for u-direction, and after rotating the mount and CCD camera by 90°, for v-direction,
3. Calculation of the u_0 and v_0 in-plane displacement fields,
4. Annealing of the sample,
5. Mounting the sample on the rotary mount in the same position as in point 1. This operation is possible due to the three point support system used,
6. Registration and calculation of the u_1 and v_1 displacement fields,
7. Calculation of the displacement values $u=u_1-u_0$ and $v=v_1-v_0$. In this step the initial distortion of the specimen grating and the interferometer imperfections are removed,
8. Calculation of the strains: ε_x, ε_y, and γ_{xy}.

In the case of RS analysis by hole drilling method, a special arrangement of three-mirror grating interferometer combined with the accurate hole drilling device was used in the experiment [97]

7.4. Experimental results and discussion

7.4.1. Residual strain determination for railway rails

The full-field information about the residual displacement and strains in rails was obtained by the analysis of the annealed slice of rail. The interefrograms obtained after stress releasing are shown in Fig. 7.5 together with the displacement fields (u, v), which according to the methodology described in Section 7.3, were obtained by subtraction of the final and initial displacement fields ($u=u_1-u_0$, $v=v_1-v_0$). The experimental strain fields calculated from u and v displacements are shown in Fig. 7.6a-c and they are compared with their numerical predictions (Fig 7.6d-f) calculated according to the analytical model described by Magiera and Orkisz [91].

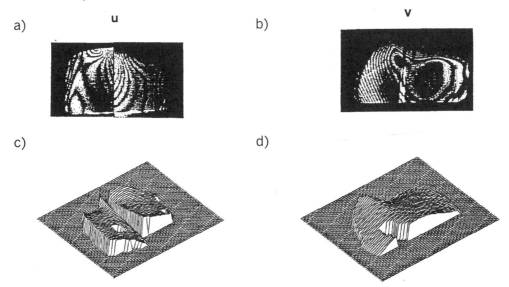

Fig. 7.5, Interferograms representing in-plane u_1 (a) and v_1 b) displacement fields in YM slice of rail after the residual stress relieving by annealing and 3D plots of the calibrated displacement fields $u=u_1-u_0$ (c) and $v=v_1-v_0$ (d)

For the comparison, the interferogram obtained for u displacement map at the rail slice with the stress relieved by cutting is presented in Fig. 7.7. The interferogram consists of several separated fringe patterns and additionally the component fringe patterns may have severe disturbances at the edges due to the cutting process. Here the modified software enables, by a semi-manual procedure, to put a mask on the distorted areas. Further processing was performed on modified interferograms and average strain values ε_x and ε_y were calculated for each component interferogram.

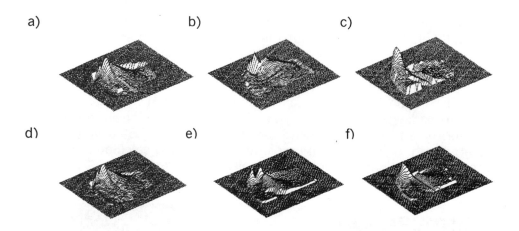

Fig. 7.6, 3D plots of experimental strain distributions calculated from u and v displacement fields: ε_x (a), ε_y (b), γ_{xy} (c) and numerically calculated strains ε_{xt} (d), ε_{yt} (e), γ_{xyt} (f) within the same area as experimental data

Fig. 7.7, The rail slice with the stress released by cutting together with the interferograms representing

The theoretical and experimental ε_x strain distribution determined by strain gauges (YM) and moiré interferometer methods, while stress was relieved by cutting (IM·C) and by annealing (IM·W), in the crossections B-B and C-C (as marked in Fig. 7) are shown in Fig. 7.8. The character of the strain distributions for THEOR and IM·W for the crossection (C-C) were very similar, however the experimental values were lower. This may be caused by nonfully performed process of stress relieving by annealing [94]. The pair of strain distribution in C-C obtained for YM and IM·C methods at the samples with the stress relieved by cutting have similar character and values, however they differ from THEOR and IM·W. This proves the statement that the physical nature of stress relieving influence the values of stress determined.

The crossection B-B, which is placed closer to the high plasticity region produced by the contact area rail/wheel indicates significant differences between the strain determined theoretically and experimentally. The theoretical predictions in this area may be incorrect due to non proper model within plasticity region and wrong local material constant values applied during calculations. Therefore the experimental data especially in this region are strongly required.

Introducing higher magnification of imaging optics in the system presented in Fig. 7.4 enables the analysis within the area A (indicated in Fig. 7.1). The interferogram, u-displacement and ε_x strain within the region close to the contact area rail/wheel are shown in Fig. 7.9. These experimental data allow proper modelling of the rail behaviour in the contact area.

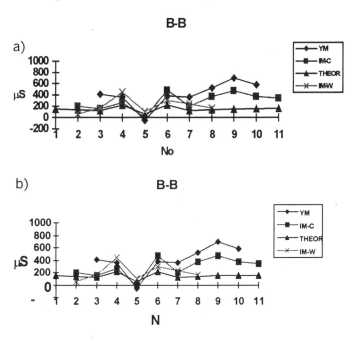

Fig. 7.8, The strains ε_x distributions obtained by various stress relieving and measurement methods: stress relieving by cutting and measurements by strain gauges (YM) and grating interferometry (IM-C); stress relieving by annealing and measurements by grating interferometry (IM-W); theoretical predictions (THEOR)

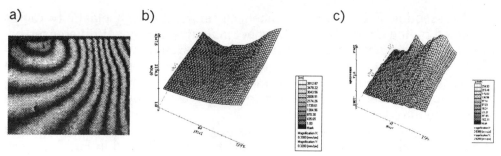

Fig. 7.9, The analysis of wheel/rail contact region A: the interferogram representing u-displacement (a), 3D plot of u-displacement (b) and 3D plot of ε_x strain (c)

4.2. Residual stress determination in laser beam welds

As indicated in Section 7.2, three types of welded specimen were considered with different way of residual stresses relieving.
The measurements of ferritic steel specimen enable to determine displacement/strain fields at the weld and heat affected zone Fig. 7.10.

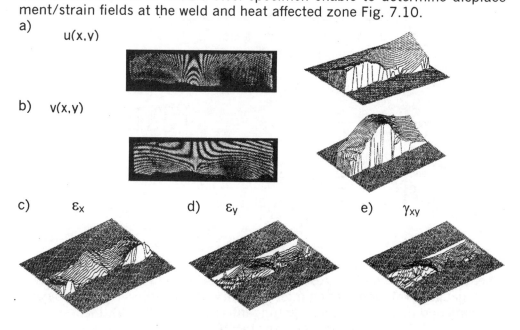

Fig. 7.10, The residual stress related in-plane displacements and strains in ferritic steel specimen. Sequential pairs of images show interferogram and 3D plot of the respective u, and v, displacement maps; ε_x (c), ε_y (d) and λ_{xy} (e) strains

Analysing the u and v displacement in the weld it can be seen that u has zero value at the center with decreasing and increasing values (±3.5 µm) at both sides of the weld. The v displacement reaches maximum (18.3 µm) at the weld center. This result confirms with the experience from laser beam welded thick plates (t=12 mm), that the biggest displacements occure inside the 600° isotherm, parallel to the welding direction. The strains calculated from the relevant displacement maps reach peak-to-valley values as follows: $\varepsilon_x = 400 \cdot 10^{-6}$, $\varepsilon = 1200 \cdot 10^{-6}$ and $\gamma_{xy} = 2800 \cdot 10^{-6}$. The values of displacements and strains seem to be in the possible range as it can be evaluated today.

The structural steel specimen SS, after sectioning along the weldment indicated significant u and v displacements with the highest gradients within a few mm from the weld [98]. Fig. 7.11 shows the interferograms, displacements and strains obtained at the region S (Fig. 7.3b). The P-V values of strains equal $\varepsilon_x = 3550 \times 10^{-6}$, $\varepsilon_y = 3950 \times 10^{-6}$ in the measurement area 30x30 mm².

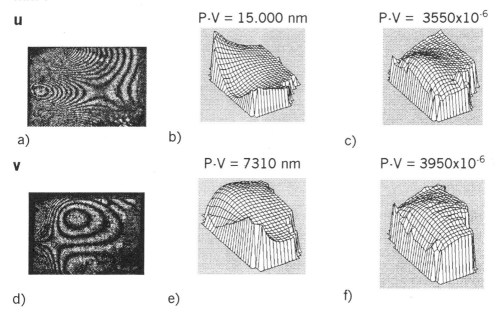

Fig. 7.11, The results of the welded specimen SS analysis after the residual stress releasing by cutting a, d) interferograms and b, e) 3D plots of u and v in-plane displacements; c, f) 3D plots of ε_x, ε_y strains respectively.

The TA6V alloy specimen was analysed for the incrementally increased depth of the drilled hole. The typical u and v fringe patterns obtained for the depths: 0,3 mm and 0,5 mm are shown in Fig. 7.12. u - displacement field in the longitudinal direction of the weld contained more fringe orders than the corresponding v - field in the transverse direction for each incremental depth and therefore only u - displacements are used in the determination of residual stresses. After determination of the fringe orders the distributions of residual stresses in depth, were obtained for regions R1, R2, R3 and R4 according to the relation:

$$N_x^i(x_k,y_k) = \sum_{j=1}^{i} 2fs\left(A^{ij}\cos\Theta_k + B^{ij}\cos\Theta_k\cos\Theta_k - C^{ij}\sin\Theta_k\sin\Theta_k\right)\sigma_{xx}^j +$$

$$\left(A^{ij}\cos\Theta_k - B^{ij}\cos2\Theta_k\cos\Theta_k + C^{ij}\sin2\Theta_k\sin\Theta_k\right)\sigma_{yy}^j + \left(2B^{ij}\sin2\Theta_k + 2C^{ij}\cos2\Theta_k\sin\Theta_k\right)\tau_{xy}^j \quad (7.2)$$

where: k=1, 2, 3; i=1, 2, ..., n; n is the total number of incremental steps; j=1, 2, ..., J; A, B, C - are the coefficients which can be determined by three dimensional FEM when two loading cases are considered:
1) $\sigma_{xx} = \sigma_{yy} = \sigma$, $\tau_{xy} = 0$; an equibiaxial residual stress field which is equivalent to a uniform pressure $p = \sigma$ acting on the hole boundary,
2) $\sigma_{xx} = -\sigma_{yy} = \sigma$, $\tau_{xy} = 0$; a pure shearing residual stress field.
The residual stress profiles in depth for regions R1 and R4 are shown in Fig. 7.13a,b.

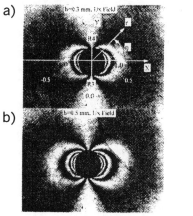
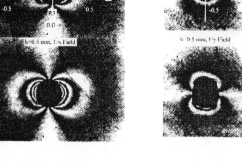

Fig. 7.12, Typical interferograms obtained for u and v- displacements after drilling 0.3 mm (a) and 0.5 mm (b) depth hole [93]

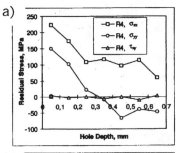
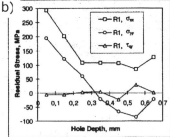

Fig. 7.13, residual stress profile in depth in region R1 (a) and R4 (b) [93]

7.5. Conclusions

The chapter summarizes the research results obtained by experimental full-field analysis of the residual stresses released by sectioning or annealing and measured by automated moiré interferometry. The results gathered till now support our knowledge about residual stresses, however at the moment they provide the information about in-plane displacement/strain fields which occur at the surface of the specimen after stress releasing. Due to high spatial resolution, high sensitivity and accuracy (when compared with strain gauges measurement) of the experimental results, it is proposed to use these data to compute the original 3D stress state [87]. It can be approached by hybrid procedure which includes:
- experiment applying oblique slicing technique, stress relieving by annealing and 2D stress state determination by moiré interferometry,
- numerical methods which compute 3D stress state through iterative calculation of mutual accumulative impacts of the out-of-plane stress components derived for each slices.

If hybrid method is applied, it will allow to neglect the second stage of Battle technique i.e. Meier sectioning of rails which through a very tedious process provides information about δ_{zz} stress state. It may be also very helpfull to compute 3D stress state in such engineering objects as thick welded elements or materials formed by explosive method.

Additionally method for residual stress determination in depth by incremental drilling /moiré technique is presented. This method connected with proper numerical analysis provides detailed information about 3D residual stress distribution in the given point of structure. Combining both methodologies may be in future an excellent basis to form a new hybrid method of 3D residual stress distribution determination.

However even without getting into the more sophisticated 3D residual stress determination the moiré interferometry has provided an excellent tool to extend knowledge about local residual stresses in weldment and heat effected zone as well as about the regions in neighbourhood of the contact wheel/rail.

8. CONCLUSIONS AND FUTURE POTENTIALS

8.1 Moiré in relation to other methods

There are a number of methods which can give the information in the same form as a moiré, e.g. holographic and speckle interferometer (including ESPI). They have a number of arrangements that provide sensitivity vectors in a variety of different directions. The ones that are responsive essentially to in-plane displacement have a two-beam set-up that is akin to moiré interferometry. They have a great attraction that there is no surface treat-

ment required (no specimen grid). They also can examine arbitrary large ares with surface curvature (provided it is taken into account in the fringe analysis).
Although all these statements are true, moiré and especially moiré interferometry stays a very attractive technique in experimental mechanics and material engineering. The main reasons for it are listed below [12]

- freedom from positional restriction

Because the surface is regular, these is no difficulty in responding the interrogating system to correlate with the deformed specimen surface. There is no particular limit to the displacement which may occur between one side of the specimen and the other before correlation is lost; 20 per cent strain and beyond is possible.

- freedom form speckle

Possibly the most significant generic advantage is the reduction of speckle effect, which inevitably accompanies observation of a diffuse surface illuminated coherently. With geometric moiré it is totally absent and in moiré interferometer, where coherent light is used, the specular nature of the imposed grid can produce, ideally, speckle-free diffraction (although in practise there is always some measure of optical noise, essp. for higher magnifications).

- level of detail

The relatively low speckle noise effectively enables the elucidation of fine detail on a strain field. This capability more than any other lead to the more widespread use of moiré and moiré interferometry in recent years. As the local high quality information is avaliable, it can be, with high confidence, inserted as material constants distribution, boundary conditions etc. into FEM programs to obtain more exact modelling of the most difficult mechanics problems and design of structures.

- dynamic range

The low noise is also conductive to a wide dynamic range, since the fringes may be very closely packed before they become degraded. Using typical CCD detector with resolution 512x512 pixels, a dynamic range of 10^4 on strain may be achieved.

- tolerance to plasticity and temperature

The grid, as typically applied in resin, has the ability to continue to diffract efficiently even as the surface below changes its nature, due, for example, to plastic strain creating a quite different diffuse surface, which would run out of correlation using non-grid methods.

Further, the high-temperature grating survive and diffracts at temperatures at which the natural surface would in many materials be oxidized to a totally different condition from the original, making correlation impossible.

8.2 Direction of development and future potentials

The main directions of development of grid based technique concentrates on:

- improving opto-mechanical and analysis systems to implement an active approach to design output fringe pattern to facilitate its analysis for measurements in unstable environment (including high temperature)and dynamic events (e.g. fracture mechanics)
- development of commercial automatic grating interferometry systems
- development of easily transferable high-frequency gratings for the commercial GI systems
- combining global and local approaches of measurement and numerical analysis
- introducing new concepts for fringe pattern analysis and postprocessing of displacement fields, including regularization methods, wavelent transform, sophisticated image processing algorithms etc.
- combining the analysis system of moiré and GI with FEM packages to form full hybrid experimental-numerical systems.
- extending the area of applications of visible light and electron beam moiré interferometry.

REFERENCES

1. Patorski K.: The Handbook of the Moiré Technique, Elsevier, Oxford, (1993)
2. Cloud, G.: Optical Methods of Engineering Analysis, Cambridge Univ. Press, (1995)
3. Kobayashi, A.S., ed: Handbook of Experimental Mechanics, Prentice Hall, Inc. (1987)
4. Theocaris, P.S.: Moiré Fringes in Strain Analysis, Pergamon Press, Oxford, (1969)
5. Durelli, A.J. and V.J. Parks: Moiré Analysis of Strain, Prentice-Hall, Englewood Cliffs, New Jersey, (1970)
6. Bach, C. and R. Bauman: Elastikität und Festigkeit, Springer, Berlin, (1922)
7. Bell, J.F.: Diffraction grating strain gauge, Proc. SEM, 17(2), Bethel, (1960)
8. Backes, P.G. and W.M. Stevenson: High accuracy image centroid position determination with matrix sensors: An experimental comparison of methods, Proc. of Fifth Int. Congress on Applications of Lasers and Elektro-Optics, Arlington, (1986)

9. Van der Heijden, F.: Image based measurement systems, J.Wiley and Sons Ltd, Chichester and New York, (1994)
10. Born, A. and E. Wolf: Principles of Optics, New York, Oxford, Pergamon Press, (1959)
11. Sciammarella, C.A.: Basic optical law in the interpretation of moiré patterns applied to the analysis of strains · Part I., Exp. Mechanics, 5, (1965), 154-160
12. McKelvie, J.: Moiré strain analysis: an introduction, review and critique, including related techniques and future potentials, The Journal of Strain Analysis, 33, (1998), 137-152
13. Guild, J.: The interference System of Crosed Diffraction Gratings: Theory of Moiré Fringes, Clarendon Press, Oxford, (1956)
14. Weller, R. and B.M. Shepard: Displacement measurement by mechanical interferometry, Proc. Soc. Exp. Stress Analysis, 6(1), (1948), 35-38
15. Weissman, E.M. and D. Post: Moiré intrferometry near the theoretical limit, Applied Opt., 21(9), (1982), 1621-1623
16. Sevenhuisen, P.J.: Grid methods · a new future, in Proc. of SEM Spring Conf. on Experimental Mechanics, (1989), 445-450
17. Post, D. and B. Han, P.Ifju: High Sensitivity Moiré Interferometry, Springer-Verlag, Berlin, (1994)
18. Naumann, J.: Grundlagen and Anwendung des In-plane-Moiréverfahrens in der experimentallen festkörpermechanik, Mechanik/Bruchmechanik, 110, VDI Verlag, (1992)
19. Sciammarella, C.A.: Moiré fringe multiplication by means of filtering and wave front reconstruction process, Exp. Mech., 9, (1969), 179-185
20. Post, D.: Moire fringe multiplication with a nonsymmetrical, doubly blazed reference grating, Appl. Optics, 10, (1971), 901-907
21. Huntley, M.C.: Diffraction gratings, Academic Press, (1982)
22. Wiliams, D.C. (ed): Optical methods in engineering metrology, Chapman & Hall, London, (1993)
23. Burch, J.M. and C. Forno: A high-sensitivity moiré grid technique for studying deformation in large objects, Opt. Eng., 14, (1975), 175-185
24. Forno, C.: Deformation measurement using high resolution moiré photography, Opt. Lasers Eng., 8, (1988), 189-212
25. McKelvie, J. and C.A. Walker: A practical multiplied moiré-fringe technique, Exp. Mechanics, 18, (1978), 316-320
26. Ifju, P. and D. Post: Zero thickness specimen gratings for moiré intrferometry, Exp. Techqs., 15(2), (1991), 45-47
27. Kujawinska, M. and J.R. Pryputniewicz: Micromeasurement: a challenge for photomechanics, Proc. SPIE, 2782 (1996), 15-24
28. Kearney, A. and C. Forno: High temperature resistant gratings for moiré interferometry, Exp. Techqs. 17(6), (1993), 9-12

29. Dally, J.W. and D.T. Read: Electron-beam moiré, Exp. Mechanics, 33, (1993), 270-277
30. Dally, J.W. and D.T. Read, E.S. Drexler: Transitioning from optical to electronic moiré, Exp. Mechanics, Allison I.M. (ed), Balkema, Rotterdam, (1998), 437-447
31. Han, B.: Higher sensitivity moiré interferometry for micromechanics studies, Opt. Eng., 31, (1992), 1517-1525
32. Yatagai, T.: Intensity based analysis methods In Interferogram Analysis, D.W. Robinson and G.T. Reid (eds), Institute of Physics, Bristol, (1993)
33. Osten, W. and W. Jüptner: Digital processing of fringe patterns in optical metrology in Optical Measurement Techniques & Applications, P.K. Rastogi Artech House, Boston, (1997)
34. Creath, K: Temporal phase measurement methods, in Interferogram Analysis, D.W.Robinson, G.T.Reid (eds), Institute of Physics, Bristol, (1993)
35. Kujawinska, M: Spatial phase measurement methods, in Interferogram Analysis, D.W.Robinson, G.T.Reid (eds), Institute of Physics, Bristol, (1993)
36. Takeda, M. and H. Ina, S. Kobayashi: Fourier transform method of fringe pattern analysis for computer – based topography and intrferometry, J.Opt. Soc. Am., 72, (1982), 156-160
37. Bruning, J.H. et al.: Digital wavefront measuring interferometer for testing optical surfaces and lenses, Appl. Opt. 13, (1974), 2693-2703
38. Surrel, Y.: Design of algorithms for phase measurements by the use of phase stepping, Appl. Opt., 35, (1996), 51-60
39. Stetson, K.A. and W.R. Brohinsky: Electrooptic holography and its application to hologram interferometer, Appl. Opt., 24, (1985), 3632-3637
40. Surrel, Y.: Additive noise effect in digital phase detection, Appl. Opt., 36, (1997), 271-276
41. Huntley, J.M.: Automated fringe pattern analysis in experimental mechanics: a review, J.Strain Analysis, 33 (1998), 105-125
42. Kujawiska, M. and J. Wójciak: Spatial-carrier phase-shifting technique of fringe pattern analysis, Proc. SPIE, 1508, (1991), 61-67
43. Creath, K. and J. Schmit: N-point spatial phase measurement technique for nondestructive testing, Opt. Lasers Eng., 24, (1996), 365-379
44. Pirga, M and M. Kujawinska: Two-directional spatial-carrier phase shifting method for analysis of crossed and closed fringe pattren, Opt. Eng., 34, (1995) 2459-2466
45. Burton, D.R. and M.J. Lalor: Multichannel Fourier fringe analysis as and aid to automatic phase unwrapping, Appl. Opt., 33, (1994), 2939-2948
46. Osten, W. and W. Nadeborn, P. Andrä: General hierarchical approach in absolute phase measurement, Proc. SPIE, 2860, (1996) 2-13

47. Takeda, M: Recent progress in phase-unwrapping techniques, Proc. SPIE, 2782, (1996), 334-343
48. Huntley, J.M.: New methods for unwrapping noisy phase maps, Proc.SPIE, 2340, (1994), 110-123
49. Bone, D.J: Fourier fringe analysis: the two-dimensional phase unwrapping problem, Appl.Opt., 30, (1991), 3627-3632
50. Towers, D.P and T.R. Judge, P.J. Bryanston-Cross: Automatic interferogram analysis techniques applied to quasi-heterodyne holography and ESPI, Opt. Lasers Eng., 14, (1991), 239-281
51. Ghiglia, D.G. and G.A.Mastin, L.A. Romero: Cellular automata method for phase unwrapping, J.Opt.Soc.Am. A, 4, (1987), 267-280
52. Servin, M. and R. Rodriguez-Vera, A.J. Moore: A robust cellular processor for phase unwrapping, J.Mod.Opt., 41, (1994), 119-127
53. Huntley, J.M. and H. Saldner: Temporal phase unwrapping algorithm for automated interferogram analysis, Appl. Opt. 32, (1993), 3047-3052
54. Vrooman, H.A. and A.A. Mass: New image processing algorithms for the analysis of speckle interference patterns, Proc. SPIE, 1163, (1989), 51-61
55. Czarnek, R.: High sensitivity modre interferometer with compact achromatic head, Opt. Lasers Eng., 13, (1990), 93-101
56. Epstein, J.: Moiré interferometry: past achievements and present directions, Opt. Lasers Eng., 12, (1990), 77-79
57. McKelvie, J. and C.A. Walker, P.M. MacKenzie: A workaday more interferometer: conceptual and design considerations: operation; applications; variations; limitations., Proc. SPIE, 814, (1987), 464-474
58. McKelvie, J. and K. Patorski: Influence of the slopes of the specimen grating surface on out-of-plane displacements by moiré interferometry, Appl. Opt., 27, (1988), 4603-4605
59. Kujawinska, M. and L.Salbut: Recent development in instrumentation of automated grating interferometry", Optica Applicata, 25, (1995), 211-232
60. Czarnek, R.: Three-mirror four-beam interferometer and its capabilities, Opt. Lasers Eng., 15, (1991),93-101
61. Poon, C.Y. and M. Kujawinska M., C. Ruiz: Spatial carrier phase-shifting method of fringe pattern analysis for moire interferometer, J. of Strain Analysis, 28, (1993), 79-88
62. Salbut L., Kujawinska M., Dymny G., „Portable, automatic grating interferometer for laboratory and field studies of material and mechanical elements", Proc. SPIE, 2342, (1994), 58-65
63. Kozlowska, A. and M. Kujawinska, Ch. Górecki: Grating interferometer with a semiconductor light source, Appl. Opt., 36, (1997), 8116-8120

64. Han, B. and D. Post: Immersion interferometry for microscopic moiré interferometry, Exp. Mechanics, 32, (1992), 38-41
65. Salbut, L. and M. Kujawinska: Grating microinterferometer for local in-plane displacement/strain fields analysis, Proc. SPIE, 3407, (1998) in press
66. Salbut, L. and K. Patorski, M. Kujawinska: Polarization approach to high sensitivity moiré interferometry, Opt. Eng., 31, (1992), 434-439
67. Kujawinska, M. and L. Salbut, P. Czarnocki: Materials studies of composites by automatic grating interferometer, Proc. SPIE, 2004, (1993), 282-288
68. Kosinski, C. and A. Olszak, M. Kujawinska: Adaptive system for smart fringe image processing, Graphics and Machine Vision, 5, (1996), 245-256
69. Pryputniewicz, R.J.: A hybrid approach to deformation analysis, Proc. SPIE, 2342 (1994) 282-296
70. Brown, G.C. and R.J.Pryputniewicz: Experimental and computational determination of dynamic characteristic of microbeam sensors, Proc.SPIE, 2545, (1995), 108-119
71. Olszak, A. and K.Patorski: Modified electronic speckle pattern interferometer with reduced number of elements for vibration analysis, 138, Opt.Comm., (1997), 265-269
72. Nakadate, S. and T.Yatagai, H.Saito: Digital speckle pattern shearing interferometry, Appl.Opt., 19, (1980), 4241-4246
73. Wyant, J.C.: Computerized interferometric measurement of surface microstructure, Proc.SPIE, 2576, (1995), 122-130
74. Kujawinska M.: Micromechanics: New challenges for photonics, Proc. ATEM'97, (1997), 367-372
75. Salbut, L. and M. Kujawinska, G.Dymny: Polycrystalline material studies by automatic grating interferometry, Proc. SPIE, 2782, (1996), 513-521
76. McKelvie, J. and P.M.Mac Kenzie, A. McDonach, C.A. Walker: Strain distribution measurement in a coarse-grained titanium alloy, Exp. Mechanics, 33, (1993), 320-325
77. Poon, C.Y. and M.Kujawinska, C.Ruiz: Strain measurement of composite using an automated moiré interferometry, Measurement, 11, (1993), 45-57
78. Salbut, L. and M.Kujawinska: Novel material studies by automatic grating interferometry, Proc.SPIE, 2861, (1996), 212-219
79. Han, B. and Guo Y., Lim C.K.: Application of interferometric techniques to verification of numerical model for microelectronics packaging design", EEP 10-2, Advances in Electronic Packaging, ASME (1995), 1187-1194

80. Kujawińska, M. and T. Tkaczyk, R. Pryputniewicz: Computational and experimental hybrid study of deformations in a microelectronic connector, Proc. SPIE, 2545, (1995), 54-70
81. Jüptner, W. and M. Kujawinska, W. Osten, L. Sa☐but, S. Seebacher: Combinative measurement of silicon microbeams by grating interferometer and digital holography, Proc. SPIE, 3407, (1998) in press
82. Salbut, L. and M. Kujawinska: Moire interferometry/thermovision method for electronic packaging testing, Proc. SPIE, 3098,(1997), 10-17
83. Guo, Y.: Experimental determination of effective coefficients of thermal expansion in electronic packaging", EEP 10-2, Advances in Electronic Packaging, ASME (1995), 1253-1258
84. Selverian, J.H and S. Kang: Ceramic–to-metal joints: Part II – performance and strength preditions, American Ceramic Society Bulletin, 71, No 10, (1992)
85. Salbut, L. and M.Kujawinska, J.Bulhak: Ceramic-to-metal joint testing by automated grating interferometer, Experimental Mechanics, Allison (ed) Balkema, Rotterdam, (1998), 633-638
86. Rowlands, R.E.: Residual stress in SEM Handbook of Experimental Mechanics, A.S. Kobayasahi, Edd., Prentice-Hall, Englewood Cliffs, New York, (1987)
87. Kujawinska, M.: Experimental-numerical analysis of 3D residual stress state in engineering objects, Akademie Verlag Series in Optical Metrology, 3, (1996), 151-158
88. Swiderski, Z. and A. Wójtowicz: Plans and progress of controlled experiments on rail residual stress using the EMS-60 machine, in Residual Stress in Rails, 1, Kluver Academic Publ., (1992), 57-66,
89. Groom, J.J: Determination of residual stresses in rails, Final Report for US DOT No DOT/FRA/ORD-83/05, (1983)
90. Orkisz, J. et al: *Discrete analysis of actual residual stress resulting from cyclic loading*, Computers & Structures, 35, (1990), 397-412
91. Magiera, J. and J. Orkisz: Experimental-numerical analysis of 3D residual stress state in railroad rails by means of oblique slicing technique, Proc. SPIE, 2342, (1994), 314-325,
92. Gordon, R.: Residual stress and distortion in welded structure – an overwiev of current, U.S.Research Initiatives, IIW-Doc. XV, (1995), 878-95
93. Wu, Z. and J.Lu: Residual stress by moiré interferometry and incremental hole drilling Exp. Mechanics, I.M.Allison (ed) Balkena, Rotterdam, 2, (1998), 1319-1324
94. Kujawinska, M. and L. Salbut, A. Olszak, C. Forno: Automatic analysis of residual stresses in rails using grating interferometry in Recent Advances in Exp.-Mech., S.Gomez et al. (eds) Balkena Rotterdam, (1994), 699-704

95. Kujawinska, M.: The architecture of a multipurpose fringe pattern analysis system, Opt. Lasers Eng., 19, (1993), 261-268
96. Schwider, J. et al: Digital wave-front measuring interferometry: some systematic errors sources, App. Optics, 22, (1993), 3421-3432
97. Wu, Z. and J.Lu, P.Jouland: Study of residual stress distribution by moiré interferometry incremental drilling method, The Fifth Int. Conf. on Residual Stresses, Linkoping, Sweden, (1997)
98. Kujawinska, M.and L. Salbut, S. Weise, W. Jüptner: Determination of laser beam weldment properties by grating interefrometry method, Proc. SPIE, 2782, (1996), 224-232
99. Salbut, L. and M. Kujawinska, D. Holstein, W. Jüptner: Comparative analysis of laser weldment properties by grating interferometer and digital speckle photography, Exp. Mechanics, Allison I.M. (ed), Balkema, Rotterdam, (1998), 1331-1337

CHAPTER IV

INTERFEROMETRIC METHODS

W. Jüptner
BIAS, Bremen, Germany

ABSTRACT

Interference is one of the most fundamental phenomena of the light as an electromagnetic wave. Huygens postulated that one can define the intensity in any given point anywhere in space by superposing the elementary wave coming from a surface with known electromagnetic excitation. The Huygens principle involves the interference otherwise only the intensities would be summed up.

The phenomena of interference is a result of the linearity of the wave equation. This equation is similar for all kinds of waves. It is a differential equation which contains at least a term with the second derivative after space and one after time, respectively. With any found solution the sum of the solutions is a solution, too. This consequence is the basis for most of the coherent-optical metrology methods.

Holography is based on the interference of an object wave and an reference wave forming the hologram. The reconstruction of the hologram results in the original object wave which can interfere with the wave from a changed object state. The interference pattern is a measure for the displacement of surface points: the principle of holographic interferometry in all kind and of shearography. In Speckle photography the recorded Speckle fields before and after displacement interfere forming Young's fringes.

1. INTRODUCTION

1.1 Historic Remarks

Some historical remarks may be given at the beginning of the lectures on interferometric methods since some of the backgrounds needed a long time of recognition of physical dependencies in the past. Two main inputs were required for the understanding and the use of interferometry: Once the description of the nature of the light, especially the behavior as an electromagnetic wave, had to be developed in order to understand the interference phenomena. But this knowledge could only be used in the way of today by the invention of the laser.

The modern optics started with the work of Huygens who postulated the light to be a wave [1]. He explained the excitation in any point of the space by the thesis of elementary waves coming from a known area. This theory was improved e.g. by Young and Fresnel describing the diffraction and interference of light. Fresnel`s zone plate is still interesting in the field of modern optical systems. It was first realized by Lord Rayleigh who performed it experimentally with a mirror and the rounded tip of a pin in its middle. The pin served as a point source and the mirror reflected the reference wave: the zone plate is a hologram of a point source. Later on Gabor invented the holography [2], which can be thought to be the sum of zone plates. The laser enabled the introduction of off-axis holography by Leith and Upathics [3] and the holographic interferometry, first published by Stetson and Powell [4].

The fundamental discoveries and developments in physics which lead to the invention of the laser are made in the beginning of this century: In the time from 1910 to 1920 physicists like Planck, Einstein, and others could explain the interaction between light and material, the fundamental input to the invention of the laser many years later. Especially Einstein´s theory of absorption and emission of light is the basis for all later developments [5]. The theory was proven by measuring the change of the absorption behavior of materials when it is illuminated by high intensities through works e.g. by Ladenburg [6]. First approaches to realize the laser were made in Russia by Fabrikant with his proposal to amplify light in a material with a negative absorption coefficient. Main progresses were made in the time from 1950 to 1960. Scientist like e.g. Prokhorov, Bloembergen, Schawlow, Townes, Weber, and others developed the microwave amplification in gases and the required knowledge about the laser [e.g.: 7]. 1960 Maiman build the first laser, a pulsed Ruby laser [8]. After this breakthrough the development of lasers of all kinds speeded up and until to the end of the sixties most of the laser types were invented.

Historical Overview

Year	Scientist	Contribution

Laser Metrology

Year	Scientist	Contribution
~ 1690	Huygens, Ch.	Wave nature and propagation of light
~ 1800	Young, T.	Diffraction and interference of light
~ 1820	Fresnel, A.J.	Diffraction of light, Fresnel's zone plate
1871	Lord Rayleigh	Experimental realization of zone plate, own notices
1870	Abbe, E.	Theory of forming taking the diffraction into account
1948	Gabor, D.	*A New Microscopic Principle, Nature 161 (1948)* First description of holographic imaging
1963	Leith, E.N., Upathnieks, J.	*Off-axis holography;* Use of laser for holography
1965	Stetson, K.A.; Powell,	*Interferometric hologram evaluation and real-time vibration analysis of diffuse objects, JOSA 55 (1965)*

Laser

Year	Scientist	Contribution
1873	Maxwell, J.C.	Theory of Electrodynamics; Maxwell's Equations
1917	Einstein, A.	*Quantum mechanics of radiation* Basic theoretical treatment of the interaction between radiation and matter: stimulated emission of light
1928	Ladenburg, R.; Kopfermann, H.	*Experimental prove of negative dispersion* Z. Phys. Chemie Abt.A 139, S. 375ff Experimental prove of stimulated emission of radiation by the measurement of a negative dispersion in gases
1939	Fabrikant, V.A.	Dissertation, Lebedev-Institut, UdSSR Proposal to amplify electromagnetic waves in a gas discharge *with negative absorption coefficient*
1950	Purcell, E.M.; Pound, R.V.	Experimental *prove of inversion* in Li_2F with look-out to microwave amplification in gases

Year	Scientist	Contribution
1951	Fabrikant, V.A.	Patent: Proposal to amplify electromagnetic radiation by the transmission through a material with more higher states of energy compared to the thermal equilibrium; patent was nearly unknown up 1959
1951	Townes, C.H.	Discussion of realization of such an amplifier
1953 - 1955	Weber, J. Basov, N.G.	Publication of proposals for realization of an amplifier according to the patent of Fabrikant.
1954	Townes, C.H. Gordon, J.P. Zeiger, H.J.	*Maser* Prove of amplification by stimulated emission of micro waves in ammonia as oscillator medium
1956	Bloembergen	Introduction of 3-level-method in solid state materials with the use of an independent pump frequency
1956	Basov, N.G. Prokhorov, A.M.	Introduction of 3-level-method in *gases*
1957	Scovil Feher, Seidel	Experimantal prove of Bloembergen's theory by the construction of a Maser
1958	Scovil	*Ruby-Maser* The Ruby-Maser was developed to a product as an extremely low noise amplifier
1958	Schawlow, Townes, C.H.	Expansion of the Maser idea into the optical regime
1959	Javan	First proposal to build a He-Ne laser
1960	Maiman,	*First laser.* Pulsed Ruby laser
1960	Schawlow, Collins	Ruby laser emission
1960		*He-Ne laser:* first cw laser
1962		Proposal for semiconductor laser
1964	Townes, C.H., Basov, N.G., Prokhorov, A.M.	Nobel price

1.2 Properties of Electromagnetic Waves

Light can described as an electromagnetic wave. The electromagnetic field follows physical laws given by the six Maxwell equations [9]: The first two equations concern the sources of the electric and the magnetic field. The second pair of equations concerns to the rotation of the electric and the magnetic field, and by this the interaction between the two fields. It leads as a special solution to the propagation of an electromagnetic wave. The third pair of equations is related to the interaction with the material; they may be neglected here. The Maxwell equations are:

$$div \, \vec{E} = \frac{\rho}{\varepsilon \varepsilon_0} \tag{1.2.1}$$

$$div \, \vec{H} = 0 \tag{1.2.2}$$

$$rot \, \vec{E} = \frac{1}{\varepsilon \varepsilon_0 c^2} \frac{\partial \vec{H}}{\partial t} \tag{1.2.3}$$

$$rot \, \vec{H} = \varepsilon \varepsilon_0 \frac{\partial \vec{E}}{\partial t} + \sigma \vec{E} \tag{1.2.4}$$

with $\vec{E}$: electric field vector
$\vec{H}$: magnetic field vector
ρ : charge density
σ : electrical conductivity
ε : dielectric constant; index 0 for the vacuum

For the following an electrical neutral matter with no free charges is assumed, i.e. $\rho = 0$, $\sigma = 0$. By forming the rotation of the rotation it can be derived.

$$rot(rot \, \vec{E}) = grad(div \, \vec{E}) - div^2 \, \vec{E} \tag{1.2.5}$$

and with Equ.(1.2.5) the wave equation, e.g. for the electric field vector:

$$\frac{\partial^2 \vec{E}}{\partial r^2} - \frac{\varepsilon}{c^2} \frac{\partial \vec{E}}{\partial t} = 0 \tag{1.2.6}$$

The magnetic field may be treated in a similar way with a comparable result. The solution of this equation is any function $f = f(r - ct)$ which has the second derivatives after space and time. In common the exponential wave function

$$\vec{E} = \vec{E}_0 \, exp\left[i(\phi(\vec{r}) - \omega t)\right] \tag{1.2.7}$$

with $\vec{E}_0$: electric field vector amplitude
$\phi(\vec{r})$: space dependent phase factor
ωt : time dependent phase factor

and $\omega = 2\pi f$
   f : Frequency of the wave

is named as main solution. This is allowed since the wave equation (1.2.6) is a linear equation. The description with the Equ.(1.2.7) is a common simplification: Fourier synthesis allows to generate any wave form. However, only the real valued terms of the complex function are physically meaningful.

Equ.(1.2.7) contains three quantities to define the wave:

- The amplitude of the electric vector E_0. This vector oscillates in a plan perpendicular to wave propagation direction for all cases to be regarded. The magnetic vector oscillates perpendicular to the electric vector and to the wave propagation since the electromagnetic wave is a transversal wave. These waves can be polarized and the polarization behavior is defined by the oscillation of the electric (or magnetic) vector:

  = The vector oscillates in a constant direction: linear polarization
  = The oscillation direction turns constantly around: circular polarization
  = The oscillation direction changes randomly: no polarization

- The space dependent phase term $\phi(\vec{r})$ defines the locations of constant phase. It depends on the wave propagation conditions. Two typical examples are:

  = A sphere wave has spheres as locations of constant phase. This means that e.g. for a coordinate system with the 0-point in the middle of the source the phase function $\phi(\vec{r})$ can be written as

  $$\phi(\vec{r}) = k\, r \quad ; \quad k = |\vec{k}| \,, \quad r = |\vec{r}| \tag{1.2.8}$$

  with $\vec{k} = \dfrac{2\pi}{\lambda} \vec{e}$: wave vector

  = A plane wave has planes perpendicular to the wave propagation as locations of constant phase. $\phi(\vec{r})$ can be written as

  $$\phi(\vec{r}) = \vec{k}\, \vec{r} \tag{1.2.9}$$

- The time dependent term contains with ω the frequency of the wave. In light waves the electric vector oscillates so fast, that it cannot be measured directly. All detectors react on the intensity of the light, which is related to the electric vector of the wave by

$$I = \varepsilon \varepsilon_0\, c\, |\vec{E}_0|^2 = \varepsilon \varepsilon_0\, c\, \vec{E}_0\, \vec{E}_0^{\,*} \tag{1.2.10}$$

1.3 Interference of Waves

Interference is the ability of waves to extinguish themselves. Interference appears when two or more waves are superimposed: The sum of single solutions of the wave Equ.(1.2.6) is a solution, too, since the wave equation is a linear differential equation. Regarding monochromatic light waves the electric field vector $E(r,t)$ of the resulting field at given time t and location r is the sum of the contributing single waves:

$$\vec{E}(\vec{r},t) = \sum_i \vec{E}_i(\vec{r},t) \quad ; \quad i = 1,2,... \tag{1.3.1}$$

The case of the interference of two waves shall be regarded as an example which can be extended easily to the common case of more than two waves. The two waves can be written as:

$$\vec{E}_1(\vec{r},t) = \vec{E}_{10} \exp\{i[\phi_1(\vec{r}) - \omega_1 t]\} \tag{1.3.2a}$$

$$\vec{E}_2(\vec{r},t) = \vec{E}_{20} \exp\{i[\phi_2(\vec{r}) - \omega_2 t]\} \tag{1.3.2b}$$

The electrical vector cannot be detected directly; any detector is sensitive for the intensity of the light which is proportional to the squared amplitude of the vector:

$$I \propto |\vec{E}_r|^2 = |\vec{E}_1 + \vec{E}_2|^2 = (\vec{E}_1 + \vec{E}_2) \cdot (\vec{E}_1 + \vec{E}_2)^* \\ = \vec{E}_1 \vec{E}_1^* + \vec{E}_2 \vec{E}_2^* + \vec{E}_1 \vec{E}_2^* + \vec{E}_1^* \vec{E}_2 \tag{1.3.3}$$

The first term on the right side of Equ.(1.3.3) represents the intensity I_1 of the wave $\vec{E}_1(\vec{r},t)$, the second term represents I_2 of wave $\vec{E}_2(\vec{r},t)$. The sum of these terms is the addition of the two waves without interference, i.e. what will be measured with thermal light. Taking this into account Equ.(1.2.3) results in:

$$I \propto I_1 + I_2 + \\ + \vec{E}_{10} \vec{E}_{20}^* \exp\{i[(\phi_1(\vec{r}) - \phi_2(\vec{r})) - (\omega_1 t - \omega_2 t)]\} + \\ + \vec{E}_{10}^* \vec{E}_{20} \exp\{-i[(\phi_1(\vec{r}) - \phi_2(\vec{r})) - (\omega_1 t - \omega_2 t)]\} \tag{1.3.4}$$

The fourth term on the right side of Equ.(1.3.4) is the complex conjugated term to the third. The sum of those terms results in:

$$\exp(ix) + \exp(-ix) = 2\cos(x) \tag{1.3.5}$$

Additionally $\vec{E}_{10} \vec{E}_{20}^*$ and $\vec{E}_{10}^* \vec{E}_{20}$, respectively, can be combined to

$$\vec{E}_{10} \vec{E}_{20}^* = \vec{E}_{10}^* \vec{E}_{20} = \sqrt{I_1 I_2} \tag{1.3.6}$$

From the Equ.(1.3.4), (1.3.5) and (1.3.6) the superposition of two waves results:

$$I = I_1 + I_2 + 2\sqrt{I_1 I_2}\,\cos\left[\left(\phi_1(\vec{r}) - \phi_2(\vec{r})\right) - \left(\omega_1 t - \omega_2 t\right)\right] \quad (1.3.7)$$

The third term on the right side is the interference term. This term has a maximum for all values of the argument Φ with:

$$\Phi = \left(\phi_1(\vec{r}) - \phi_2(\vec{r})\right) - \left(\omega_1 t - \omega_2 t\right) = 2n\pi \quad ; \quad n = 0,1,2,\ldots \quad (1.3.8)$$

This term has a minimum for all values of the argument Φ with:

$$\Phi = \left(\phi_1(\vec{r}) - \phi_2(\vec{r})\right) - \left(\omega_1 t - \omega_2 t\right) = (2n+1)\pi \quad ; \quad n = 0,1,2,\ldots \quad (1.3.9)$$

The argument Φ contains a time dependent term: the second summand of the argument. For common free running light waves the difference between the frequencies is so large and fluctuating, that no interference can be observed. For small and stable differences ($\Delta\omega \ll \omega_1, \omega_2$) this term is a time modulation of the interference resulting from the space dependent summand. This fact is used in *heterodyne interferometry* to compare the phase difference between an arbitrary point P($\vec{r}$) to a reference point R($\vec{r}$) electronically.

The time dependent term disappears if two light waves with the same frequency interfere. The remaining interference equation is given by:

$$I = I_1 + I_2 + 2\sqrt{I_1 I_2}\,\cos\left[\left(\phi_1(\vec{r}) - \phi_2(\vec{r})\right)\right] \quad (1.3.10)$$

The intensity maximum is achieved for $\cos(\Phi) = 1$, the minimum for $\cos(\Phi) = -1$

$$I = I_1 + I_2 + 2\sqrt{I_1 I_2} \qquad \text{maximum intensity} \qquad (1.3.11)$$
$$I = I_1 + I_2 - 2\sqrt{I_1 I_2} \qquad \text{minimum intensity} \qquad (1.3.12)$$

Using Equ. (1.3.11) and (1.3.12) the contrast K can be defined as:

$$K = \frac{I_{max} - I_{min}}{I_{max} + I_{min}} \quad (1.3.13)$$

For two light waves according to Equ. (1.3.2a) or (1.3.2b) this results:

$$K = \frac{2\sqrt{I_1 I_2}}{I_1 + I_2} \quad (1.3.14)$$

The contrast has a maximum for $I_1 = I_2$ and a minimum when $\sqrt{I_1 I_2}$ disappears, i.e. when no interference can be observed. The ability of light to interfere is called coherence. The before mentioned treatment of interference assumed a fixed phase relation in space and time for the interacting waves. For thermal light sources this is valid only under very limited conditions in contrary to the light of lasers.

1.4 Diffraction of Light

Conventional photography uses an optical system, e.g. a lens, to image the object on to a light sensitive medium. In this process the phase gets lost since all media are only sensitive for the intensity. One step into the direction to conserve the phase, too, was made by Abbé with his work on the forming process of an image in a microscope: The generation of the image is a two step process, Fig.1.1.: First a diffraction image ("first image") is generated in the plane E by the interference of all diffraction orders coming from the object, which is thought to be the superposition of gratings. This first image is transformed into the second image in the image plane. Filtering may be performed in the first image plane in order to accentuate part of the image.

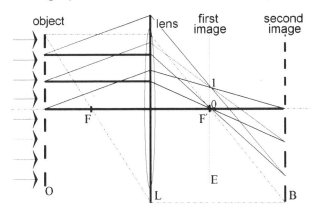

Fig.1.1: Image formation with diffraction

Linear gratings with a binary transmission not only generate the zero and the first orders of diffraction but higher orders, too. The positive and the negative orders are symmetrically to the center axis of the grating and the angle of the maximum intensity of the nth order is given by, Fig.1.2a:

$$sin\alpha_n = n\lambda/d \quad ; \quad n = 1, 2, ... \tag{1.4.1}$$

with α_n : angle of diffraction order n
  n : order of diffraction
  l : wave length
  d : grating constant, distance between middle of two slots

Higher order diffraction can be avoided by gratings with sinusoidal transmission. In this case only the zero and the two first orders are generated. Additionally, one can change the grating constant with the distance from the center. If the distance d between two neighbored transmission maxima m and m+1 is inverse proportional to the square root of the maximum number,

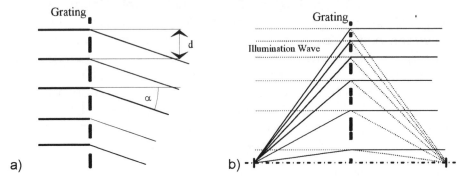

Fig.1.2: Diffraction by a grating, a) constant d, b) d proportional to $1/\sqrt{m}$

counting the center maximum as 1, then the grating will diffract a plane incident wave once in a line in the distance f from the grating. This is the minus first order diffraction. The first order diffraction forms a divergent wave which seems to come from a line in the distance -f before the grating, Fig.1.2b. Furthermore, if one takes a grating with rings instead of lines the minus first order diffraction of a plane incident wave forms a point in the distance f. The first order diffraction forms a spherical divergent wave seemingly coming from a point in the distance -f before the grating, Fig. 1.3. Such a grating is an imaging system, it is called after its inventor *Fresnel zone plate*.

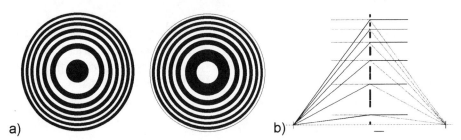

Fig.1.3: Fresnel zone plate, a) transmission b) imaging characteristic

A simple procedure to create a Fresnel zone plate was given by Lord Rayleigh: He proposed to illuminate a photographic plate with a plane wave and a spherical wave coming from a source point in the distance f from the plate. Both waves should have parallel axis in his set-up. After the development of the photo plate a system of concentric rings appeared where the radii followed the before mentioned square root law[1]. The imaging laws of such a zone plate can be defined easily as:

$$\frac{1}{a}+\frac{1}{b}=\frac{m\lambda}{r_m^2}=\frac{1}{f} \qquad (1.4.2)$$

with   a : distance of a object point (source point) to the plate
       b : distance of the image point from the plate

The focal length can be defined as in Equ.(1.4.2):

$$f=\frac{r_m^2}{m\lambda} \qquad (1.4.3)$$

This results in the rule for the radii:

$$r_m=\sqrt{f\lambda m} \qquad (1.4.4)$$

Gabor published in 1949 a new technique to image an object with the preservation of the phase calling it *holography* since he stored amplitude and phase in the photographic plate. For demonstration he took a slide with points as object. The points served as point sources while the unaffected incident light served as reference wave, Fig.1.4. The set-up is comparable to that of Lord Rayleigh, however, for several source points instead of one. In this sense the Fresnel zone plate is the hologram of a single point source.

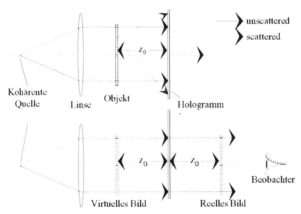

Fig.1.4: Gabor set-up for generating a hologram

---

[1] Lord Rayleigh used the rounded end of a pin in the middle of a mirror to generate the two waves out of a plane incident wave.

1.5 References

1. Geerthsen, Chr., v.Vogel, H.: Physik, Springer Verlag 1997
2. Gabor, D.: Microscopy by reconstructed wave fronts, Proc. Royal Soc. A, 197, (1949), p.454-487
3. Leith, E.N.; Upathnics, J.: Wavefront reconstruction with diffused illumination and three-dimensional objects, Journ. Opt. Soc. Amer., 54 (1965)
4. Powell, R.L.; Stetson, K.A.: Interferometric vibration analysis by wave front reconstruction, Journ. Opt. Soc. Amer., 55, 1965, p.1593-1608
5. Einstein, A.: Zur Quantentheorie der Strahlung (About the quantum theory of radiation), Phys. Z. 18 (1917), S. 121 - 128
6. Ladenburg, R.; Kopfermann, H.: Experimenteller Nachweis der negativen Dispersion (Experimental proof of the negative dispersion), Z. Phys. Chemie Abt. A 139 (1928), S. 375 - 385
7. Bloembergen, N.: Proposal for a new type solid state maser, Phys. Rev. 104, 2 (1956), p. 324 - 327
8. Maiman, T. H.: Stimulated optical radiation in ruby, Nature 187 (1960), p. 493f
9. Born, M.; Wolf, E.: Principle of Optics, Pergamon Press, Oxford, London, Edinburgh 1980

2. HOLOGRAPHIC INTERFEROMETRY

2.1 Holography

2.1.1 Fundamentals of Holography

Holography was first introduced by Gabor in 1949 [1]. His fundamental idea was to combine the light coming from an coherently illuminated object - the object wave - with an information free wave - the reference wave, chap. 1.4. Both waves interfere and the micro-interferences are stored in a photographic plate. After the development of the plate, now called a *hologram*, the original wave can be reconstructed by illuminating the plate with the reference wave. A break-through came with the laser and with the idea of Leith and Upathniks to combine the two waves under a certain angle. However, holography is always a two step method, Fig.2.1.

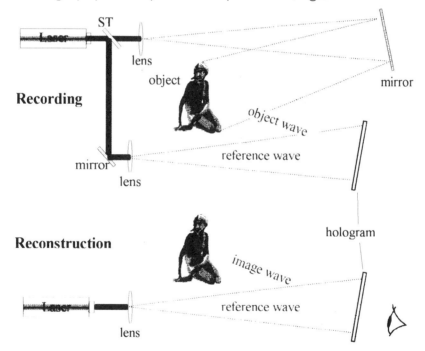

Fig.2.1: Holography as two step imaging method

The mathematical description of a hologram, e.g. taken in a set-up according to Fig.2.2, can be formulated in the following way [2]: The object wave in the plane of the plate may be expressed by, see chap. 1.3:

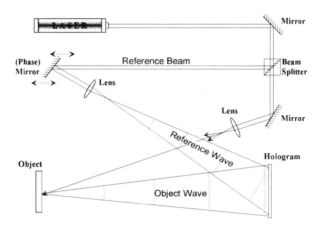

Fig.2.2: Holographic recording set-up, schematically

$$\vec{E}_O = \vec{E}_{OO}\, exp[i(\phi_O(\vec{r}) - \omega_O t)] \tag{2.1.1}$$

with $\vec{E}_O$: electric field vector of the object wave
$\vec{E}_{OO}$: amplitude of the electric field vector of the object wave
ϕ : spatial phase of the object wave
r : point vector
ω : circular frequency of the wave

The object wave interferes with a reference wave according to

$$\vec{E}_R = \vec{E}_{R0}\, exp[i(\phi_R(\vec{r}) - \omega_R t)] \tag{2.1.2}$$

with the nomination of the variables according to that of the object wave. ϕ_R is in common a slowly in space varying function with no information. However, this is not a requirement as will be shown later.

One can prove easily, that two waves with different circular frequencies ω_O and ω_R result in a flat gray transmission of the plate [3]. So, in the following it is assumed ω_O and ω_R to be equal. This leads to an intensity in the hologram plane as follows, chap. 1.3:

$$I \approx (\vec{E}_O + \vec{E}_R)\cdot(\vec{E}_O + \vec{E}_R)^* \\ = \vec{E}_{OO}^2 + \vec{E}_{R0}^2 + \vec{E}_{OO}\vec{E}_{R0}^*\, exp[i(\phi_O(\vec{r}) - \phi_R(\vec{r}))] + \vec{E}_{OO}^*\vec{E}_{R0}\, exp[i(\phi_O(\vec{r}) - \phi_R(\vec{r}))] \tag{2.1.3}$$

Equ.(2.1.3) describes a time independent intensity function in the space conserving the information about the object in the phase ϕ_O and about the reference in the term ϕ_R. The intensity distribution can be recorded by any light sensitive sensor with the required spatial resolution: the distance between two interference fringes is in the magnitude of a few wave length or less.

The transmission of the photographic plate after a correct exposure and after the development is linear to the incident intensity:

$$T = T_0 + \beta I \tag{2.1.4}$$

with T : transmission of the hologram
      T_0 : transmission of the unexposed photo plate
      β : gradation of the photo plate

After the replacement into the set-up the developed hologram is illuminated with the reference wave. The intensity distribution behind the hologram is the one of the reference wave but altered by the local transmission:

$$\begin{aligned}\vec{E}_R T &= \vec{E}_R T_0 + \vec{E}_R \beta \vec{E}_{O0}^2 + \vec{E}_R \beta \vec{E}_{R0}^2 \\ &+ \beta \vec{E}_{O0} \vec{E}_{R0} \vec{E}_{R0}^* \exp\left[i(\phi_o(\vec{r}) - \omega t)\right] \\ &+ \beta \vec{E}_{O0}^* \vec{E}_{R0}^* \vec{E}_{R0} \exp\left[-i((\phi_o(\vec{r}) - 2\phi_R(\vec{r}) - \omega t))\right]\end{aligned} \tag{2.1.5}$$

This Equ.(2.1.5) may be written in a slightly changed form:

$$\begin{aligned}\vec{E}_R T &= \left(T_0 + \beta \vec{E}_{O0}^2 + \beta \vec{E}_{R0}^2\right)\vec{E}_R & &\text{reference wave} \\ &+ \beta \vec{E}_{R0}^2 \vec{E}_O & &\text{object wave} \\ &+ \beta \vec{E}_{O0}^* \vec{E}_{R0}^* \vec{E}_{R0} \exp\left[-i(\phi_o(\vec{r}) - \omega t)\right]\exp(2\phi_R(\vec{r})) & &\text{real image wave}\end{aligned} \tag{2.1.6}$$

This means that Equ.(2.1.6) describes three waves:

- The first term, the zeroth diffraction order, is the reference wave, only changed in the amplitude,

- the second term, the first diffraction order, describes the original object wave, only changed in its amplitude

- the third term, the minus first diffraction order, describes a real image wave, but mirrored at the reference wave axis according to the second exponential function

The equation system Equ.(2.1.1) to (2.1.6) is very powerful [4]. By changing the boundary conditions one can derive the different applications of holography. Only some remarks on this:

- Holography is a real 3D imaging method: The object wave is reconstructed with all the phase - i.e. depth - information about the object, but windowed by the hologram as an aperture.
- The description of the object and the reference wave only differs in the indices, which are not relevant from the physical point of view. This means, that the object wave may be used as reference wave resulting in the reconstruction of the reference wave. This can serve for the correlation between two objects.
- According to the before mentioned holograms can be generated in the computer and used as matched filters.
- Furthermore, the reference has not to be an information free wave, e.g. it can be the object wave itself but slightly changed. This leads to the modern technology of shearing interferometry.

2.1.2 Amplitude and Phase Holograms

Holograms are in principle diffraction gratings [5]. Gratings can be generated by a periodical or quasi-periodical change of the transmission or the diffraction index. Holograms with a changing amplitude of the transmission are amplitude holograms since they modify primarily the amplitude of the transmitted wave, Fig.2.3. Holograms changing primarily the optical path length of the transmitted wave are called phase holograms. Both types of holograms are able to reproduce whole the object wave. However, phase holograms do not change the intensity of the reference wave. Therefore they may have the higher efficiency of diffraction.

Both types of holograms can be generated as transmissive or as reflective holograms, which means that it is possible to have transmission or reflection holograms.

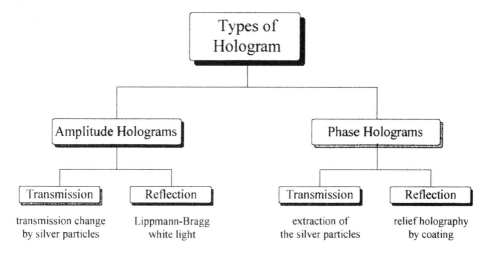

Fig.2.3: Transmission and phase holograms

2.1.3 Types of Holograms

Holograms differ in the manner of the incident object wave: the object may be imaged onto the hologram - the object seems to be in the hologram plane -, the object may be close to the hologram with any object point as a source of a spherical wave, and the object may be far away with any object point as a source of a plane wave. These differences require different spatial resolution of the photo material and different stability conditions of the set-up.

Image plane hologram

An image plane hologram can be achieved, when the object is imaged by a lens into the hologram plane, Fig.2.4. This means, that any object point refers to one point in the hologram. The magnification of the image system has to be adapted to the size of the object and the hologram area, otherwise information will be lost. The reference wave is added as a parallel wave in common but not necessarily. The way of generating this hologram makes it comparable to a slide with the wave phase coded in micro-interferences.

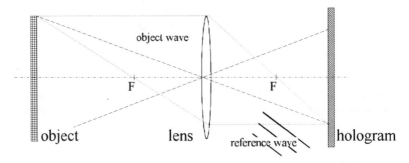

Fig.2.4: Image plane hologram

The advantage of the hologram is the low requirement of spatial resolution and stability of the set-up. Therefore it fits the boundary conditions of TV holography - or ESPI, DSPI, EOH - very well, chap 5.

The disadvantage is the sensitivity against damage, which is in its main applications no problem, and its requirements on the contrast transfer.

Fresnel hologram

A hologram taken from an object in the neighborhood to the hologram is called a Fresnel hologram, Fig.2.5. Any point of the surface can be thought to be the source of an elementary sphere wave. These waves interfere in the hologram plan superposed with the reference wave.

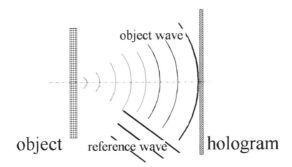

Fig.2.5: Fresnel hologram

The advantage of this kind of hologram is the distribution of the information over all the hologram surface. So, the hologram and by this the stored information is insensitive against damage.

The disadvantage is the high demand on the spatial resolution of the storing material and on the stability of the set-up.

Fraunhofer or Fourier transform hologram

The hologram is called a Fraunhofer or Fourier transform hologram, when the object is very far away from the hologram, which is comparable with a set-up with the object in the focal plane of a lens, Fig.2.6. Then the object is focused into infinity and the waves incident onto the hologram are plane waves. The recorded wave front is proportional to the Fourier spectrum of the object surface. Fourier holograms may have a high contrast, since every point of the object contributes to any point of the hologram. They fit well the requirements of filtering.

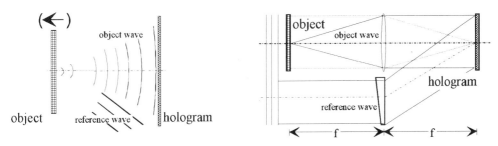

Fig.2.6: Fraunhofer or Fourier transform hologram

2.1.4 Stability Requirements for Recording Holograms

The recording of a hologram means the superposition of two interfering waves. In the theoretical treatment of the recording stable light paths during recording were assumed without mentioning it: the equations for the waves

$$\vec{E}_o(\vec{r},t) = \vec{E}_{oo} \exp[i(\phi_o(\vec{r}) - \omega t)] \tag{2.1.7}$$

$$\vec{E}_R(\vec{r},t) = \vec{E}_{Ro} \exp[i(\phi_R(\vec{r}) - \omega t)] \tag{2.1.8}$$

(variables see chap. 2.1.1)

include spatial phase terms which were assumed to be time independent and noise free. In reality this is not valid in this strong formulation. For the treatment of a real recording situation both terms ϕ_o and ϕ_R must be extended by a noise component. This noise term is assumed to be randomly distributed and they do not depend on the phase itself. Then the phases may complemented as follows [6]:

$$\phi_o(\vec{r},t) \Rightarrow \phi_o(\vec{r}) + \phi_{so}(t) \tag{2.1.9}$$

$$\phi_R(\vec{r},t) \Rightarrow \phi_R(\vec{r}) + \phi_{SR}(t) \tag{2.1.10}$$

$$\vec{E}_o(\vec{r},t) = \vec{E}_{oo} \exp[i(\phi_o(\vec{r}) + \phi_{so}(t) - \omega t)] \tag{2.1.11}$$

$$\vec{E}_R(\vec{r},t) = \vec{E}_{Ro} \exp[i(\phi_R(\vec{r}) + \phi_{SR}(t) - \omega t)] \tag{2.1.12}$$

The interference between the two waves is transferred to

$$\begin{aligned} I &\approx (\vec{E}_o + \vec{E}_R) \cdot (\vec{E}_o + \vec{E}_R)^* \\ &= \vec{E}_{oo}^2 + \vec{E}_{Ro}^2 + \\ &\quad + \vec{E}_{oo}\vec{E}_{Ro}^* \exp[i((\phi_o(\vec{r}) - \phi_R(\vec{r})) + (\phi_{so}(t) - \phi_{SR}(t)))] \\ &\quad + \vec{E}_{oo}^*\vec{E}_{Ro} \exp[-i((\phi_o(\vec{r}) - \phi_R(\vec{r})) + (\phi_{so}(t) - \phi_{SR}(t)))] \end{aligned} \tag{2.1.13}$$

A stable interference requires the time dependent noise term to be small against 2π

$$\phi_S(t) = |\phi_{so}(t) - \phi_{SR}(t)| \ll 2\pi \quad \text{for} \quad t \leq t_a \tag{2.1.14}$$

The phase terms can be expressed by changes of the light paths by any disturbance ls:

$$\phi_{SO}(t) = \frac{2\pi l_{SO}(t)}{\lambda} \qquad (2.1.15)$$

$$\phi_{RO}(t) = \frac{2\pi l_{RO}(t)}{\lambda} \qquad (2.1.16)$$

The stability condition Equ.(2.1.14) is then transferred to

$$|l_{SO}(t) - l_{SR}(t)| \ll \lambda \quad \text{for} \quad t \leq t_a \qquad (2.1.17)$$

The light paths may have different reasons:

- the object moves during the recording time,
- optical components of the set-up are not stable,
- the environment is changing, e.g. the temperature or humidity with its influence on the diffraction index or on components of the set-up.

The stability condition Equ.(2.1.17) includes different opportunities of a solution:

- The trivial one is to guaranty the stability of the set-up itself. This is the standard under laboratory conditions with a vibration isolation. This may be improved by air conditioning for constant temperature and humidity in long-term or other critical measurement situations. This means to minimize the disturbance terms, Equ.(2.1.15) and (2.1.16), separately.

- The recording time can be minimized by using pulsed lasers. This results in small disturbance terms during the exposure time even when the object or other parts of the set-up are moving with a final speed v. Problems may arise in the case of using the hologram for holographic interferometry measurements: Short pulse recording ensures to record the two holograms. However, the reconstructed objects waves must not necessarily interfere if the rigid body motion is too large between the two hologram recordings.

- The difference between the object wave and the reference wave can be held constant even with changing single light paths. One easy way is the use of a phase reference mirror on the object, i.e. the reference wave is guided to the hologram via a mirror on the object: Whenever the object is dislocated the reference wave follows it and the difference in Equ.(2.1.17) is as small as required. There are comparable approaches of the use of an object controlled reference wave, shearography is the most advanced one [7].

2.2 Holographic Interferometry

2.2.1 Fundamentals of Holographic Interferometry

The reconstructed hologram contains all optical information about the recorded object. If this holographic image is superposed with the object wave from the same object, but slightly changed, e.g. by a displacement or a deformation, the two waves will interfere [8]. The simplest way to do this is to make a first exposure from the object in a reference state, to change the state of the object and to make a second exposure, Fig.2.7. This is called a double exposure hologram.

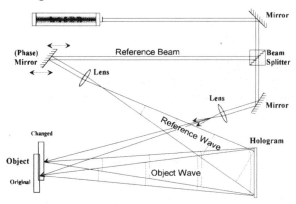

Fig.2.7: Recording of a double exposure hologram

After the development of the photographic plate, the hologram will be illuminated with the reference wave; thus leading to the reconstruction of both states of the object at the same time in the original position, Fig.2.8.

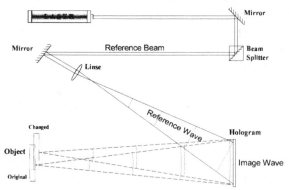

Fig.2.8: Reconstruction of a double exposure hologram

The two images cannot be discriminated by the human eye from each other if the displacement is small enough. However, the two reconstructed image waves interfere forming a fringe pattern: According to Stetson and Powell [9] generates a change of the light path from the illumination source point Q via the regarded object point P to the observation point B an intensity change of the image of this point P, Fig.2.9. The interference will result in a minimum of the intensity when the optical path change by an odd multiple of λ/2 - λ being the wave length of the used laser -, and it will result in a maximum of the intensity for a change of an integer multiple of λ.

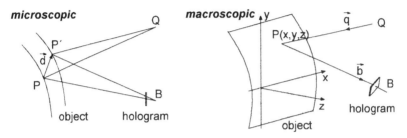

Fig.2.9: Interference by displacement

The interference between the two object waves can be derived as mentioned in former chapters. With $\phi_1(x,y,z)$ as the spatial phase term of object state 1 and $\phi_2(x,y,z)$ as the corresponding phase term 2 the interference is given by:

$$\begin{aligned}
I &\approx (\vec{E}_1 + \vec{E}_2) \cdot (\vec{E}_1 + \vec{E}_2)^* \\
&= \vec{E}_{01}^2 + \vec{E}_{02}^2 + \\
&\quad + \vec{E}_{01}\vec{E}_{02}^* \exp[i(\phi_1(\vec{r}) - \phi_2(\vec{r}))] \\
&\quad + \vec{E}_{01}^*\vec{E}_{02} \exp[-i(\phi_1(\vec{r}) - \phi_2(\vec{r}))]
\end{aligned} \qquad (2.2.1)$$

or in a similar way as in chap. 1.3 with the intensities:

$$I = I_1 + I_2 + 2\sqrt{I_1 I_2} \cos(\phi_1(\vec{r}) - \phi_2(\vec{r})) \qquad (2.2.2)$$

The two phase terms can be combined to $\phi(x,y,z)$. This phase difference between the spatial phase terms of the two object waves is according to the optical path difference:

$$\phi(x,y,z) = \frac{2\pi}{\lambda} \Delta l \qquad (2.2.3)$$

The optical path difference is the geometrical one but projected onto the sensitivity vector, which is the vector of maximum change in light path. According to Abramson is this vector the normal of the ellipses (with Q and B as focal points) in the regarded point P [10]. The sensitivity vector is also the vector given by the difference of the unity vector from the illumination source to the point P and the unity vector from P to the observation point:

$$\Delta l = \vec{d} \cdot (\vec{b} - \vec{q}) \tag{2.2.4}$$

With the Equ.(2.2.2) - (2.2.4) the interference equation gets:

$$I = I_B + I_I \cos\left[\frac{2\pi}{\lambda} \vec{d} \cdot (\vec{b} - \vec{q})\right] \tag{2.2.5}$$

Since the phase term is only the projection of the displacement vector onto one sensitivity vector one needs for a 3D-measurement of the vector three sensitivity vectors. The three vectors may be introduced by changing the observation direction. However, the most adapted way is to change the illumination direction.

2.2.2 Evaluation of Interference Pattern

The evaluation of the interference pattern will be reported by W. Osten in section VI, *Digital Processing and Evaluation of Fringe Pattern in Optical Metrology*. Therefore only a few remarks shall be given here.

The evaluation can be performed in two different ways:

- The fringe pattern can be evaluated qualitatively. This is common in non-destructive testing [11]. Usually the tester evaluates the fringe pattern or an enhanced image visually and qualifies areas with defects by his knowledge about interference fringes and his experience with the test of the special material or component.

  Modern methods of evaluation have been developed meanwhile which are based on knowledge assisted evaluations.
  - = One of these methods uses a catalogue of defect indicating fringe patterns. Each of these patterns has special qualities for the defect indication which can be found automatically. Since the number of defect indicating fringe pattern is limited [12], a computer can be programmed to look for these patterns.
  - = Another method is based on the evaluation by a neural network. The input into the network will be quantities about the slope of the phase differences [13]. Choosing the appropriate structure and teaching the network with defect indicating fringe patterns as well as with patterns of sound structures will result in a fast evaluation of the fringe patterns of components under test.

- The fringe pattern can be evaluated quantitatively [14]. This is a procedure of several steps [15]. The first step is usually a preprocessing of the pattern in order to reduce the coherent and electronic noise. This step is followed by the determination of the phase map. Three main methods are used for this:
  - = Skeletonizing from one single interferogram, a method which is close to the fringe counting of former days
  - = Phase shifting or stepping, a methods which needs the acquisition of at least three interferograms with shifted phases
  - = Fourier Transform Evaluation from one single interferogram. The interferogram is Fourier transformed, filtered unsymmetrically, and retransformed. Out of this complex intensity the phase can be reconstructed

  The phase field has to be transformed into the field of dislocation or diffraction index change using the knowledge about the set-up.

2.3 Applications of Holographic Interferometry

2.3.1 Non-destructive Evaluation by Holographic Interferometry

Non-destructive testing (NDT) is a necessary tool to improve the quality of a product in an economic production. In order to guaranty the properties of a component to match the required ones a non-destructive test can help to find bad parts. However, the improvement of the production needs the additional valuation of the faults and their influence on the component. This task is dependent on the material under test and on the complexity of the component, two parameters with increasing importance in modern constructions.

The basic idea in HNDT is that a component with a defect will react in a different manner compared to a sound structure, since the stiffness or the heat conductivity will be modified locally or globally. This results in a locally or globally changed fringe system [11].

Standard materials as metals or homogenous plastics can be tested easily with standard methods. Materials like fiber reinforced plastics (FRP) can include faults hardly to be detected with these methods by physical reasons [16]. In this case an additional or alternative method has to be applied more related to the behavior of the component under load. The experience of the last years points out holographic interferometry (HI, or related methods like ESPI, electronic holography, shearography) to be such a method. Even more, holographic interferometry in combination with FEM can help to quantify the faults and to understand the influence of the fault on the components behavior under load [17]. The combination of holography with the mathematical prediction of the FEM to a hybrid method is a powerful tool for NDT and for non-destructive evaluation (NDE) [18]. As an example for this the deformation of a fault in a bonded specimen under thermal load will be demonstrated.

Fundamentals of Non-Destructive Testing NDT

NDT is based on the comparison of the tested components behavior under any load with the reaction of a master component under the same load [13]. In practice the master can be a real part or its behavior may be calculated by computer methods like Computer Aided Design - CAD - combined with Finite Element Methods -FEM. For the comparison an energy of any kind is emitted into the component and the reaction is measured with a detector. So each NDT method consists of four basic elements, Fig.2.10:

– The sender (emitter) emits any energy into the component under test. The energy can be in the form of ultrasonic in ultrasonic testing, heat in thermographic or in holographic nondestructive testing (HNDT), mechanical load in acoustic emission or again HNDT etc.

- The test object transfers the energy to any surface where it can be detected. The transfer of the energy - the attenuation, the scattering, the transformation or else - depends on the materials properties and on the geometry of the component.
- The sensor or detector records the transferred energy and converts it to a detectable signal. It should be pointed out, that the detected signal must not be of the same physical quality as the emitted one, e.g. the load might be heat and the detected signal might be deformation.
- The evaluation in HNDT is usually done by visual inspection of the interferogram, although a computer is used for an image preprocessing. But a testing person has to decide about the soundness of the component. Some approaches to solve this problem will be reported in section VI.

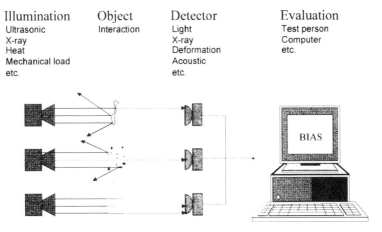

Fig.2.10: Components of a NDE system

Before starting the NDT of a given component it must be known from research and development, how the change of the material or the geometry (any fault may regarded as one of these two changes) influences the answer to the load. In practice there are two ways of performing the loading:

- The load may be adapted to the fault without any relation to the operational load. If during the test a change of the energy flux compared to the master component is observed, the tester relates it to known fault indications. This NDT procedure cannot classify into "good" or "bad" by itself. The tester has to qualify the significance of the detected signal by his experience. Method like this are „relative" ones. An example is the thermal loading.

- One can apply the operational load to the component and detect unallowed deformations. Then the real behavior during the operation can be predicted: The component can be classified to be "usable" or "not usable" regardless of the existence of faults. Such method are called "absolute" ones. HNDT with the operational load is an absolute method.

Faults in GRP Tubes

GRP (glass fiber reinforced plastic) is a modern material. By its complex structure it is not easy to be tested nondestructively. The first step must be to study the answer of the material to the test load. Two approaches are made:

- The different faults are produced artificially and there behavior is measured in advance to the test. This method was applied to the NDT of (GRP) tubes in a round robin test [19].
- The behavior of the fault and its influence on the surface deformation is determined by FEM calculations. The advantage is the easy way to change the parameters and to perform parametric studies. It is even possible to optimize the set-up and the test parameters [20].

Both methods were applied to the same types of GRP tubes:

The tubes were about 1m in length, 0.1 m in width and the wall thickness was 5 mm. Tey were loaded by internal pressure in several steps. The maximum load did not exceed 1/10 of the operational load.

The holographic set-up was a standard one. The illumination beam and the observation direction were symmetrically to the normal of the tube surface in the middle of the tube. This resulted in a main sensitivity direction out of the surface of the tube.

The evaluation of the interference pattern was performed by visual inspection. Although no training of the test personal with tubes of known defects was done before the test itself, nearly all defects could be detected. Examples are given for two types of faults [19], Fig.2.11.

reference

local

global

Fig.2.11: Examples of HNDT of glass fiber reinforced plastic tubes

Evaluation in NDT

In HNDT the question is to detect fault indicating fringe irregularities and to classify them; but not any inhomogeneity indicates a defect. Usually a test person is needed to evaluate the fringe pattern. Several methods were proposed to overcome this severe restriction to an extended use in production line: The standard methods of full field of deformations measurements or statistical methods do not match industrial needs of reliability. The fact, that test persons can detect fault indicating fringe systems changes even in complex structures lead to the approach to apply knowledge to the evaluation of the fringes. Two of the methods have been proven to be appropriate for this task:

– Neural networks "simulate" the human ability to learn knowledge by structuring a system of so-called neurons and their connections in a learning phase. In the working phase these structures are able to recognize new inputs and classify them.

The ability of the network to detect even small defects was tested with a simulated interferogram [21], Fig.2.12. One small crack in a bended plate was simulated for this test. Even for a test person it is hard to find this fault indicating fringe pattern inhomogeneity. However, the network indicated it correctly. However, up to now the network only indicates the appearance of a defect anywhere in the component. It is not able to show the location or to classify the criticality. Future work must be done to improve the performance to more information in the output.

Fig.2.12: Smallest detectable defect by neural networks or knowledge based systems

- Knowledge based systems store the experience of experts in facts and rules of evaluation. A new input to the system in the work phase is compared to the stored facts by the given rules and classified.

   A system based on the knowledge about defect indicating fringe pattern was generated under the assumption, that the number of these patterns is limited [21]. The computer was programmed to search for some specific quantities in a fringe pattern to be evaluated. The evaluation of the before mentioned fringe pattern resulted in the correct indication of the defect, but in this case with the determination of the defect location.

The comparison of the two systems showed the neural network system to be the much faster one. In contrary, the knowledge based system was able to determine the location of the defect and involved the potential of a defect classification. The combination of the two systems - the neural network for the screening and the knowledge based system for the evaluation - seems to be a good approach for the automatic evaluation of HNDT in the future.

2.3.2 Application of Holographic Interferometry: Fracture Mechanics

Theoretical Remarks on Fracture Mechanics

The determination of the structural safety of components and constructions is based not only on the detection of existing faults, such as cracks, but more importantly on the assessment of the potential risk associated in the presence of these faults. For this purpose a stress intensity factor, K_I, can be determined. The derived critical stress intensity factor K_{IC} then characterizes the fracture toughness of a material [22]. When this value is reached under an applied load, a critical zone of plastic deformation exists at the crack tip. The size of the zone is directly related to the start of crack propagation. The stress and strain distributions are described in the theory of fracture mechanics for well defined conditions [23]: The linear-elastic fracture mechanics theory is based on an infinite specimen with (mainly) elastic behavior. In the neighborhood of a crack tip - and the crack is the most severe defect - the stresses are described by the Sneddon or the Williams-Irwin equations, Fig.2.13:

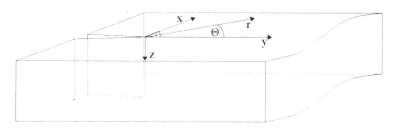

Fig.2.13: Coordinate system for the Sneddon equations

$$\sigma_{xx} = \frac{K}{\sqrt{2\pi}} \frac{1}{\sqrt{r}} \cos\frac{\Theta}{2} \left(1 + \sin\frac{\Theta}{2} \sin\frac{3\Theta}{2}\right) \quad (2.3.1)$$

$$\sigma_{yy} = \frac{K}{\sqrt{2\pi}} \frac{1}{\sqrt{r}} \cos\frac{\Theta}{2} \left(1 - \sin\frac{\Theta}{2} \sin\frac{3\Theta}{2}\right) \quad (2.3.2)$$

$$\tau_{yz} = \frac{K}{\sqrt{2\pi}} \frac{1}{\sqrt{r}} \cos\frac{\Theta}{2} \sin\frac{\Theta}{2} \sin\frac{3\Theta}{2} \quad (2.3.3)$$

with $\quad K = K_1 = \sigma_n \sqrt{2\pi a} \quad$ for normal stress load $\quad (2.3.4)$
$\quad\quad$ a : half length of the crack

As proven by experiments, the material behavior can be calculated taking into account the equations for plane strain and $\tau_{yz}=0$. Stresses can only be measured by the related deformations. According to Hooke's law the stresses are determined by the strains. The strains in the x-direction e.g. are given by (E: Young's modulus, v: Poisson ratio):

$$\varepsilon_{xx} = 1/E \left(\sigma_{xx} - v \sigma_{yy} \right) \quad (2.3.5)$$

The common way to determine the K_I-value is to measure the deformation field and to extract the strains as the first derivative of the displacement. This method introduces a significant amount of errors due to different reasons, e.g. coherent noise. These errors are magnified by the derivation operation. A better way to gain K_I out of the deformation field is to use the displacements directly. This requires to integrate the Equ.(2.3.5). However, one can even simplify this problem - and increase the accuracy - by integrating the equation in a special direction, e.g. in the x-direction with the strains ε_{xx}. This means [24]:

$$\Theta = 90°, \quad r = x, \quad dr = dx$$

$$t_x = \int_0^x \varepsilon_{xx} \, dx \quad (2.3.6)$$

with ε_{xx} according to Equ. (2.3.5). The result is given by

$$t_x = \frac{K_I}{\sqrt{2 \pi} \, E} \sqrt{x} \cos(45°) (3 - v) \quad (2.3.7)$$

$$K_I = \frac{2 \sqrt{\pi} \, E \, t_x}{(3 - v) \sqrt{x}} \quad (2.3.8)$$

By means of Equ.(2.3.8) the K_I-value can be calculated by the displacement t_x at any given point x on the line perpendicular to the crack through the crack tip. Furthermore, there is a tool to improve the accuracy of the results by fitting the t_x-x-function, which has to be a square-root-function.

Experimental Results

The experimental investigations were based on a three-point-bending test of a notched specimen with a crack induced by dynamic loading. During the bending the deformation were measured by means of holographic interferometry [24]. To facilitate the splitting of the displacement components into parallel and perpendicular ones with respect to the specimens surface, two interference fringe patterns were simultaneously recorded from two different observation points, Fig.2.14. In this way the light paths of the object waves run from source A via the object O to either observation point B_1 or B_2. The two observation point are symmetrical to the middle of the specimen. The recording of the two interferograms yields two symmetrical patterns from which one is shown in Fig.2.15. Due to the special set-up, the in-plane deformation is proportional to the difference of the two interference phase fields. By means of the standard equations for the relation between phase or fringe order, sensitivity vector, wave length of the laser, and the dislocation one can determine t_x as a function of the location x. The results of this evaluation yields the stress intensity factor K_I for each point x, Fig. 2.16.

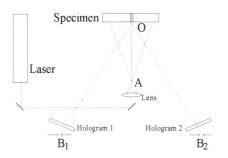 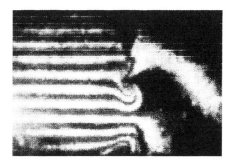

Fig.2.14: Holographic set-up   Fig.2.15: Example of a fringe pattern

The plotted results show a constant value with a low variation in distances of more than 2 mm from the crack tip. This constant value corresponds to the one measured with strain gauges. Closer to the crack the measured K_I-value decreases. This is due to the change from a plain strain to the plain stress deformation of the material.

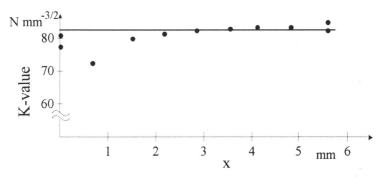

Fig.2.16: Stress intensity factor K_I vs location x

An analysis of the error of this measurements established a deviation of less than 10 % compared with strain gauge based methods. However, the variation of the measured K_I-values is up to 10% with standard methods and less than 5 % with holographic interferometry. By this result, the optical metrology performed in the described manner of application seems to be more reliable.

2.3.3 Application of Holographic Interferometry: Vibration Analysis

The dynamic behavior of objects or components is an intensive field of application in production related fields. Resonant vibrations are the sources of noise. In order to understand the reasons for the noise and the countermeasures one has to determine the deformations behavior beside the measurement of the resonance frequencies.

One of the first applications was the vibration investigation of a disk brake. The first disk brakes of VW (and other car manufacturers) happened to squeak. The common approach was to assume the disk itself to act as a membrane vibrating in resonant frequencies. Felske investigated not only the disks but the system of the disk in combination with the caliper [25], Fig.2.17. By the evaluation of the vibration deformation, he could demonstrate, that the vibrating caliper was the main reason for the squeaking of the brake system, Fig.2.18.

Fig.2.17: Disk Brake   Fig.2.18: Vibration mode, frequ. 4.8 kHz

Füzessy also worked on this topic [26]. One example is the international cooperation on dynamic behavior of the motor of the "Wartburg", a car produced in the former GDR, Fig.2.19. There are different problems to be solved. One is the noise emission of the motor. This is depending on the amplitude of the vibration as well as its transfer to the car body. With the knowledge of the vibrating parts of the system countermeasures can be planned and investigated, Fig.2.20. Another point is the control of the knocking of the motor. Today cars have knocking sensors. However, the efficiency of the control system depends on the position of the sensor: The best point is the anti-nodal point, i.e. the area of the amplitude maximum. By the visualization of the vibration modes one can avoid errors in positioning, e.g. in a nodal line. In this case the evaluation can be performed by visual inspection.

Fig.2.19: Wartburg car with optical set-up   Fig.2.20: Vibration mode

However, the dynamic behavior is also of interest for micro components. A better understanding of the components behavior is given by the comparison to theoretical calculations of the different modes. Pryputniewicz [27] as well as Höfling [28] could prove that not all modes seeming to be fundamental ones are really pure modes. Sometimes two modes with nearly the same frequency are excited at the same time forming a connected mode. By the calculation of the fundamental modes by FEM one can understand this combined mode.

2.4 References

1. Gabor, D.: Microscopy by reconstructed wave fronts, Proc. Royal Soc. A, 197, (1949), p.454-487
2. Jüptner, W.: Holographic techniques, NATO ASI Series F, Vol. 43, *Sensors and Sensory Systems*, Springer-Verlag Berlin Heidelberg 1988, p. 279-294
3. Mandel L.; Wolf, E.: Optical Coherence and Quantum Optics, Cambridge University Press, New York 1995
4. Wernicke, G.; Osten, W.: Holografische Interferometrie (Holographic Interferometry), Physik-Verlag, Weinheim 1982
5. Denisyuk, Y.N.: Fundamentals of Holography, Mir Publishers, Moscow 1978
6. Jüptner, W., Kreitlow, H.: Holografische Aufnahme von nicht-schwingungsgeschützten Objekten (Holographic recording of non-vibration-isolated objects), Poc. *Laser 77*, IPC, Guildford, GB, 1977, p. 420-429
7. Hung, Y.Y.: Shearography: a new optical method for strain measurement and nondestructive testing, Optical Engineering 21, 1982, p. 391-395
8. Ostrowski, J.I.: Holografie - Grundlagen, Experimente und Anwendungen, BSB B.G. Teubner Verlagsgesellschaft, Leipzig 1987
9. Powell, R.L.; Stetson, K.A.: Interferometric vibration analysis by wave front reconstruction, Journ. Opt. Soc. Amer., 55, 1965, p.1593-1608
10. Abramson, N.: The holo-diagram: A practical device for making and evaluating holograms, Appl. Opt., 8, 1969, p.1235-1240
11. Jüptner, W.P.O.; Kreis, Th.: Holographic NDT and Visual Inspection in Production Line Applications. SPIE Proc. 604, Los Angeles 1986
12. Mieth, U.: Musterbasierte Erkennung von Materialfehlern mit holografischen Interferogrammen (Pattern based recognition of material defects by means of holographic interferograms), PhD thesis, University Bremen, Bremen 1998, to be published
13. Jüptner, W.: Nondestructive Testing with Interferometry, Proc. *Fringe '93*, Akademie Verlag GmbH, Berlin 1993, p. 315-324
14. Kreis, Th.: Holographic Interferometry, Akademie Verlag, Berlin 1996
15. Osten, W.: Digitale Verarbeitung und Auswertung von Interferenzbildern (Digital processing and evaluation of fringe pattern), Akademie Verlag GmbH, Berlin 1991

16. Jüptner, W.; Stadler, H.-J.; Kleberger, W.A.: Investigations in Non-Destructive Testing of GRP Tubes and Project Management of the Experimental Studies. BMVg Report Nr. T/RF52/RF520/42016, Bonn 1977 (German)

17. Aswendt, P.; Höfling, R.: Interferometrische Dehnungsmessung - Aufbau und Anwendung eines DSPI Systems (Interferometric Strain Measurement - Set-up and Application of a DSPI System. LASER 93, Munich 1993

18. Jüptner, W.: Defect Quantification by a Hybrid Finite Element Method. Proc. 9th Int. UFEM Symposium, Worcester 1987

19. Steinbichler, H.: Prüfung von GFK-Hochdruckrohren mittels Holografie (Testing of GRP high pressure tubes by means of holography). BMVg Report Nr. T/RF52/RF520/42016/03 Bonn 1976

20. Höfling, R. Aswendt, P.; The influence of defects on the deformation behavior of GRP materials calculated by a specialized FEM program. Private communication on the BMFT project HOLOMETEC

21. Jüptner, W.; Kreis, Th.; Mieth, U.; Osten, W.: Application of Neural Networks and Knowledge Based Systems for Automatic Identification of Fault Indicating Fringe Patterns, Proc. SPIE, Vol. 2342, 1994, p. 16-26

22. Irvin, G.R.; McClintock: F. *ASTM*, STP 381 (1965)

23. Kerkhoff, F.: Einführung in die Bruchmechanik (Introduction into fracture mechanics), Freiburg 1969

24. Meyer, LW., Jüptner, W.; Steffens, H.-D.: Fracture Toughness Investigations Using Holographic Interferometry. Proc. Laser 75, München, 1975

25. A. Felske, "Holographic analysis of oscillations in squealing disk brakes," *Proc. SPIE*, Vol. 136, 148-155 (1977)

26. Z. Füzessy, "Applications of Double-Pulse Holography for the Investigations of Machines and Systems," *Application of Metrological Methods in Machines and Systems*, G. Frankowski, N. Abramson, Z. Füzessy, 75-108, Akademie Verlag Berlin (1991)

27. Pryputniewicz, R.; Grabbe, D.: Developments in micromechanics through analysis and experimentation. Proc. 10th Int. UFEM Symp., Worcester, 1991

28. R. Höfling, "Combined theoretical and experimental methods in materials mechanics," FRINGE '97 *Automatic Processing of Fringe Patterns*, ed. W. Jüptner, W. Osten, 379-386, Akademie Verlag Berlin (1997)

3. Digital Holography

3.1 Fundamentals of Digital Holography

3.1.1 Principle of Digital Holography

Digital Holography is a new technique of hologram recording [1] and reconstruction. The Fresnel hologram from an object is generated directly on a CCD-target, stored electronically and reconstructed numerically, Fig.3.1. In contrast to well known methods of electronic speckle interferometry [2,3,4] no imaging optics are needed to project the image of the object onto the target of the CCD-camera and no artificial aperture is introduced to increase

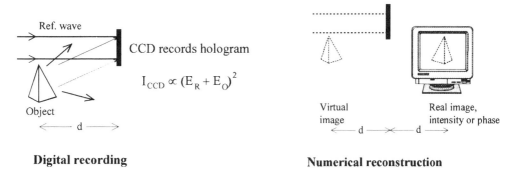

Digital recording **Numerical reconstruction**

Fig.3.1: Recording of a Digital Hologram

the Speckle size, Fig.3.2. The reference wave can be added as an in-axis wave - according to the Gabor set-up[5] - or off-axis wave - according to the Leith/Upathnics [6] set-up. The intensity distribution on the target is given by the well-known relation:

$$I \approx \left(\vec{E}_O + \vec{E}_R\right)\cdot\left(\vec{E}_O + \vec{E}_R\right)^* = \vec{E}_O\vec{E}_O^* + \vec{E}_R\vec{E}_R^* + \vec{E}_O\vec{E}_R^* + \vec{E}_O^*\vec{E}_R \tag{3.1.1}$$

which includes the information about the phase of the object wave in the third and the fourth term, whereas the first and the second term describe the intensity of the object wave and the reference wave, respectively. The object wave is defined by its objective Speckle field. Therefor the interfering intensity field seams to be an random distribution and the information cannot be detected by an human observer. The intensity distribution is digitized and stored in the memory of the computer.

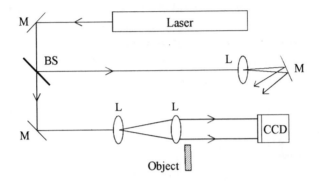

Fig.3.2: General set-up for digital recording of holograms (M: mirror, BS: beam splitter, L: lens)

3.1.2 Recording of Digital Holograms

Recording a Digital Hologram one has to recognize that the spatial resolution of a CCD target is much lower than that of a holographic plate. This means that the sampling theorem limits the spatial resolution of the intensity distribution to be stored. The maximum spatial frequency f_{max}, to be resolved by the recording medium is determined by the maximum angle α_{max} between the reference and the object wave[7]:

$$f_{max} = \frac{2}{\lambda} sin\left(\frac{\alpha_{max}}{2}\right) \qquad (3.1.2)$$

or for small angles

$$f_{max} \lambda = \alpha_{max} \qquad (3.1.3)$$

where λ denotes the wavelength. Emulsions of silver halide, the standard holographic recording media, have resolutions as high as 5000 lines/mm. With these emulsions, there is no limitation for the angle between the reference wave and the object wave. CCD cameras have pixel sizes of about 7 µm, i.e. 50-100 lines/mm. For that reason the maximum angle between the interfering waves is limited to a few degrees, Fig.3.3. Without any optical imaging, small objects or objects in a large distance from the CCD target may be recorded. There are two different ways out of this restriction. One solution is the optical magnification of the micro-interference pattern as proposed by Schnars [8]: The micro-interference pattern may be generated not directly on the CCD target but on a transparent screen. The hologram on the screen can be magnified and imaged onto the CCD. Another solution is to use a divergent lens in a Fresnel set-up, Fig.3.4. This reduces the angle under which the object light incidents the target. The effect is a significant reduction of the spatial frequency spectrum to be resolved. The (negative) focal length can be chosen by the well-known optical law of imaging according to the requirements given by the object and the set-up.

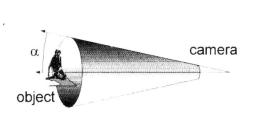

Fig.3.3: Possible position for objects

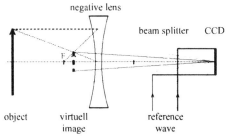

Fig.3.4: Angle transformation by a lens

3.1.3 Reconstruction of Digital Holograms

Optical reconstruction of holograms recorded on a photographic film or photo-thermoplast is performed by illumination of the hologram with the reference wave. The reference wave is diffracted by the gratings formed by the micro-interferences due to the mixture of the object wave with the reference wave. The laws of diffraction are well known from the fundamental formulation from Huygens, Kirchhoff, Fresnel, Fraunhofer and others [9]. The theory is based on the introduction of elementary sphere waves coming from each point of an excited area. The reconstruction of the stored object wave can therefore be performed numerically in Digital Holography: The recorded and stored intensity $h(\xi,\eta)$ is multiplied with numerical model of the reference wave $r(\xi,\eta)$ (according to the reconstruction illumination) and the mathematical expression of a spherical wave

$$f(\rho) = \frac{1}{\rho} exp(i k \rho) \qquad (3.1.4)$$

$$\rho = \sqrt{d'^2 + (\xi - x')^2 + (\eta - y')^2} \qquad (3.1.5)$$

The excitation in any point (x′,y′) in a plane outside the hologram is given by the integral of all source points (ξ,η) in the hologram plane, which is the known diffraction integral [7]:

$$\Gamma(x',y') = \frac{1}{i\lambda} \iint h(\xi,\eta) r(\xi,\eta) \frac{1}{\rho} exp(i k \rho) d\xi\, d\eta \qquad (3.1.6)$$

The calculation of the integral in a straight-on-forward way is even with modern computers a time consuming task. Therefor several approximation were formulated in the past. The most severe approximation was given by Fraunhofer, assuming, that the distance between the hologram and the imaging plane is very large against the dimensions of the hologram and the reconstructed image of the object. In this case the integral can be replaced by the Fourier transform of the hologram multiplied by the reference wave with an arbitrary amplitude factor [9].

Solution by Fresnel Approximation

The mostly used approximation is the Fresnel approximation: If the distance d between the object and the hologram plane, and equivalently d` = d between the hologram and the image plane is large compared to (ξ-x′) and (η-y′), then the ρ in the denominator of Equ.(3.1.4) can be replaced by d′. For the ρ in the exponential function of the numerator the binomial expansion of the square root might be used to obtain

Interferometric Methods

$$\rho \approx d'\left[1 + \frac{1}{2}\left(\frac{\xi - x'}{d'}\right)^2 + \frac{1}{2}\left(\frac{\eta - y'}{d'}\right)^2\right] \tag{3.1.7}$$

With this approximation the diffraction integral becomes

$$\Gamma(\nu,\mu) = exp\{i\pi d'\lambda(\nu^2 + \mu^2)\}\iint h(\xi,\eta)\,r(\xi,\eta)\,exp\left(\frac{i\pi}{d'\lambda}(\xi^2 + \eta^2)\right)exp\{-2i\pi(\xi\nu + \eta\mu)\}d\xi d\eta \tag{3.1.8}$$

$$\nu = \frac{x'}{d'\lambda} \quad ; \quad \mu = \frac{y'}{d'\lambda} \tag{3.1.9}$$

where the constant factor $exp(ikd')/(i\lambda d')$ has been omitted for clarity.

The stored Digital Hologram consists of N x M discrete values, each recorded by a pixel of size $\Delta\xi$ x $\Delta\eta$. So number and size of the elements of the CCD array define the discretization

$$\Gamma(n,m) = exp\left\{i\pi d'\lambda\left(\left(\frac{n}{N\Delta\xi}\right)^2 + \left(\frac{m}{M\Delta\eta}\right)^2\right)\right\}\sum_{k=0}^{N-1}\sum_{l=0}^{M-1}h(k\Delta\xi,l\Delta\eta)\,r(k\Delta\xi,l\Delta\eta)$$

$$\cdot exp\left\{\frac{i\pi}{d'\lambda}(k^2\Delta\xi^2 + l^2\Delta\eta^2)\right\}\cdot exp\left\{2i\pi\left(\frac{kn}{N} + \frac{lm}{M}\right)\right\} \tag{3.1.10}$$

The field reconstructed by this numerical Fresnel transform, Equ.(3.1.10), is complex. Intensity and phase can be calculated by

$$I(n,m) = \Gamma'(n,m)\cdot\Gamma'^*(n,m) \quad \text{and} \quad \phi(n,m) = arctan\frac{Im[\Gamma'(n,m)]}{Re[\Gamma'(n,m)]} \tag{3.1.11}$$

The pixel spacing - or equivalently the pixel size - in the reconstructed field is

$$\Delta x' = \frac{d'\lambda}{N\Delta\xi} \quad \text{and} \quad \Delta y = \frac{d'\lambda}{M\Delta\eta} \tag{3.1.12}$$

Solution by the Convolution Approach

The diffraction integral Equ.(3.1.5) is a superposition integral

$$\Gamma(x',y') = \frac{1}{i\lambda}\iint h(\xi,\eta)\,r(\xi,\eta)\,f(x',y',\xi,\eta)\,d\xi\,d\eta \tag{3.1.13}$$

with the impulse response f(x´,y´,ξ,η) given by the function of a sperical wave

$$f(x',y',\xi,\eta) = \frac{1}{i\lambda} \frac{exp\{ik\sqrt{d'^2 + (\xi-x')^2 + (\eta-y')^2}\}}{\sqrt{d'^2 + (\xi-x')^2 + (\eta-y')^2}} \qquad (3.1.14)$$

The linear system characterized by $f(x',y',\xi,\eta) = f(x'-\xi,y'-\eta)$ is space-invariant, so the superposition integral Equ.(3.1.13) is a convolution. For the determination of $\Gamma(x',y')$ the convolution theorem can be invoked, which states the convolution of the product hr with the impulse response f to be the inverse Fourier transform of the product of the individual Fourier transforms of hr and f:

$$\Gamma = F^{-1}\{F(h \cdot r) \cdot F(f)\} \qquad (3.1.15)$$

The calculation of the Fourier transform is performed effectively by the FFT algorithm. The numerical realization of the impulse response for free space propagation is

$$f(k,l) = \frac{1}{i\lambda} \frac{exp\{ik\sqrt{d'^2 + (k-N/2)^2 \Delta\xi^2 + (l-M/2)^2 \Delta\eta^2}\}}{\sqrt{d'^2 + (k-N/2)^2 \Delta\xi^2 + (l-M/2)^2 \Delta\eta^2}} \qquad (3.1.16)$$

The coordinate shifts by N/2 and M/2 are on symmetry reasons.

The Fourier transform of the impulse response f(k,l) is the transfer function F(n,m) of free space propagation, which can be calculated from f as indicated in Equ.(3.1.16) or may be defined directly by

$$F(n,m) = exp\left\{\frac{2\pi i d'}{\lambda}\sqrt{1 - \frac{\lambda^2\left(n + \frac{N^2\Delta\xi^2}{2d'\lambda}\right)}{N^2\Delta\xi^2} - \frac{\lambda^2\left(m + \frac{M^2\Delta\eta^2}{2d'\lambda}\right)}{M^2\Delta\eta^2}}\right\} \qquad (3.1.17)$$

This allows to save one Fourier transform

$$\Gamma = F^{-1}\{F(h \cdot r) \cdot F(n,m)\} \qquad (3.1.18)$$

By this mathematical treatment, the diffraction integral is determined exactly as long as the conditions for the Huygens-Kirchhoff theorem is fulfilled[10,11].

Comparison of the Reconstruction Methods

There is a conceptual difference between Fresnel transform reconstruction and the convolution methods. If the (ξ,η)-plane of the Digital Hologram is interpreted as the spatial domain, the procedure Equ.(3.1.10) gives a result in the spatial frequency domain due to the single Fourier transform. Thus the pixel values in the (x´,y´)-plane - the image plane - are spatial frequencies, their spacing varies with distance and wavelength.

The algorithms based on the convolution theorem first transform into the frequency domain. There the multiplication with the transfer function is performed. The product is Fourier transformed back into the spatial domain. As a consequence the pixel size in the reconstructed image is now $\Delta x´=\Delta\xi$ and $\Delta y´=\Delta\eta$. The field reconstructed by the convolution approach represents only a part of the scene that would be reconstructed by the Fresnel transform but in size independently from the distance of reconstruction plane. It is possible to reconstruct shifted parts of the image plane and to join them together as in a mosaic[10,11].

3.1.4 Elimination of the Zero Order Diffraction Wave

The reconstructed wave field of the hologram

$$E_R \cdot H = E_R(E_O E_O^* + E_R E_R^*) + E_R E_R^* E_O + E_R E_R E_O^* \qquad (3.1.19)$$

contains in the first product on the right side the reference wave multiplied by the intensity of the sum of the reference and the object wave. This term can be interpreted as the zero order diffraction term. This reference wave is disturbing the reconstruction since it is a bright light term in optical reconstruction as well as in the numerical one. In practice, this term is modified by the aperture function of the CCD target. In optical reconstruction there is no way around this bright noise in the wave field behind the hologram. However, in Digital Holography one can subtract the well known field after the hologram. Then the zero order diffraction or better the undiffracted reference wave disappears [12], Fig. 3.5.

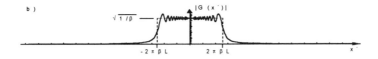

Fig.3.5: Suppression of the DC-term

3.2 Applications of Digital Holography

3.2.1 Light-in-Flight Measurements

Holographic recording of Light-in-Flight (LiF) was first proposed in 1978 by Abramson [13]. He pointed out that a hologram can only image the distances in space where the optical way of the reference wave matches the one of the object wave. The basic idea of LiF consists of recording a hologram of a plane object with streaking illumination by a short coherence length laser. For this purpose he used a cw Ar-Ion laser without intracavity etalon having a coherence length in the range of few millimeters or less and a picosecond pulsed dye laser for displaying LiF [14]. The reference wave was guided nearly parallel to the holographic plate. In this way, only those parts of the object are recorded (and later reconstructed) for which the optical path difference (OPD) between object and reference wave is smaller than the coherence length of the light source. By changing the observation point in the developed plate, the above condition is met for different parts of the object, thus allowing observation of a wavefront as it evolves over the object.

In Digital Holography a Fresnel hologram is stored in a digital image processing system and the real image of the object as well as the virtual one can be numerically reconstructed by calculating the diffraction of the reconstructing wave at the microstructure of the hologram. The advantage of this method is to avoid an imaging of the object onto the target by a lens. Therefore several holograms can be taken simultaneously in different parts of the target which all allow to reconstruct whole the object. This method has been applied to LiF recordings by using reference waves with different optical paths[15].

The general setup for Digital Holography can be used for recording Light-in-Flight if a short coherence length light source is used for illuminating the object. The source used in this investigation was an Ar-Ion laser pumped cw dye laser (Rhodamine 6G, λ = 574 nm). No frequency selecting elements were installed in the laser resonator. Therefore the output spectrum consists of many oscillating modes, resulting in a short coherence length, which was determined by the LiF experiments to be 2.3 mm.

The laser beam was divided into a plane reference wave illuminating the CCD array and into a diverging wave illuminating the object, Fig.3.6. The path differences were provided by glass plates with different but known thickness. The object consisted of a plane aluminum plate of 2 cm × 2 cm area, the distance between object and CCD sensor was set to 1.67 m, the angle of illumination α (referred to the normal of the object) was about 80 degrees, a wavelength of λ = 574nm was used and the maximum angle between object and reference wave has been α_{max} = 2°. In this experiment a KODAK MEGAPLUS 4.2 camera has been used. The matrix consists of 2048 x 2048 light sensitive elements.

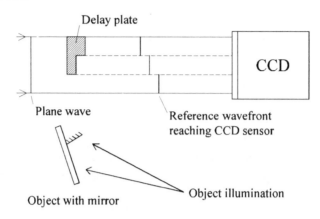

Fig.3.6: Optical set-up with delay lines by glass plates

The first experiment was performed with the set-up described before. Since the object was illuminated under an angle only a part of the object could be illuminated in coherence to the reference wave. The numerical reconstruction of such a digitally recorded hologram shows, as expected, a bright stripe representing the wavefront, Fig.3.7. It has to be pointed out, that the shown image is the direct result of one hologram.

The reconstructed image is available in a digital form and further processing is easy to be done. The coherence length e.g. can be calculated from this image, Fig.3.7. The width of the bright stripe (wavefront) seen at the object is determined from both, the coherence length of the light source, K, and the geometrical conditions of the holographic setup. However, since a plane wave was used as reference and the angle between the interfering waves was small - α_{max} = 2° -, only changes in the optical path due to the illumination beam have to be considered, Fig. 3.8. If the laser has a coherence time t, (coherence length K = c t, being c the speed of light in the corresponding medium) and the object is illuminated at an angle a with respect to its normal, thus the bright zone at the object has a width w given by:

$$w = \frac{c\,t}{\sin\alpha} = \frac{K}{\sin\alpha} \qquad (3.2.1)$$

Interferometric Methods

Fig.3.7: Numerically reconstructed wavefront

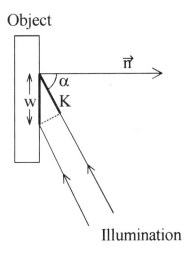

Fig.3.8: Geometrical considerations for calculating the coherence length

After measuring the width of the bright stripe representing the wavefront w, Equ.(3.2.1) can be used to calculate the coherence length of the light source using the known angle α of the incident wave. Since the measurements of the width are disturbed by electronic and coherent noise, direct measurement of the intensity profile lead to errors. A good result can be achieved by low-pass filtering of the image and by applying the autocorrelation function to the intensity profile line. For the experimental conditions:

$\Delta\xi = 9$ μm, $d = 1.67$ m, $\lambda = 574$ nm, $\alpha = 80°$

this results to a coherence length, Equ. (3.2.1):

$K = \Delta\xi \cdot 45 \cdot sin80° = 2.3$ mm , $\Delta x = 52$ mm (3.2.2)

It is also possible to apply Digital Holography to follow the evolution of a wavefront in its "flight" over an object, as proposed in the original work of Abramson [13]. However, because of the reduced size of the CCD target (18mm × 18mm) and the lower resolution compared to a holographic plate, only slightly different points of view of the wavefront can be registered in each hologram. Some view-time expansion is also needed. A possible setup for this purpose, using a skewer reference wave, has been proposed by Pettersson et.al. [16]. However, this solution cannot be applied here because the high spatial frequencies that would be produced at the sensor are not resolvable. A solution to this problem to record a hologram introducing different phase delays in different parts of the plane reference wave. That can be achieved e.g. by introducing plane-parallel plates of different thickness in the plane wave illuminating the CCD sensor, Figure 3.6. A plate of thickness p and refraction index n will produce a delay Δt with respect to air (or the vacuum with light speed c) given by:

$$\Delta t = (n-1)\frac{p}{c} \qquad (3.2.3)$$

In this way it is possible to record at one time several holograms of the object using a corresponding number of reference waves delayed with respect to each other. The numerical reconstruction can be done for each part of the CCD array in which the phase of the reference wave has a particular delay, giving rise to the desired "times of evolution" of the wavefront illuminating the object. This is equivalent to choose another observation point in the original LiF experiment, or to change the position of the delay line in the DSPI setup of LiF. In this sense, the phase delays introduced in the different parts of the reference wave can be interpreted as artificial extensions of the CCD sensor and allows a better visualization of the phenomenon.

The number of different "times of evolution" that can be recorded simultaneously is limited by the minimum number of pixels required for a part of the hologram to be successfully reconstructed. Furthermore, diffraction effects due to the borders of the plates introduced in the reference wave distort the information at the CCD sensor originating dark zones which cannot be used in the numerical reconstruction process.

In the experiments 6 mm thick PMMA plates (refraction index n ~ 1.5) were used to produce the phase delays in the reference wave, Fig.3.6. One third of the original plane reference wave does not travel through PMMA, the second third, illuminating the sensor in the middle, travels through 6 mm PMMA (representing 10 ps delay with respect to air) and the last third travels through 18mm PMMA (30ps delay with respect to air). The object as seen from the CCD sensor is schematically sketched for better recognition of the results, Fig.3.9. It consists of a 3 cm × 3 cm plane aluminum plate, which was matt white painted for better light scattering. A small plane mirror (1cm x 1cm area) is attached to the plate, perpendicular to its surface and at an angle of about 10 degrees away from the vertical.

Interferometric Methods

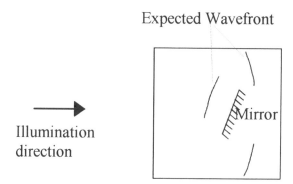

Fig.3.9: Object used for displaying the temporal evolution of a wavefront as seen from the CCD sensor

The three reconstructed stripes of the hologram, corresponding to three different times of evolution of the wavefront illuminating the object are shown in Fig.3.10. One part of the wavefront is reflected by the mirror, the other part is traveling in the original direction. The three pictures can be interpreted as a slow-motion shot of the wavefront. As demonstrated before, quantitative results can be derived from these images, e.g. the speed of light.

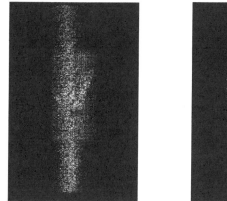 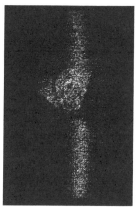 

Fig.3.10: The wavefront at three different times, reconstructed from one single holographic recording.
Left: No delay, wavefront just reaching the mirror.
Middle: 10 ps delay, the mirror reflects one part of the wavefront.
Right: 30 ps delay with respect to the left recording, one part is reflected into the opposite direction, the other part is traveling in the original direction.

3.2.2 Deformation Measurement

The method of holographic interferometry is a well proven and widely used technique for deformation measurement of diffusely reflecting objects. The quantitative material analysis and determination of its properties is an accepted application. Because of the considerable progress in micro-system manufacturing a measurement technique is required which works fast, robust, and contactless with objects having lateral extensions of about 1mm to 10mm. Digital Holography presents itself as a versatile tool for this purpose.

In conventional holographic interferometry the analysis of the fringes and the transformation into an absolute phase value or fringe order is a complicated and imperfect procedure. To increase the reliability and accuracy several phase reconstruction methods have been developed[17]. Digital Holography overcomes several drawbacks of conventional holography by storing the holograms directly in the computer. As an important consequence Digital Holography allows the direct access to both the phase and the intensity of the object. Once the holograms of the two (or more) states of the object are stored in the computer, they can be reconstructed numerically. The reconstructed picture of the object contains the complex amplitude of the object wave on the object surface from which intensity and phase can be calculated. Doing digital holographic interferometry (DHI) means to subtract the two reconstructed phases. Consequently the mod2π phase distribution can be derived without any additional effort as e.g. phase shifting, Fig.3.11.

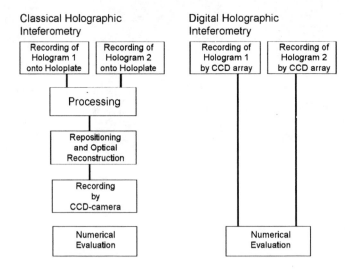

Fig.3.11: Differences between classical and digital holographic interferometry

The deformation measurement of a cantilever beam under a magnetic load shall serve as example for the application of Digital Holography to materials properties evaluation[18]. The experiments were performed with a full 3D-analysis although in this special case the deformation field is only a two dimensional one. The beam was bent in a homogeneous magnetic field. It was illuminated for this measurement from four different directions and the interferograms were taken according to the before mentioned procedure. Glass fibers were used to build-up a small and compact set-up, Fig.3.12. The main components beside the fiber optical system were a He-Ne-laser with an output power of 20 mW, a high resolution camera (2043x2024 pixels, pixel size 9x9 µm²) and a lens as well as a beam splitter cube for the creation of a parallel reference wave. All the components were mounted together on a X95-profile, Fig.3.13.

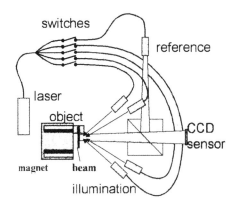

Fig.3.12: Set-up for Digital Holography   Fig.3.13: Measurement set-up

The set-up allows a fast access of the four different interferograms in a series by switching mechanically the four illumination fibers to the laser output for the holograms in each state of load. The calculation of the mod-2π phase distribution is performed according to the before mentioned subtraction of the phases, Fig.3.14. After this, the phase maps have to be unwrapped. The unwrapping procedure can de done easily in this case since the zero order fringe is known to be at the fixture at the top of the beam. Furthermore, the change of the phase must be monotonically according to the way of loading. With the unwrapped phase maps the three deformation components of can be calculated by the standard evaluation of holographic interferometry:

$$\lambda N_i = S_i d \quad ; \quad i = 1, 2, 3, 4 \qquad (3.2.4)$$

with λ : wave length of the laser, here: l = 633 nm
 N_i : order of the interference fringe in the i-th pattern
 S_i : sensitivity vector for the ith pattern according to set-up geometry
 d : dislocation

The deformation behavior of the beam shows the expected result that mainly the z-component (component parallel to the load direction) is different from zero, Fig.3.15. If the deformation and loading force is known, the material constants can be derived. As an example, Young's modulus is calculated from the shown deformation results: The measured results can be compared with an analytical solution along a line parallel to the y-direction. The analytical solution for the bending of a thin beam is given by:

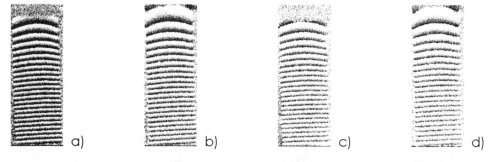

Fig.3.14: Interferograms, illumination: a) top left, b) bottom left, c) top right, d) bottom right

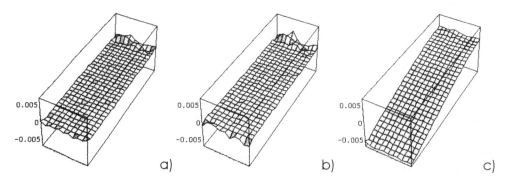

Fig.3.15: Deformation components a) x-component, b) y-component, c) z-component

$$w(x) = \frac{q l^4}{24 E I_y}\left(3 - 4\frac{x}{l} + \left(\frac{x}{l}\right)^4\right) \tag{3.2.5}$$

with w(x) the amount of bending, q the bending force per meter, E the Young's modulus, I_y the y-component of the momentum of inertia, l the length and x the distance from the attachment. The comparison fitted well between theory and experiment.

3.2.3 Shape Measurement

Shape measurement by means of optical methods get increasing significance in the industry. Digital holography can be used for two-wavelength contouring as follows. Two holograms are recorded with two different wavelengths λ_1 and λ_2. The reconstructed phases are subtracted and one gets an interference pattern with mod-2π fringes. These fringes correspond to contour lines on the surface of the object. The height difference between two fringes depends on the difference of the two wavelengths which forms a synthetic wave length λ_{syn}:

$$\lambda_{syn} = \frac{\lambda_1 \lambda_2}{\lambda_1 - \lambda_2} = \lambda_1 \frac{\lambda_2}{\Delta\lambda} \approx \frac{\lambda_i^2}{\Delta\lambda} \quad i = 1,2 \tag{3.2.6}$$

The larger $\Delta\lambda$ is, the smaller are the height differences of two fringes, which means a greater resolution. In conventional holographic interferometry the wavelength differences are limited by the aberrations in the imaging due to the different magnification in lateral and transversal direction when changing the wave length during reconstruction. In Digital Holography there is no restriction in the wavelength difference since the aberrations by wavelength transformation can be corrected in the computer directly.

Experiments were performed on a small cylindrical lens. The lens was holographically imaged as described before with a wavelength difference of 25 nm. From the two hologram an interferogram was calculated, Fig.3.16. The visible part has an extension of about 3mm x 2mm. The curvature height is only 30µm and can resolved precisely, Fig.3.17.

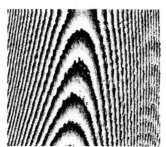
Fig.3.16: Two wavelength contour mapping: interferogram

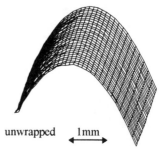
Fig.3.17: Evaluated shape curves

3.2.4 Nondestructive Evaluation

One of the most important applications of holographic interferometry is nondestructive testing (NDT) or nondestructive evaluation (NDE). It can be used wherever the presence of a structural weakness results in a changed deformation of the surface of the stressed component. The load can be realized by the application of a mechanical force or by a change in pressure or temperature [19]. Holographic nondestructive testing (HNDT) indicates deformations down to the submicrometer range, so loading amplitudes far below any damage threshold are sufficient to produce detectable fringe patterns]20].

The object in the reported investigations is a pressure vessel with a diameter of 0.5 meter, Fig.3.18. Such vessels are used as gas tanks in satellites. The thickness of the wall is only about one millimeter, the internal pressure is more than 10 bar. Up to now conventional holographic interferometry is used to find possible existing irregularities such as cracks or weaknesses in the wall. The faults become visible qualitatively as typical patterns or disturbances in an interference pattern. Here we use the method of digital holography to measure the defect induced deformation quantitatively.

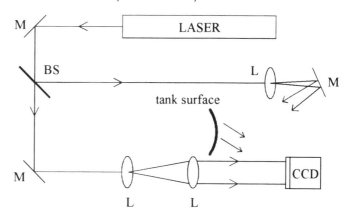

Fig.3.18: Satellite tank

Fig.3.19: Experimental set-up; L: lens, M: mirror, BS: beam splitter

The experimental set-up is shown in Fig.3.19. The surface of the tank is divided into segments of 5×5 cm², which are tested separately. The light from an Ar-ion laser is divided into a plane reference wave and an object illuminating wave by means of a beam splitter. The reference wave is expanded with a telescope and illuminates the CCD-target, which is placed in a distance of ~

2 m from the tank. At this distance the angle between object and reference wave is small enough for direct recording of the resulting spatial frequency spectrum: In the experiments a KODAK MEGAPLUS 4.2 CCD camera with pixel sizes of 9 × 9 microns was used.

For each segment, a series of holograms is recorded and stored in a digital image processing system. Between the exposures, the pressure inside the tank is varied by a few hundred hPa. In the numerical reconstruction process, the phase difference between any pair of exposures can be calculated. Even if the total deformation between the first and the last hologram is too high for direct evaluation, the total phase difference can be calculated step by step as the sum of the individual phase changes. With conventional holographic interferometry using photographic plates, this would cause an unresolveable fringe density.

As a typical result the interference phase between one pair of holograms is shown in Fig.3.20. The disturbance in the middle of this phase distribution is an indication for a fault in the wall of the tank. The interference phase can be converted into a continuous phase map by unwrapping the mod2π phase map, Fig.3.21. This phase map represents the out-of-plane deformation field of the surface, since the sensitivity vector is nearly constant and perpendicular to the surface. The global deformation field is superimposed by a local peak in the center, which corresponds to the defect. The total amplitude of the defect induced deformation is approximately 1.0 μm, compared with ~ 0.6 μm in the surrounding.

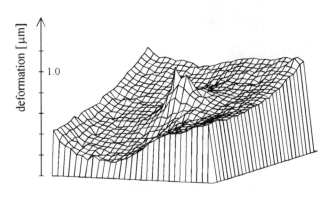

Fig.3.20: Interference phase modulo 2π

Fig. 3.21 Deformation field

3.3 References

1. Schnars, U.: Direct phase determination in hologram interferometry with use of digitally recorded holograms, JOSA A 11(7) (1994), p. 2011-2015

2. Lokberg, O.: Electronic Speckle Pattern Interferometry, Phys. Technol. 11, (1980) p. 16-22

3. Creath, K.: Phase-shifting Speckle interferometry, Appl. Opt. 24(18), (1985) p. 3053-3058

4. Pryputniewicz, R.J.; Stetson, K.A.: Measurement of vibration patterns using Electro-Optic Holography, Proc. SPIE, Vol. 1162, (1989) p. 456-467

5. Gabor, D.: Microscopy by reconstructed wavefronts, Proc. Royal Soc. A, 197, (1949) p.454-487

6. Leith, E.N.; Upathnics, J.: Wavefront reconstruction with diffused illumination and three-dimensional objects, J. Opt. Soc. Amer., 54 (1965)

7. Goodman, J.W.: *Introduction to Fourier Optics*, McGraw-Hill Companies Inc., New York, 1996, (example)

8. U. Schnars, *Digitale Aufzeichnung und mathematische Rekonstruktion von Hologrammen in der Interferometrie*, VDI-Verlag, Reihe 8, Nr. 378, Düsseldorf (1994), pp. 9-12

9. Born, M.; Wolf, E.: Principles of optics. Electromagnetic theory of propagation, interference and diffraction of light, Pergamon Press, Oxford, London, Edinburgh (1980)

10. Kreis, Th.; Adams, M.; Jüptner, W.: Methods of Digital Holography: A Comparison, Conf. On Optical Inspection and Micromeasurements II, Proc. SPIE vol. 3098, 1997, p. 224-233

11. Kreis, Th.; Jüptner, W.: Principles of Digital Holography, Proc. of Fringe 97, 3rd Intern. Workshop on Automatic Processing of Fringe Patterns, 1997, p. 353-363

12. Kreis, Th.; Jüptner, W.: The DC-term in digital holography and its suppression, 1997, p.353-363

13. Abramson, N.: Light-in-Flight recording by holography", Opt. Lett. 3, 121-123 (1978).

14. Abramson, N.: Light-in-flight recording : high speed holographic motion pictures of ultrafast phenomena", Appl. Opt. 22, 215-232 (1983).

15. Jüptner, W.; Pomarico, J.; Schnars, U.: Light-in-Flight measurements by Digital Holography", SPIE Proc. Vol. (1996), p.

16. Pettersson, S.-G.; Bergstrom, H.; Abramson, N.: Light-in-flight recording 6: Experiment with view-time expansion using a skew reference wave", Appl. Opt. 28, 766-770 (1989)

17. Osten, W.: *Digitale Verarbeitung und Auswertung von Interferenzbildern (Digital Processing and Evaluation of Fringe Patterns)*, Akademie Verlag Berlin (1991)

18. Jüptner, W.; Osten, W.; Seebacher, S.: Measurement of Materials Behaviour in Microstructures by Means of Digital Holography, Proc. LANE 97,

19. Hariharan, P.:*Optical Holography*, Cambridge University Press, Cambridge, pp. 257-260 (1984)

20. Kreis, Th.; Jüptner, W.; Biedermann, R.: Neural network approach to holographic non-destructive testing, Appl. Opt., Vol. 34, No. 8, pp. 1407-1415, March 1995

4. Speckle Photography

4.1 Fundamentals of Speckles

4.1.1 Origin of Speckles

Speckles occur when the surface of a rough object is illuminated by coherent light. The surface of the object seems to be covered with a granular structure called *Speckle* [1], Fig.4.1. Currently after its discovery the effect was connected to the roughness of the surface: When the typical rough surface is

Fig. 4.1: Part of the image of a rough surface
λ = 514 nm, diameter: 10mm, distance: 930 mm

illuminated with coherent light, the reflected waves in any regarded point P in the space before the object will interfere with each other. The intensity in this point P is the sum of numerous elementary waves which are not phase-correlated but coherent [2]. Since the roughness of the surface is given by statistically distributed peaks and valleys, the intensity and the phase of the resulting wave in different points P_i will be distributed randomly. These intensity fluctuations are the so-called *Speckles*. They are called *objective* Speckles if no imaging system is used [3], Fig.4.2.

The Speckles are called *subjective* if they are the result of an imaging process: If the object is observed through an imaging system one has to take into account the diffraction by the optics, even when the system is free of aberration [4]. The optical system has a finite aperture, so the spatial frequencies are limited. The image of a point is not any longer a point but convoluted with the point spread function of the system, Fig.4.3.

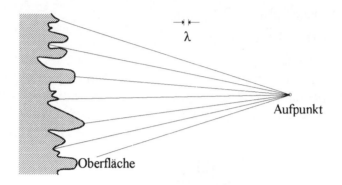

Fig.4.2: Origin of Speckles for arbitrary points

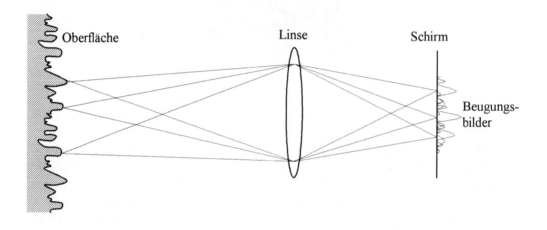

Fig.4.3: Origin of Speckles with imaging systems

Speckles will be generated in any observation of an object with or without an imaging system.

4.1.2 Statistics of Speckles

The statistics of Speckles will be described under the assumption that a monochromatic, linear polarized wave can be regarded[1], expressed by:

$$\vec{E}(\vec{r},t) = \vec{E}_0(\vec{r})\exp(i\omega t) \tag{4.1.1}$$

where ω is the circular frequency and $\vec{E}_0$ describes the complex amplitude with a polarization factor $\vec{E}_{00}$ and a complex location depending phase term

$$\vec{E}_0(\vec{r}) = \vec{E}_{00}(\vec{r})\exp(i\phi(\vec{r})t) \tag{4.1.2}$$

The intensity of the wave is given by the time average of the squared amplitude:

$$I(\vec{r}) = \langle \vec{E}_0(\vec{r})^2 \rangle = \lim_{T\to\infty} \int_{-T/2}^{T/2} \vec{E}(\vec{r},t)\,dt \tag{4.1.3}$$

The amplitude of the electric vector at the regarded point $P(r)$ is the sum of numerous non-phase correlated contributions from different scattering areas of the surface. Therefor the amplitude can be written as

$$\vec{E}_0(\vec{r}) = \sum_{k=1}^{N} \frac{1}{\sqrt{N}} \vec{e}_k(\vec{r}) = \frac{1}{\sqrt{N}} \sum_{k=1}^{N} |\vec{e}_k| \exp(i\phi_k) \tag{4.1.4}$$

With the assumptions:
- the amplitude e_k and the phase ϕ_k are statistically independent from each other and from any other elementary wave,
- the phases ϕ_k are statistically even distributed over the interval $(-\pi,\pi)$, which means that the surface roughness has peak-to-valley values large against the wavelength of the used light,
- the surface will not depolarize the wave and is ideally rough,

the statistics of the Speckle pattern can be derived.

Statistics of the complex amplitude

For the real part and the imaginary part of the amplitude one can write:

$$\vec{E}_0^r = Re(\vec{E}_0) = \frac{1}{\sqrt{N}} \sum_{k=1}^{N} |\vec{e}_k| \cos\phi_k \tag{4.1.5}$$

$$\vec{E}_0^i = Im(\vec{E}_0) = \frac{1}{\sqrt{N}} \sum_{k=1}^{N} |\vec{e}_k| \sin\phi_k \tag{4.1.6}$$

---

[1] Otherwise the components of the wave are regarded separately.

The averages of $\vec{E}_0^r$ and $\vec{E}_0^i$ over an entity of macroscopic similar but microscopic different surface areas - taking into account the a.m. assumptions - is given by

$$\langle \vec{E}_0^r \rangle = \frac{1}{\sqrt{N}} \sum_{k=1}^{N} \langle |\vec{e}_k| \cos\phi_k \rangle = \frac{1}{\sqrt{N}} \sum_{k=1}^{N} \langle |\vec{e}_k| \rangle \langle \cos\phi_k \rangle = 0 \tag{4.1.7}$$

$$\langle \vec{E}_0^i \rangle = \frac{1}{\sqrt{N}} \sum_{k=1}^{N} \langle |\vec{e}_k| \sin\phi_k \rangle = \frac{1}{\sqrt{N}} \sum_{k=1}^{N} \langle |\vec{e}_k| \rangle \langle \sin\phi_k \rangle = 0 \tag{4.1.8}$$

In similar way one can derive

$$\langle [\vec{E}_0^r]^2 \rangle = \frac{1}{\sqrt{N}} \sum_{k=1}^{N} \sum_{m=1}^{M} \langle |\vec{e}_k \vec{e}_m| \rangle \langle \cos\phi_k \cos\phi_m \rangle = \frac{1}{2\sqrt{N}} \sum_{k=1}^{N} \langle |\vec{e}_k|^2 \rangle \tag{4.1.9}$$

$$\langle [\vec{E}_0^i]^2 \rangle = \frac{1}{\sqrt{N}} \sum_{k=1}^{N} \sum_{m=1}^{M} \langle |\vec{e}_k \vec{e}_m| \rangle \langle \sin\phi_k \sin\phi_m \rangle = \frac{1}{2\sqrt{N}} \sum_{k=1}^{N} \langle |\vec{e}_k|^2 \rangle \tag{4.1.10}$$

$$\langle \vec{E}_0^r \vec{E}_0^i \rangle = \frac{1}{\sqrt{N}} \sum_{k=1}^{N} \sum_{m=1}^{M} \langle |\vec{e}_k \vec{e}_m| \rangle \langle \cos\phi_k \sin\phi_m \rangle = 0 \tag{4.1.11}$$

where the well known averages for the trigonometric functions were used.

The result of Equ.(4.1.9) - (4.1.11) is that the real part and the imaginary part of the complex electromagnetic field amplitude is zero. They have the same variance and they are uncorrelated. For $N \to \infty$ the sums can be replaced by the Gaussian distribution functions and it can be written:

$$p_{r,i}(\vec{E}_0^r, \vec{E}_0^i) = \frac{1}{2\pi\sigma^2} \exp\left[-\frac{(\vec{E}_0^r)^2 + (\vec{E}_0^i)^2}{2\sigma^2}\right] \tag{4.1.12}$$

with $\quad \sigma^2 = \lim_{N \to \infty} \frac{1}{2N} \sum_{k=1}^{N} \langle |\vec{e}_k|^2 \rangle$ (4.1.13)

Statistics of the Intensity and the Phase

The measured quantity in optics is the intensity instead of the amplitude of the wave. Therefor the statistics of the intensity has to be derived from the statistics of the amplitude. Both are connected by:

$$\vec{E}_0^r = \sqrt{I} \cos\phi \tag{4.1.14}$$
$$\vec{E}_0^i = \sqrt{I} \sin\phi \tag{4.1.15}$$
$$I = (\vec{E}_0^r)^2 + (\vec{E}_0^i)^2 \tag{4.1.16}$$
$$\phi = \arctan(\vec{E}_0^r / \vec{E}_0^i) \tag{4.1.17}$$

Interferometric Methods

After the transformation of the variables it follows from the probability function of the complex amplitude Equ.(4.1.12) with (4.1.13):

$$p(I,\phi) = \begin{cases} \dfrac{1}{4\pi\sigma^2} \exp\left[-\dfrac{I}{2\sigma^2}\right] & \text{for} \quad \begin{cases} I \geq 0 \\ -\pi \leq \phi \leq \pi \end{cases} \\ 0 & \text{else} \end{cases} \qquad (4.1.18)$$

The probability density function for the intensity is given by the integration of the function Equ.(4.18) over ϕ:

$$p(I) = \int_{-\pi}^{\pi} p(I,\phi)\,d\phi = \begin{cases} \dfrac{1}{2\sigma^2} \exp\left[-\dfrac{I}{2\sigma^2}\right] & \text{for} \quad \begin{cases} I \geq 0 \\ -\pi \leq \phi \leq \pi \end{cases} \\ 0 & \text{else} \end{cases} \qquad (4.1.19)$$

The same procedure can be performed for the phase to get:

$$p(\phi) = \int_{-\infty}^{\infty} p(I,\phi)\,dI = \begin{cases} \dfrac{1}{2\pi} & \text{for} \quad \begin{cases} I \geq 0 \\ -\pi \leq \phi \leq \pi \end{cases} \\ 0 & \text{else} \end{cases} \qquad (4.1.20)$$

Since the two variables intensity and phase are independent one can write

$$p(I,\phi) = p(I)\,p(\phi) \qquad (4.1.21)$$

Equ.(4.1.19) and (4.1.20) indicate, that the probability density follows an exponential decreasing function and the probability for all phases is the same.

4.1.3 Size of Speckles

The mean size of Speckles depends whether they are objective - no imaging optics between rough surface - or subjective ones after imaging optics.

Objective Speckles

In the case of objective Speckles all surface points contribute to the interference at the regarded point: The larger the object, i.e. the reflecting surface, the larger is the number of statistically intensifying and extinguishing components. This means, that the Speckle size will decrease. A mathematical treatment can be done by an autocorrelation. However, this requires assumptions about the roughness of the surface [5]. On the other hand the mean Speckle size may be derived from the laws of diffraction, when an optimal rough surface is assumed. The Speckle at a given point may be regarded as an aperture with the Speckle size as diameter, Fig.4.4. By the law of inverse

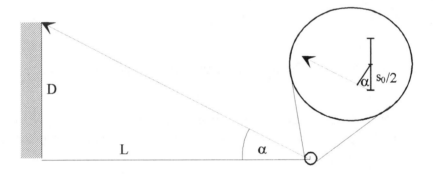

Fig.4.4: Model for the estimation of the mean Speckle size

light paths one can regard the object surface as illuminated through this aperture. The first minimum point of the diffracted intensity must be outside the edge of the object since in contrary all parts of the object are assumed to contribute to the intensity at the given point. This means (quantities: Fig. 4.4):

$tan\alpha = \lambda / s_0$ (4.1.22)

and by geometrical reasons, see Fig. 4.4:

$tan\alpha = D/L$ (4.1.23)

This leads to the mean Speckle size:

$$s_0 \approx \frac{\lambda L}{D} \qquad (4.1.24)$$

The autocorrelation results in

$$s_0 \approx 1{,}2\,\frac{\lambda L}{D} \qquad (4.1.25)$$

The small difference between the two results are due to the different assumptions about the intensity at the edge of the object.

Subjective Speckles

The mean size of subjective Speckles can be derived from the minimum point size of an imaging system. For this size the aperture of the lens is the crucial quantity:

$$s_0 \approx 1{,}2\,\frac{\lambda f}{D} \qquad (4.1.26)$$

with   f   : focal length of the lens
       D   : diameter of the lens

4.2 Speckle Photography

4.2.1 Principle of Speckle Photography

Speckles are strongly related to the surface of an object and its roughness. A simple method to measure dislocations or deformations of object surfaces is the Speckle photography or in modern words Speckle correlation: The object is imaged in a reference load state with a set-up comparable to a common photo camera, Fig.4.5. The use of an aperture introduces Speckles to the image. The size of the Speckle has to be chosen appropriate, chap. 4.1.3: The mean Speckle diameter should be smaller than the expected dislocation of the surface points. However, when the size is too small, the visibility of the later generated interference fringes is low. Afterwards the objects is loaded. The load may vary over the observed surface. However, the areas of nearly constant dislocation should cover several Speckles. Then a second image is taken onto the same photo plate, which developed afterwards [6].

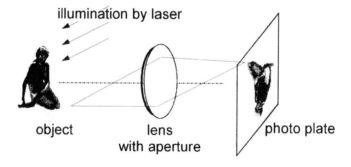

Fig.4.5: Recording of a Speckle photo

The Speckle photos are illuminated with a laser beam, Fig.4.6. The diameter of laser beam must be small enough to cover only areas of nearly the same dislocation. The Speckle pattern scatters the incident laser beam into an angle, which is comparable to the recording aperture. The scatter cone is called *halo*. If the two Speckle pattern are dislocated by a certain distance d, the two patterns serve as two apertures generating interference fringes called the Specklegram. The parallel equidistant fringes are called Young´s fringes. The direction of the fringes is perpendicular to the dislocation direction and their distance d_s is a measure for the dislocation d:

$$d = \frac{\lambda L}{M d_s} \tag{4.2.1}$$

with $\bar{d}$: dislocation of illuminated area
      M : magnification of the image
      L : distance of observation plane
      d_s : distance of fringes

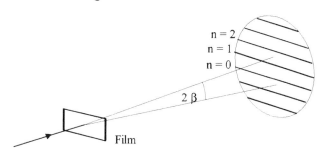

Fig.4.6: Evaluation of a Specklegram

The interference pattern can be evaluated after storing it in a computer by means of a CCD camera and a frame grabber. A Fourier transform results two peaks since the fringes are equidistant. The position of the peaks delivers as well the amount as the direction of the dislocation. The full field of dislocations is evaluated by a point wise evaluation in the before mentioned manner.

4.2.2 Speckle Photography with Spatial Filtering

The Fourier transform can be performed optically. Then it is possible to recognize lines of equal dislocation: The double-exposed Specklegram is illuminated by a parallel light wave, Fig.4.7. By means of a lens the transmitted light focused into the Fourier plane, where a pin hole in a point (β,δ) serves as a filter [7]. The angles β and δ are the angle coordinates according to the aperture distance from the axis and the angle in a polar coordinate system in the aperture plane. δ also represents the angle of the dislocation. The relation between β and the amount of the dislocation is given by

$$d_\delta \sin\beta = n\lambda$$

Behind the aperture the filtered image is reconstructed by a second lens. In the filtered image only those parts are reconstructed with a not vanishing intensity which were dislocated according to the aperture hole in the Fourier plane, i.e. dislocated in a according direction and distance, Fig.4.8.

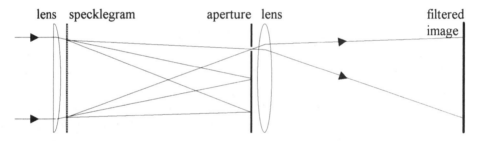

Fig.4.7: Recording of a filtered Speckle photo

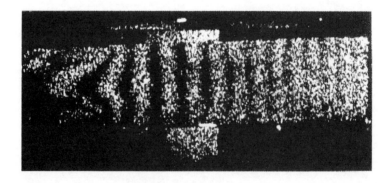

Fig.4.8: Filtered Image

4.3 Digital Speckle Photography DSP

4.3.1 Principle of Digital Speckle Photography

Digital Speckle Photography (DSP) is based on the well-known Speckle photography and provides fieldwise in-plane displacements and strains, respectively [8]. However, in contrast to classical Speckle Photography the Speckle images are stored and correlated electronically [9]. Owing to the numerical procedure, the evaluation effort of the Digital Speckle Photography is low in comparison to the pointwise probing of a double exposed photographic film, which is the classical evaluation procedure. Generally, the DSP has the potential to measure under dynamic testing conditions, because a single photo of the load state of interest is sufficient for the evaluation. Furthermore, low level demands to vibration isolation turn this optical contact free method into an attractive tool for measurements under workshop conditions.

The sample under investigation is coherently illuminated by means of an expanded laser beam as described in chap. 4.2.1. A Speckle pattern of the reference state and a Speckle pattern of the considered load state are digitized by means of a high resolution CCD camera. The evaluation of the displacements occurring between these two load states is done numerically. At first the whole available Speckle images with a size of 2024 x 2043 Pixels² are divided into subimages, Fig. 4.9. The sizes of these subimages are 64 x 64 Pixel² or 32 x 32 Pixel² in common. The calculation of the local displacement vectors at each subimage is performed by cross correlation of the subimages at the reference state (I) and the subimage at the load state (I'):

$$R_{II'}(\tau) = \int_{-\infty}^{\infty} I(u) I'(u+\tau) du \qquad (4.3.1)$$

which may be calculated by two Fourier transformations and one inverse Fourier transformation:

$$R_{II'}(\tau) = F^{-1}\{F[I(u)] F[I'(u+\tau)]\} \qquad (4.3.2)$$

where $R_{II'}$ is the cross correlation function, F the Fourier transform, F^{-1} the inverse Fourier transform and * the complex conjugation. This refers to the classical evaluation technique, where double exposed Speckle photos are locally illuminated. The mean displacement vector of the evaluated subimage is given by the location of the peak of the cross correlation function, Fig.4.9. After the evaluation of all subimages the full in-plane displacement map of the monitored area is available.

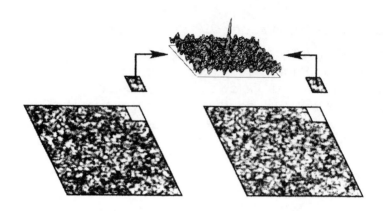

Fig.4.9: Cross correlation of subimages

The numerical evaluation of the mean translation (u_x, u_y) between the subimages is calculated in integer pixels in this first step. For some applications, e.g. determination of qualitative displacement fields, this discrete evaluation is sufficient. But usually the peak of the correlation function $R_{II'}$ is located between two pixels. Consequently, the error of the displacement evaluation can rise up to 0.5 Pixel, if only the integer pixels are taken into account. Therefore, experimental strain analyses with DSP cannot be performed by means of this first evaluation step. In order to determine the plane strains, the differences of the displacements are evaluated:

$$\varepsilon_{xx} = \frac{\Delta u_x}{\Delta x} \quad \varepsilon_{yy} = \frac{\Delta u_y}{\Delta y} \quad \varepsilon_{xy} = \frac{1}{2}\left(\frac{\Delta u_y}{\Delta x} + \frac{\Delta u_x}{\Delta y}\right) \tag{4.3.3}$$

Thus, a more sophisticated second evaluation step is necessary, where the floating point values of the in-plane displacements u_x and u_y are calculated on the basis of a subpixel algorithm. One possible way to calculate the floating point values of the peak location with acceptable calculation effort is the determination of the "center of gravity". In this analogy the location of the "center of gravity":

$$u_x = \frac{\sum_i u_{x,i} G_i}{\sum_i G_i} \quad u_y = \frac{\sum_i u_{y,i} G_i}{\sum_i G_i} \tag{4.3.4}$$

is calculated by means of the coordinates $u_{x,i}$ and $u_{y,i}$ of the i considered pixels and their respective gray levels G_i. The calculated coordinates are not only the "center of gravity", but also the location of the peak of the correlation function $R_{II'}$ and therefore the mean translation of the considered subimage. Furthermore, this simple and fast algorithm has the advantage that the information of a few pixels around the peak is sufficient for the subpixel evaluation.

4.3.2 Application of DSP: Deformation of a Laser Weld Seam

Laser beam welding provides a high process speed and a low energy input into the sample in comparison to conventional welding [10]. Owing to these features, the resulting laser welds have very narrow dimensions of the weld material and the heat affected zone in comparison to conventional welds, Fig.4.10. Furthermore, the various zones of the laser weld provide very different properties [8]. The introduction of laser beam welding into industry requires a profound knowledge about the mechanic-technological properties of the weld seam. The yield and the tensile strength of each zone of the laser weld must be available [9]. Due to steep gradients of the strains in the laser weld, the characterization by means of conventional measurement with strain gauges is usually not sufficient. The lack of local resolution is the main reason for the inability of strain gauges to provide the required information for the heat affected zone and the weld material itself. Thus, the application of an adapted measurement technique with a high lateral resolution is necessary. Ideally, the applied technique is able to record fieldwise strains with high accuracy and in a range up to the failure of the investigated sample.

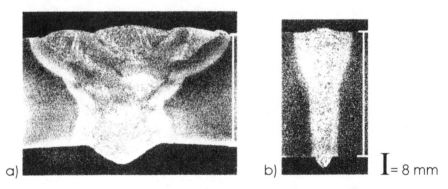

Fig.4.10: Comparison of weld seams a) conventional weld, b) laser weld

Experimental Set-up

Experimental determinations of local stress/strain diagrams of laser beam welded samples were made under workshop conditions on a hydraulic tensile test machine. The simple experimental set-up was directly mounted onto the tensile test machine, Fig.4.11. A frequency doubled Nd-YAG laser served as the light source in this set-up. The 500 mW power of the Nd-YAG laser were needed, owing to the used apertures, which provided speckles of the required quality.

Fig.4.11: Set-up for tensile tests of laser weld seams

The sample were cutted from 6 mm thick laser welded steel plates, which were welded with a power of 7.2 kW and a speed of 1.6 m/min by means of a CO_2 laser. The geometry of the samples was chosen according to the standard DIN 50120. Two different base materials with very different material properties were investigated, table 4.1.

Table 4.1: Material properties of the base material

No.	base material	yield strength [Mpa]	tensile strength [Mpa]	maximum strain
1	S355J2G3 (St 52-3 N)	355	490 - 630	22
2	S690Q (StE 690)	≥ 690	≥ 790	15

The testing procedure of the uniaxial quasistatic tensile tests ($\dot{\varepsilon} \approx 7 \cdot 10^{-5}$) was the same for both investigated samples. In the elastic range of the samples the path controlled load steps were in a distance of 0.1 mm. When the yield strength was exceeded, the distances risen up to 0.5 mm. This procedure provided between 25 and 38 load steps till the samples failed. Consequently, the same number of points in the stress/strain diagrams were obtained.

The monitored area on the samples had a size of approximately 15 x 15 mm². At the beginning of the tests the monitored area was centered in horizontal and vertical direction on the weld.

Experimental Results

Specimen 1

Specimen 1 was experimentally investigated in the tensile test machine owing 38 load steps up to failure. As an example, the horizontal displacement in the monitored area is shown, Fig.4.12. The influence of the constriction is evident in the upper part, where the heat affected zone and the base material are located in the lower part is almost no constriction visible. At this

Fig.4.12: Horizontal displacement field u_x at σ = 514 MPa

load state near the tensile strength the weld material is located in the lower part. Obviously, the strength of the weld material is higher than that of the base material, so that almost no deformation occurs in this zone. Furthermore, the weld material supports the heat affected zone (HAZ) and parts of the base material. With increasing distance from the weld material the horizontal deformation increases as well. An analysis of the strains in one point of the weld material and in one point of the heat affected zone close to the base material confirms the higher strength of the of the weld material, Fig.4.13.

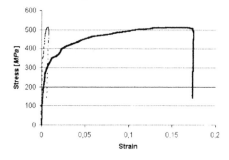

Fig.4.13: Stress-strain curves of HAZ and weld material

Specimen 2

Further measurements were performed with laser welds in the steel S690Q. With these specimen only 25 steps were needed until failure since the maximum strain was only about 65 % of that of the steel S355J2G3. A higher strength of the weld material in comparison to the heat affected zone and the base material can be observed, too. In the weld material the increase of displacements in loading direction is small, Fig.4.14. Like in specimen 1, the smaller strains appear in the weld material and the large ones as well as the failure are located in the heat affected zone and the base material.

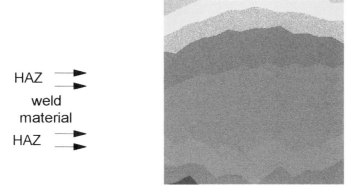

Fig.4.14: Vertical displacement field u_y at σ = 853 MPa

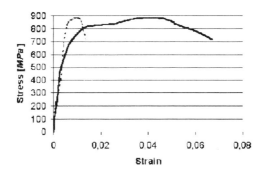

Fig.4.15: Stress-strain diagram of specimen 2

4.4 References

1. Erf, R.K. (ed.): „Speckle Metrology", Academic Press, New York (1978)
2. Osten, W.: Digitale Verarbeitung und Auswertung von Interferenzbildern (Digital processing and evaluation of interference patterns)", Akademie-Verlag GmbH, Berlin (1991)
3. Falldorf, C.: „Einfluß der Speckleinformation auf die Demodulation phasengeschobener Interferogramme (Influence of the Speckle information on the demodulation of phase-shifted interferograms)", Diplomarbeit, Universität Bremen, Bremen (1998)
4. Kolenovicz, E.: „Theoretische und experimentelle Untersuchung der Phasenverteilung in Specklefeldern (Theoretical and experimental investigation of the phase distribution in Speckle fields)", Diplomarbeit, Universität Bremen, Bremen (1998)
5. Jones, R.; Wykes, C.: „Holographic and Speckle Interferometry", Cambridge University Press, Oxford, (1989)
6. Fagan, W.F.: „Novel Speckle Camera and Analyzer", Proc. Laser 77, I.P.C. Publishing House, Guildford, Surrey (1977), p. 456-461
7. Ek, L.: „Focused Speckle Photography with Laser and White Light", Proc. *Herbstschule 77*, Hannover (1977),p. 123-125
8. Holstein, D.; Jüptner, W.; Kujawinska, M.; Salbut, L.: „Investigation of residual stresses induced by laser beam welding by means of optical methods", Proc. MECHATRONIKA, Warschau (1997)
9. Salbut, L.; Kujawinska, M.; Holstein, D.; Jüptner, W.: „Comparative analysis of laser weldment properties by grating interferometry and digital speckle photography", Proc. 11th Int. Conf. on Experimental Mechanics, Oxford (1998)
10. Seifert, K.: „Elektronenstrahlschweißen hochfester umwandlungshärtender Baustähle (Electron beam welding of high strength transition hardening construction steels)", Maschinenmarkt, 82 (1976) 45, S. 789-793

5. Electronic (Digital) Speckle Pattern Interferometry - ESPI, DSPI

5.1 Fundamentals of ESPI

Electronic Speckle Interferometry ESPI[1] is based on the idea of holographic interferometry HI. However, HI requires a lot of efforts in material and environment. Therefor it was proposed by different authors to use a TV camera as photosensitive medium [1, 2]. In DSPI the object is imaged onto the target of a camera and the reference wave is introduced parallel to this object wave, Fig.5.1.

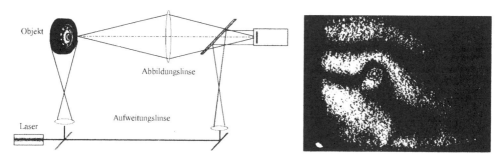

Fig. 5.1: TV-holography, ESPI: Set-up and test result at a tire

The interference on the target can be written as mentioned before - and the imaging will bring no difference in the mathematical description:

$$I_A(x,y) = I_{OO} + I_{RO} + 2\sqrt{I_{OO} I_{RO}} \cos[\phi_O(x,y) - \phi_R(x,y)] \quad (5.1.1)$$

with
$$\vec{E}_O = \vec{E}_{OO} \exp[i(\phi_O(\vec{r}) - \omega t)] \quad \text{object wave} \quad (5.1.2)$$

$$\vec{E}_R = \vec{E}_{RO} \exp[i(\phi_R(\vec{r}) - \omega t)] \quad \text{reference wave} \quad (5.1.3)$$

The cos-term describes the image, however, covered with Speckle noise: The rough surface leads to micro-interferences which are the Speckles. A deformation of the object may result in a phase difference δ(x,y) at the point P(x,y). The new object wave is then:

$$\vec{E}'_O = \vec{E}_{OO} \exp[i(\phi_O(\vec{r}) + \delta(x,y) - \omega t)] \quad (5.1.4)$$

---

[1] ESPI is named in different ways: Digital Speckle Interferometry (DSPI) or TV holography. Even Electro-Optical Holography is at least an ESPI method, but with a special evaluation.

which leads with the same reference wave to

$$I_B(x,y) = I'_{00} + I_{R0} + 2\sqrt{I'_{00} I_{R0}} \cos[\phi_o(x,y) - \phi_R(x,y) + \delta(x,y)] \qquad (5.1.5)$$

In the image processing system the intensity I_B is subtracted from I_A for each point assuming, that the object deformation changes the phase but not the amplitude of the object wave, i.e. $I'_{00} = I_{00}$. The difference is given by

$$\begin{aligned} I_B - I_B &= 2\sqrt{I_{00} I_{R0}} \{\cos[\phi_o - \phi_R] - \cos[\phi_o - \phi_R]\cos\delta + \sin[\phi_o - \phi_R]\sin\delta\} \\ &= 4\sqrt{I_{00} I_{R0}} \sin[\phi_o - \phi_R + \delta/2]\sin(\delta/2) \end{aligned} \qquad (5.1.6)$$

In order to have only positive values for the display on the monitor, the amount or the square of the result is taken. The Equ.(5.1.6) contains to sin-terms: the first one is the random Speckle noise as it is generated by the micro-interferences from the different points of the rough surface of the object. The second term describe a low spatial frequency modulation of the Speckle pattern. This is the desired information, which can be transformed in the surface deformation as described in the before mentioned chapters.

5.2 Applications of ESPI Methods

5.2.1 Deformation Analysis of Small Components

Electro-Optical Holography (EOH) is an interferometric method based on the ESPI procedure [3] with a slightly changed evaluation: The Speckle pattern interferogram is taken with phase shifts of 0, π, 2π and 3π. These four interferograms can be processed in a way that reduces the Speckle noise significantly [4]. However, the first result is an image processed Speckle interferogram. The optical configuration of the EOH system has been reported by Pryputniewicz and his coworkers several times, e.g. [5], Fig.5.2. It is the typical ESPI set-up. The system is equipped with different devices which enable analysis of the object in static and dynamic mode. Beside the basic component as beam splitter BS, spatial filters with beam expander assemblies SE1, SE2 and the mirrors M1, M2 are this especially:

- the phase stepper PS1 in the object beam, which can be driven at the same frequencies as the object excitation to provide bias modulation in dynamic mode
- the phase stepper PS2, in the reference beam, which introduces phase steps between the consecutive frames during the acquisition of the interferometric information. This enables the fringe analysis by phase stepping methods.
- the beam rotator BR, which facilitate speckle averaging and increases the quality of the image captured.

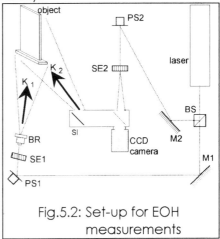

Fig.5.2: Set-up for EOH measurements

- the speckle interferometer SI combines the object and the reference beam and directs them towards the CCD chip of the camera.

The static measurements are implemented using the double exposure holographic interferometry method in which the reference frame may refer to unloaded or an arbitrary loaded state of the object. This EOH method was applied among other to the investigation and optimization of the design of a micro-connector, Fig.5.3. The experimental results were used to confirm the computation of the behaviour under compression load. The micro-connector undergoes in its operation a strong mechanical loading. Therefor the material as well as the design had to be optimized. The application of FEM modeling is the state of the art for this purpose. However, the FEM results have to be confirmed experimentally. The simplicity of the EOH method made preferable instead of holographic interferometry.

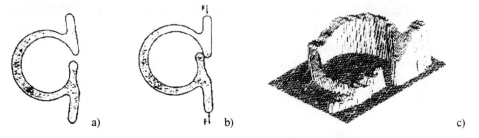

Fig.5.3: Deformation measurement of a microconnector a) unloaded, b) loaded, c) bending [3]

Arrays of these micro-connectors are meanwhile in use for the mounting of microprocessors in PCs.

5.2.2 Vibration Analysis of Small Membranes

The ESPI or DSPI method has been applied to the investigation of small membranes [6]. Höfling et.al. reported about the systematic investigation of squared diaphragm prepared by etching of silicon with dimensions of 4*4, 2*2, 1*1, and 0.45*0.45 mm². The diaphragm is fixed at all four sides. Then the natural frequency in the n-th mode is given by the following equation [7]:

$$f_n = \frac{b_n}{2\pi}\sqrt{\frac{Ed^2}{\rho a^4 (1-\upsilon^2)}} \tag{5.2.1}$$

with f : frequency
 n : mode number
 b : mode constant
 E : Young's modulus
 d : thickness of the plate
 ρ : mass density
 a : side length
 ν : Poisson's ratio

As long as the geometrical quantities are known, the equation can be used to define the the material constants. However, the Equ.(5.2.1) is based on a lot of assumptions. Therefor Höfling calculated the Eigen values of the modes and the amplitude and phase distribution by Finite Element Methods. Then he compared the results to measurements by DSPI.

The DSPI set-up was a conventional one for out-of-plane sensitivity, Fig.5.4. In

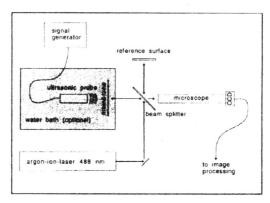

Fig.5.4: Scheme of the experimental set-up [6]

principle the set-up is a Michelson typed interferometer, but with a fine grained surface in one of the interferometer arms. An Ar-Ion laser beam has been directed via a beam splitter onto the surface of the object under test. The components were placed in a water tank for a sufficient coupling of the exciting sound wave to the diaphragms. The reflected light is combined with the reference wave and observed by a long distance microscope. The image taken by a CCD camera is stored in a PC and evaluated according to the standard methods of DSPI (ESPI). The resonant frequencies were observed and registered by viewing at the images of the of the surface amplitudes in real time, i.e. one reference image was stored and compared to the vibration images, Fig.5.5.

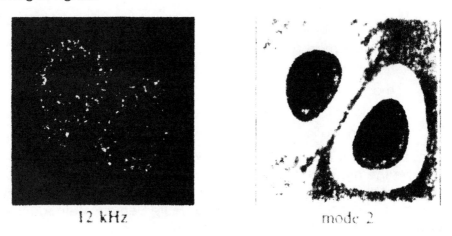

Fig.5.5: Amplitude distribution of the 2nd Eigen mode [6]

The investigation was performed as a comparison between the theoretical and the experimental frequencies. One of the new results was the behaviour of the diaphragm in certain Eigen values: The resonant frequencies were not the natural Eigen modes but combined ones, i.e. the diaphragm did not vibrate in the calculated and expected Eigen value - e.g. mode number 2 or 3 - but in a combination in the form of mode 2+3 or 2-3, taking into account the phases of the modes. This effect has been observed by Pryputniewicz and his co-workers, too [8].

5.3 References

1. Raimann, G.: "Eumig HT-10: Ein TV-Speckle Interferometer und seine praktischen Anwendungen (Eumig HAT-10: A TV-Speckle Interferometer and its practical applications)", Herbstschule '77, Hannover 1977, p. 147

2. Pedersen, H.M.; Løkberg, O.J.; Førre, B.M: "Holographic Vibration Measurement Using a TV Speckle Interferometer with Silicon Target Vidicon", Optics Communication, Vol.12, No. 4, p.421-426

3. Kujawinska, M.; Pryputniewicz, R.J.: "Micromeasurements: a Challenge for Photomechanics", In: Proc. SPIE, Vol. 2782 (1996), p. 15-24

4. Stetson, K.A.:"Theory and Application of Electronic Holography", Proc. Of the SEM Meeting *Hologram Interferometry and Speckle Metrology*, SEM, Bethel, 1990, p. 295-300

5. Pryputniewicz, R.J.; Grabbe, D.G.: "Developments in micromechanics through analysis and experimentation". In: Proc. 11th Int. Invitational UCEM Symp. 1993, Soc. Exp. Mech., Bethel, CT, p. 506-532

6. Höfling, R.; Aswendt, P.; Liebig, V.; Brückner, S.: "Synthesis of Experiment and Simulation in Speckle Interferometry: a Medical Application", In: Simulation and Experiment in Laser Metrology, eds.: W.Jüptner, W.Osten, Akademie-Verlag (1996), p. 261-268

7. Harris, C.M.: "Shock and Vibration Handbook Vol.1", McGraw Hill, 1961

8. Furlon, C.; Pryputniewicz, R.J.: "Hybrid, experimental and computational, investigation of mechanical components", SPIE Vol. 2861, 1996, p. 13-25

6. Digital Speckle Shearing Interferometry

6.1 Fundamentals of Shearography

Digital Speckle Shearography or shorter shearography is a coherent optical measurement method for deformations [1], too: the phase of the object wave is recorded additionally to the amplitude by adding a reference wave. However, in shearography the reference wave is the object wave itself, but slightly shifted in one in-plane direction through shearing, Fig.6.1.

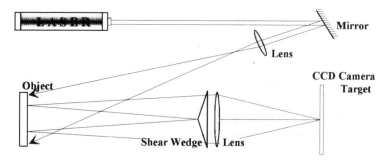

Fig.6.1: Shearography: principle set-up

The shearing of the object wave can be performed in different ways. The two methods mostly applied are the introduction of a prism, Fig.6.1, and the splitting of the object wave by a Michelson interferometer, Fig. 6.2. The Michelson typed set-up has the advantage of flexibility and evaluation comfort [2]:

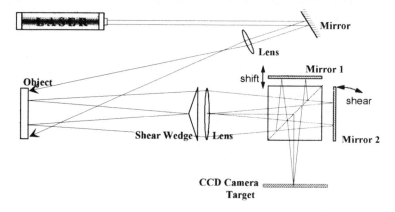

Fig.6.2: Shearing by a Michelson type set-up

Interferometric Methods

- One of the mirrors can be used as the shearing mirror: by light tilts of the mirror any shift direct can be introduced and the amount of shift can be adapted to the deformation.
- The other mirror can be used for phase shifting by moving it parallel to the mirror axis. This enables the evaluation of the fringe pattern by phase shift methods.

The interference on the target can be written as follows [3]:
The image shearing leads to two nearly collinear waves (the time dependent phase term is neglected):

$$\vec{E}_1 = \vec{E}_{10}(x,y) \exp[i\phi(x,y)] \tag{6.1.1}$$

$$\vec{E}_2 = \vec{E}_{20}(x,y) \exp[i\phi(x+\Delta x, y)] \tag{6.1.2}$$

where a shift in x-direction is assumed (without restriction of generality). The superposition of the two waves results in:

$$I_A(x,y) = I_1 + I_2 + 2\sqrt{I_1(x,y) I_2(x,y)} \cos(\Delta\phi(x,y)) \tag{6.1.3}$$

with $\quad \Delta\phi(x,y) = \phi(x,y) - \phi(x+\Delta x, y) \tag{6.1.4}$

The phase term Equ.(6.1.4) describes a stochastic distribution according to the rough surface of the object. The deformation of the object results in the two waves

$$\vec{E}_3 = \vec{E}_{30}(x,y) \exp[i(\phi(x,y) + \delta(x,y))] \tag{6.1.5}$$

$$\vec{E}_4 = \vec{E}_{40}(x,y) \exp[i(\phi(x+\Delta x, y) + \delta(x+\Delta x, y))] \tag{6.1.6}$$

The superposition of the two waves leads to

$$\begin{aligned} I_B &= I_1 + I_2 + 2\sqrt{I_1 I_2} \cos[\phi(x,y) - \phi(x+\Delta x, y) + \delta(x,y) - \delta(x+\Delta x, y)] \\ &= I_1 + I_2 + 2\sqrt{I_1 I_2} \cos[\Delta\phi(x,y) + \delta(x,y) - \delta(x+\Delta x, y)] \end{aligned} \tag{6.1.7}$$

In the image processing system the intensity I_B is subtracted from I_A for each point assuming, that the object deformation changes the phase but not the amplitude of the object wave. The difference is given by

$$\begin{aligned} I_B - I_B &= 2\sqrt{I_1 I_2} \{\cos[\Delta\phi(x,y)] - \cos[\Delta\phi(x,y) - \delta(x,y) + \delta(x+\Delta x, y)]\} \\ &= 4\sqrt{I_1 I_2} \sin\left[\Delta\phi(x,y) + \frac{\delta(x,y) - \delta(x+\Delta x, y)}{2}\right] \sin\left(\frac{\delta(x,y) - \delta(x+\Delta x, y)}{2}\right) \end{aligned} \tag{6.1.8}$$

In order to have only positive values for the display on the monitor, the amount or the square of the result is taken. The Equ.(6.1.8) contains to sin-terms: the first one is the random Speckle noise. The second term describe a low spatial frequency modulation of the Speckle pattern. This is the desired information. However, the deformation will be given by the derivative: For the derivation of this equation it assumed that the illumination vector and the observation vector will not vary over the distance of the image shift, i.e. $\vec{s}(x,y) \cong \vec{s}(x+\Delta x, y)$ and $\vec{b}(x,y) \cong \vec{b}(x+\Delta x, y)$. This assumption is in common valid, since the illumination is a slowly varying function and the observation vector is large against the shearing distance Δx. With this assumption the modulation term which contains the deformation, can be written.

$$\frac{\delta(x,y) - \delta(x+\Delta x, y)}{2} = \frac{\pi}{\lambda}\{\vec{d}(x,y)[\vec{b}(x,y) - \vec{s}(x,y)] - \vec{d}(x+\Delta, y)[\vec{b}(x\Delta x, y) - \vec{s}(x+\Delta x+, y)]\}$$

$$= \frac{\pi}{\lambda}\{[\vec{d}(x,y) - \vec{d}(x+\Delta, y)][\vec{b}(x,y) - \vec{s}(x,y)]\}$$

$$= \frac{\pi}{\lambda}\left\{\frac{\vec{d}(x,y) - \vec{d}(x+\Delta, y)}{\Delta x} \cdot [\vec{b}(x,y) - \vec{s}(x,y)]\right\} \cdot \Delta x$$

$$\approx \frac{\pi}{\lambda}\frac{\partial \vec{d}(x,y)}{\partial x}[\vec{b}(x,y) - \vec{s}(x,y)] \cdot \Delta x$$

(6.1.9)

Equ.(6.1.9) points out that the interference is proportional to the first derivative of the deformation. The shift Δx influences the sensitivity proportionally.

6.2 Applications of Shearography

6.2.1 Deformation Measurements

Digital Speckle Shearography or shorter shearography is sensitive to displacement differences instead of the displacements. This can be demonstrated by the deformation of a membrane [4]. The membrane is loaded by a static pressure load in its center. The set-up is chosen in a way that it is mainly sensitive to out-of-plane displacements, Fig.6.3.

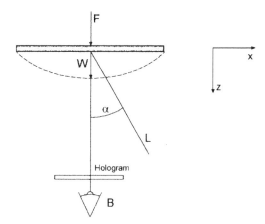

Fig.6.3: Set-up for holographic interferometry [4]

The fringe pattern for the holographic-interferometric displacement measurement is almost a system of concentric circles, Fig.6.4. This is according to the expected deformation, Fig.6.5. The appearance of the interferogram and the

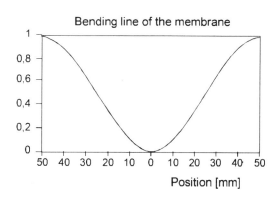

Fig.6.4: Bending of a membrane   Fig.6.5: Bending curve of a membrane [4]

result of the measurement changes significantly by applying shearography: The interferogram is now divided into two symmetrical pattern, Fig.6.6. The result of the fringe pattern analysis is given by the differential quotient of the bending curve into the direction of the shearing, Fig.6.7

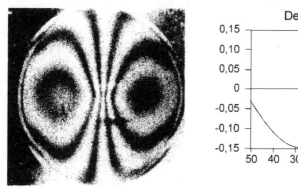
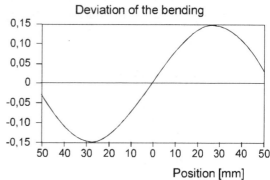

Fig.6.6: Deformation of a membrane   Fig.6.7: Deviation of the bending [4]

However, one should notice from Equ.(6.1.9) that this is not really the derivative of the displacement function since in practice the shearing has to have a nominal value to result in fringe pattern. Furthermore, the shearography does not result in strains as often mentioned. The result is the projection of the displacement difference onto the sensitivity vector, which may be significantly directed out-of-plane.

A tremendous advantage of shearography is the ability to adjust the sensitivity to the expected deformation simply by the choice of the shearing Δx. By this one can assure a reasonable fringe density.

The shearography is well suited to non-destructive testing. The method is mainly sensitive to deformations indicating local defects since in this region the deformation gradients are in common larger than in sound areas, and NDT is usually evaluated qualitatively by visual inspection. On the other hand, one has to be very careful using shearography as a quantitative method: The mathematical procedure leading to Equ.(6.1.9) is based on the assumption that one can neglect the first sin-Term of Equ.(6.1.8).

6.2.2 Non-destructive Testing

Shearography is well suited to non-destructive testing. As an example the testing of a carbon-fiber reinforced plastic tube may be cited [5]. The component is a tensile test specimen with a width of 20 mm. After the loading first delaminations appear at the edges starting from the middle of the specimen, Fig.6.8. The delaminations cause an opening of the compound with strong deformations into the out-of-plane direction which was the sensitivity direction in this test. The shearing direction had been chosen perpendicular to the specimen axis in this experiment. Therefor the derivative of the deformation into the shearing axis is shown clearly while the unavoidable rigid body displacements are suppressed. By this reason the fringe pattern indicates clearly how far the delamination is progressed.

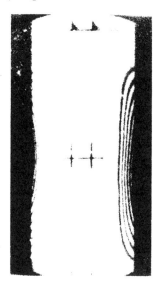

Fig.6.8: Shearogram of a delaminated specimen [5]

6.3 References

1. Hung, Y.Y.: "Shearography: anew optical method for strain measurement and nondestructive testing", Opt. Eng. 21 (1982), p. 391-395
2. Kreis, Th.: "Holographic Interferometry", Akademie Verlag, Berlin
3. Kreis, Th.: "Shearography", Anwenderforum
4. Ettemeyer, A.: „Shearografie - ein optisches Verfahren zur zerstörungsfreien Werkstoffprüfung (Shearography - an optical method for nondestructive testing)", tm Technisches Messen58 (1991) 6, p. 247 - 252
5. Klumpp, P.A.: "Delaminationsuntersuchungen an Carbonfaser/Epoxy-Verbunden mit kohärent-optischen Verfahren (Investigation of delaminations in carbon-fiber/Epoxy-resin compounds by means of coherent optical methods)", PhD thesis, University of Karlsruhe, 1990

CHAPTER V

DIGITAL PROCESSING AND EVALUATION OF FRINGE PATTERNS IN OPTICAL METROLOGY AND NON-DESTRUCTIVE TESTING

W. Osten
BIAS, Bremen, Germany

ABSTRACT

The basic principle of modern optical methods in experimental solid mechanics such as holographic interferometry, speckle metrology, fringe projection and moiré techniques consists either in a specific structuring of the illumination of the object by incoherent projection of fringe patterns onto the surface under test or by coherent superposition (interference) of light fields representing different states of the object. A common property of the methods is that they produce fringe pattern as output. In these intensity fluctuations the quantities of interest - coordinates, displacements, refractive index and others - are coded in the scale of the fringe period. Consequently the task to be solved in fringe analysis can be defined as the conversion of the fringe pattern into a continuous phase map taking into account the quasi sinusoidal character of the intensity distribution. The course starts with a physical modeling of the image content that contains the relevant disturbances in optical metrology. After that the main techniques for quantitative phase reconstruction together with the most commonly used image processing methods are presented. Because image processing is an important prerequisite for holographic non-destructive evaluation (HNDE) an overview of modern approaches in automatic flaw detection based on knowledge assisted and neural network techniques is given. Finally modern software tools for digital processing of fringe patterns are presented.

1. INTRODUCTION

Optical techniques are now becoming widely used on the production line for the automatic inspection of objects and their components, bringing significant advantages in the cost-effective improvement in the quality of products. By reason of their sensitivity, accuracy and non-contact as well as non-destructive characteristics, methods such as holographic interferometry, speckle metrology, moiré and fringe projection techniques have found an increasing interest not only for laborious investigations but also for applications on the factory floor. These applications cover such important fields as testing of lenses and mirrors with respect to optical surface forms, construction optimization of components by stress analysis under operational load, on site investigation of industrial products with diffusely reflecting surfaces for the purpose of material fault detection and last but not least optical shape recognition as an area of topical interest for the solution of problems in reverse engineering.

The basic principle of the methods considered here consists either in a specific structuring of the illumination of the object by incoherent projection of fringe patterns onto the surface under test or by coherent superposition (interference) of light fields representing different states of the object. A common property of the methods is that they produce fringe pattern as output. In these intensity fluctuations the quantities of interest - coordinates, displacements, refractive index and others - are coded in the scale of the fringe period. Using coherent methods this period is determined by the wavelength of the interfering light fields. In the case of incoherent fringe projection the spacing between two neighbouring lines of the projected transparency controls the scale of the measurement. Consequently the task to be solved in fringe analysis can be defined as the conversion of the fringe pattern into a continuous phase map taking into account the quasi sinusoidal character of the intensity distribution.

Techniques for the analysis of fringe patterns are as old as interferometric methods themselves, but until into the late 1980s routine analysis of fringe patterns for optical shop testing, experimental stress analysis and non-destructive testing was mainly performed manually. One example may illustrate this highly subjective and time consuming process: If only one minute is estimated for the manual evaluation per measuring point the total time necessary for the reconstruction of the three-dimensional displacement field of a 32x32 mesh would take almost 6 working days because at least 3 holographic interferograms with 1024 grid points each have to be evaluated. Such a procedure must be qualified as highly ineffective with respect to the requirements of modern industrial inspection, to say nothing of the limited reliability of the derived data. However, the step from a manual laborious technique only done by skilled technicians to a fully automatic procedure was strongly linked to the availability of modern computer technology. Consequently the development of automatic fringe pattern analysis closely followed the exponential growth in the power of digital computers with image processing capabilities. Since several years various commercial image processing systems with different efficiency and price level have been developed and specialized systems dedicated to the solution of complex fringe analysis problems are commercially available now (1), (2), (3). Advanced hardware and software technologies enable the use of desk-top systems with several image memories and special video processors nearby the optical set-up. This on-line connection between digital image processors and optical test equipment opens completely new approaches for optical metrology and non-destructive testing as real time techniques with high industrial relevance. In this sense automatic fringe analysis has developed mainly during the 1980s into a subject in its own right. A number of international conferences (4), (5), (6), (7) and several comprehensive publications (8), (9), (10), (11), (12), (13) are devoted specifically to this subject.

Automatic fringe analysis covers a broad area today. It would be beyond the scope of this course to discuss all methods and trends suitable. Consequently this lecture concentrates mainly on the application of digital image processing techniques for the reconstruction of continous phase fields from noisy fringe patterns. Optical and analogue electronic methods of fringe analysis are not considered, for this subject the reader is referred to (14, 15). The tasks to be solved with respect to the derivation of quantities of primary interest as coordinates, displacements, vibrational amplitudes and refractive index are described as examples of three-dimensional shape and displacement measurement. Concerning qualitative fringe interpretation some modern trends are discussed only with respect to the computer aided detection of material faults.

This course starts with the description of intensity relations in optical metrology and with the derivation of a physical model of the image content that contains the relevant disturbances in optical metrology. Such a modeling is necessary since a-priori knowledge of the image formation process is very useful for the development of algorithms. Using the image model fringe patterns for various metrological methods can be simulated and used for the development of adapted algorithms. After that the main techniques for quantitative phase reconstruction together with the most commonly used image processing methods are presented. To emphasize that the reconstruction of the continous phase distribution is only the first step in fringe pattern analysis, the computation of vector displacements using digitally measured phases is described as an example. Because image processing is an important prerequisite for holographic non-destructive evaluation (HNDE) an overview of modern approaches in automatic flaw detection based on knowledge assisted and neural network techniques is given. Finally modern software tools for digital processing of fringe patterns are presented.

2. INTENSITY MODELS, DISTURBANCES AND SIMULATIONS

Fringe patterns generated by holographic interferometry, structured illumination and other metrological methods are nothing else but the image of the object under test after a characteristic redistribution of the intensity. In the result of this process the image of the object is modified by muliplicative fringe patterns, additive random noise, and other influences. The intensity relations in optical metrology give the mathematical link between the observed intensity and the quantity to be measured. It is common for all techniques discussed here that the phase of the sinosoidal intensity modulation is the quantity of primary interest. The connection between the measured phase and the final quantitiies such as displacements or coordinates is given by so-called basic equations or geometric models.

Mathematical intensity models are quite useful for the processing and analysis of fringe patterns. They are based on the optical laws of interference and imaging. Additionally, the model includes the various disturbances such as background illumination and noise which influence both the accuracy of the reconstructed phase distribution and the choice of method to be used in its determination. Although it is quite impossible to separate the effects of different kinds of noise in a real image, a physical model describing the image formation process with all its influential contributions gives the opportunity to compute artificial images which approximate step by step the complex structure of a real image. By means of this artificial test environment the performance of algorithms and image processing tools can be studied highly effectively. Both error detection and the selection of adapted parameters become more convenient.

In this section the modeling and simulation is demonstrated especially by the example of holographic interferograms. Since this image formation process is determined by the coherent superposition of diffusely scattered wavefronts, the resulting intensity distribution is complicated and differs considerably from the theoretically assumed cosine shaped intensity modulation between bright and dark fringes of constant peak and valley values, respectively. The experiences gained with the processing of holographic interferograms can be applied with advantage to other more simple types of fringe patterns as for instance the fringes observed with the method of structured illumination.

2.1 Intensity relations in optical metrology

A general expression for the observed intensity I(x,y) and the phase difference δ(x,y) holographic interferometry and speckle metrology was introduced by STETSON (16):

$$I(x,y) = I_o(x,y) \cdot |M[\delta(x,y)]|^2 \qquad (2.1)$$

with $I_O(x,y)$ as basic intensity and M(δ) as the *characteristic function* which modulates the basic intensity depending from the used registration technique and the object movement during the exposure time t_B of the light sensitive sensor

$$|M(\delta)|^2 = \left| \frac{1}{t_b} \int_0^{t_B} e^{i\delta_t(\mathbf{r},t)} dt \right|^2 \qquad (2.2)$$

The result of this modulation is an object covered with interefence fringes. To find an analytic expression for a concrete movement it is necessary to characterize its time response during t_B. To this purpose the movement is indicated by a time-dependent displacement vector **D(r,t)** which satisfies a definite temporal function f(t):

$$\mathbf{D}(\mathbf{r},t) = \mathbf{d}(\mathbf{r}) \cdot f(t), \qquad (2.3)$$

where **d(r)** considers the vector amplitude of the movement of the measuring point P(r,t). In Equ. (2.3) was assumed that the movement can be separated into a time-dependent part of the movement of all points and their spatial position. Under that condition the time-dependent phase difference $\delta_t(\mathbf{r},t)$ can be written as

$$\delta_t(\mathbf{r},t) = \delta(\mathbf{r}) \cdot f(t) \qquad (2.4)$$

Consequently for Equ. (2.2) follows

$$M(\delta) = \frac{1}{t_b} \int_0^{t_B} e^{i\delta(\mathbf{r}) \cdot f(t)} dt \qquad (2.5)$$

For the main registration techniques in optical metrology the following wellknown characteristc functions can be derived:

a) Double-Exposure Technique
The subsequent registration of two exposures and their simultaneous reconstruction:

$$|M(\delta)|^2 = \cos^2(\delta/2) \quad , \quad \delta = N \cdot 2\pi \quad . \qquad (2.6)$$

b) Real-Time Technique
Registration of one state on the sensor and superposition of its reconstruction with the current wavefront scattered by the illuminated object:

$$|M(\delta)|^2 = \sin^2(\delta/2) \quad , \quad \bar{\delta} = N \cdot 2\pi \ . \tag{2.7}$$

c) Time-Average Technique: object moving with constant velocity
Continuous registration of a moving object and simultaneous reconstruction of all intermediate states:

$$|M(\delta)|^2 = \text{sinc}^2(\delta/2) \quad , \quad \text{sinc}\, x = \sin x / x \ . \tag{2.8}$$

d) Time-Average Technique: harmonic vibration
Continuous registration of an vibrating and simultaneous reconstruction of all intermediate states:

$$|M(\delta)|^2 = J_0^2(\delta/2) \ . \tag{2.9}$$

e) Real-Time-Averaging: harmonic vibration
Registration of one state on the sensor and superposition of its reconstruction with the current wavefront scattered by the vibrating and illuminated object:

$$|M(\delta)|^2 = 1 - J_0(\delta) \ . \tag{2.10}$$

f) Real-Time-Averaging: superposition of static displacement and harmonic vibration
Registration of one state on the sensor and superposition of its reconstruction with the current wavefront scattered by the vibrating and illuminated object:

$$|M(\delta)|^2 = 1 - \cos(\delta_1) \cdot J_0(\delta_2) \ . \tag{2.11}$$

g) Stroboscopic Technique
Registration of two discrete states of a moving object by synchronization of illumination and movement:

$$|M(\delta)|^2 = \cos^2(\delta/2) \quad , \quad \bar{\delta} = N \cdot 2\pi \ . \tag{2.12}$$

Fig. 2.1 shows the plots of all characteristic functions discussed above and Fig. 2 shows both a genuine and synthetic holographic interferogram made with double exposure technique and their intensity profiles along the dotted line.

With respect to the reconstruction of the phase distribution $\delta(\mathbf{r})$ from the measured intensity the special characteristic function M(d) has to be taken into account consitently. However, in practice the measured intensity distribution differs due to noise and other influences considerably of this one theoretically expected distributions (see Fig. 2.2).

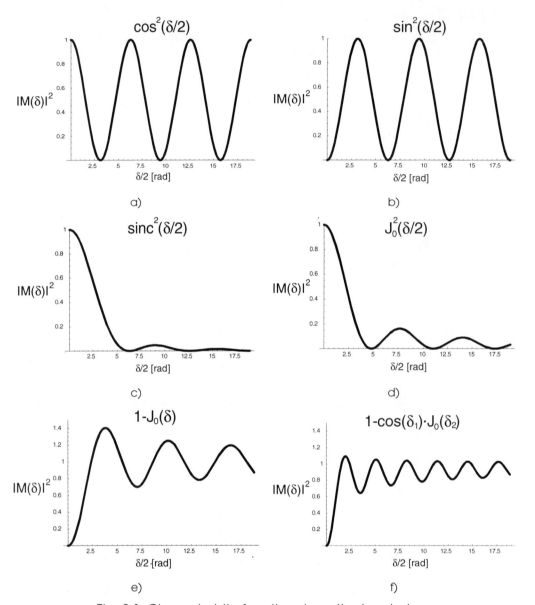

Fig. 2.1: Characteristic functions in optical metrology

Digital Processing and Evaluation of Fringe Patterns

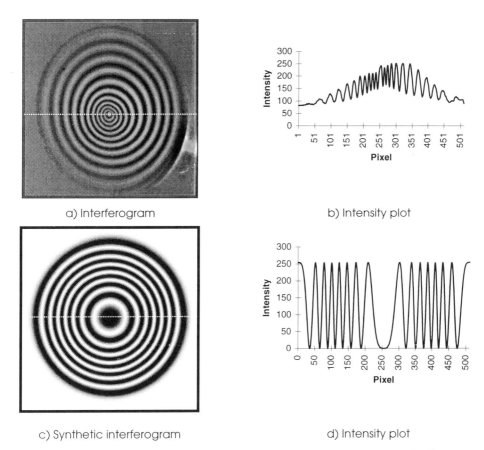

a) Interferogram

b) Intensity plot

c) Synthetic interferogram

d) Intensity plot

Fig. 2.2: Holographic interferograms of a real an simulated centrally circular plate their intensity profiles

2.2 Modeling of the image formation process in holographic interferometry

The description and interpretation of an image observed in a holographic interferometer has to consider two classes of effects that influence the image formation process, Fig. 2.3:
- the coherent superposition of diffusely reflected wavefields representing two states of an object with a rough surface and the optical imaging as well as
- the influence of the opto-electronic image processing system.

Therefore the effect of the imaging system consists not only in the transformation of the object wavefield $O(\xi,\eta)$ into an image $A(x,y)$ but also in the degradation of the original signal due to nonlinearities L of the imaging sensor and different noise components $R(x, y)$. Consequently the observed signal $A'(x,y)$ can be described as

$$A'(x, y) = L[A(x, y)] + R(x, y) = L[A(x, y)] \cdot (1 + R_1(x, y)) + R_2(x, y) \qquad (2.13)$$

with

$$A(x, y) = H[O(\xi, \eta)] \qquad (2.14)$$

At first the effects of *coherent imaging* will be investigated and the conditions of fringe pattern formation will be described. The different noise components are discussed later.

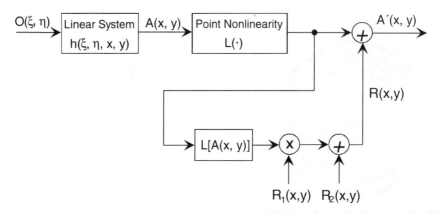

Fig. 2.3: Influence of imaging system components on the observed signal.

If linearity and space invariance can be assumed for the optical system, Fig. 2.4, the observed image $A(x,y)$ can be interpreted as the convolution of the light $O(\xi,\eta)$ coming from the object with the impulse response $h(r_i - Mr_o)$ of the imaging system with the magnification M:

$$A(x, y) = \int\int_{-\infty-\infty}^{\infty\ \infty} O(\xi, \eta) \cdot h(x - M\xi, y - M\eta) d\xi d\eta = O(\mathbf{r}_o) \otimes h(\mathbf{r}_i - M\mathbf{r}_o) \qquad (2.15)$$

In optics the impulse response is also known as the *point spread function*. For the coordinates in the image and in the object plane the symbols $r_i=(x,y)$ and $r_o=(\xi,\eta)$ are used, respectively. In comparison with classical interferometry the new quality in holographic interferometry consists in the coherent superposition of wavefronts coming from diffusely reflecting surfaces: The laser light scattered at the rough surface introduces a granular structure known as the *speckle effect*. The statistical properties of the intensity in the images of coherent illuminated objects have been investigated by LOWENTHAL and ARSENAULT (17). For a review of the statistical properties of laser speckle see GOODMAN (18). Accordingly, the complex amplitude $O(r_0)$ of the scattered light at the surface is described by

$$O(\mathbf{r}_o) = \overline{O}(\mathbf{r}_o) \cdot \rho(\mathbf{r}_o) \tag{2.16}$$

where $\overline{O}(\mathbf{r}_o)$ is the medium amplitude distribution of the object illumination and considers the macroscopic shape of the surface. The complex reflection coefficient $\rho(r_o)$ represents the influence of the *surface roughness* with its random phase variations and can be expressed as a phase function:

$$\rho(\mathbf{r}_o) = \rho_o \exp[i\phi(\mathbf{r}_o)] \tag{2.17}$$

ρ_o is a constant amplitude factor and may be neglected in the following. The coherent superposition of the light fields coming from different points of the diffusely scattering surface results in a new light field that shows speckles.

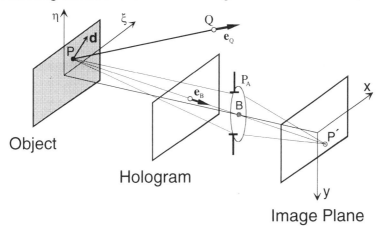

Fig. 2.4: Schematic arrangement of an holographic interferometer with the light source Q (illumination direction $\mathbf{e}_Q$), the object with the measuring point P and the displacement vector **d**, the hologram and the imaging system with the observation point B (observation direction $\mathbf{e}_B$).

DÄNDLIKER (19) has applied these results on the image formation process in holographic interferometry. Based on an analysis of the averaged interference term it can be shown that macroscopic interference fringes can be observed only if the two random phase distributions $\rho_1(r_o)$ and $\rho_2(r_o)$ of the both surfaces under test are correlated. This means that the microstructure of the rough surfaces has to be identical. Consequently, only two states of the same object can be compared interferometrically to each other - a condition that can be satisfied by holographic interferometry and speckle techniques only. For the formulation of an intensity model it is assumed that both object states differ only by a microscopic displacement $d_i=d(r_i)$ that causes a phase difference $\delta(r_i)$:

$$A_{o1}(\mathbf{r}_i) = a_o(\mathbf{r}_i) \cdot \exp[i\varphi_1(\mathbf{r}_i)] = O(\mathbf{r}_o) \otimes h(\mathbf{r}_i - M\mathbf{r}_o) \quad \text{(reference state)} \quad (2.18)$$

$$A_{o2}(\mathbf{r}_i) = A_{o1}(\mathbf{r}_i + \mathbf{d}_i) \cdot \exp[i\delta(\mathbf{r}_i)] = \exp[i\delta(\mathbf{r}_i)] \cdot [O(\mathbf{r}_o) \otimes h(\mathbf{r}_i + \mathbf{d}_i - M\mathbf{r}_o)] \quad \text{(actual state)} \quad (2.19)$$

with a_o as the amplitude of both wavefields, φ as the phase of the wavefront. The intensity relation for *two beam interference* can be found at the image point:

$$I(\mathbf{r}_i) = (A_{o1}(\mathbf{r}_i) + A_{o2}(\mathbf{r}_i)) \cdot (A_{o1}(\mathbf{r}_i) + A_{o2}(\mathbf{r}_i))^*$$
$$= A_{o1}(\mathbf{r}_i) \cdot A_{o1}^*(\mathbf{r}_i) + A_{o2}(\mathbf{r}_i) \cdot A_{o2}^*(\mathbf{r}_i) + A_{o1}(\mathbf{r}_i) \cdot A_{o2}^*(\mathbf{r}_i) + A_{o1}^*(\mathbf{r}_i) \cdot A_{o2}(\mathbf{r}_i) \quad (2.20)$$

where * denotes the complex conjugate. For low resolution objects showing a slowly varying medium local intensity distribution, the first term on the right side can be defined as the incoherent image of the object in the first state:

$$I_o(\mathbf{r}_i) = A_o(\mathbf{r}_i) \cdot A_o^*(\mathbf{r}_i) = \int\int_{-\infty}^{\infty} |\overline{O}(\mathbf{r}_o)|^2 \cdot \rho(\mathbf{r}_o) \cdot \rho^*(\mathbf{r}_o) \cdot |h(\mathbf{r}_i - M\mathbf{r}_o)|^2 d\xi d\eta = |\overline{O}(\mathbf{r}_o)|^2 \quad (2.21)$$

Since A_{o1} and A_{o2} differ only in a microscopic displacement their contributions to the intensity are equal with negligible differences. The *interference term* is given by:

$$A_{o1}(\mathbf{r}_i) \cdot A_{o2}^*(\mathbf{r}_i) = \int\int_{-\infty}^{\infty} |\overline{O}(\mathbf{r}_o)|^2 \cdot \exp[i\delta(\mathbf{r}_i)] \cdot \rho_1(\mathbf{r}_o) \cdot \rho_2^*(\mathbf{r}_o + \mathbf{d}_o) \cdot h(\mathbf{r}_i - M\mathbf{r}_o)h(\mathbf{r}_i + \mathbf{d}_i - M\mathbf{r}_o)d\xi d\eta$$

$$= |\overline{O}(\mathbf{r}_o)|^2 \cdot \exp[i\delta(\mathbf{r}_i)] \cdot \int\int_{-\infty}^{\infty} \rho_1(\mathbf{r}_o) \cdot \rho_2^*(\mathbf{r}_o + \mathbf{d}_o) \cdot h(\mathbf{r}_i - M\mathbf{r}_o)h(\mathbf{r}_i + \mathbf{d}_i - M\mathbf{r}_o)d\xi d\eta$$

$$= I_o(\mathbf{r}_i) \cdot \exp[i\delta(\mathbf{r}_i)] \cdot C_h(\mathbf{d}_i)$$

$$(2.22)$$

and for the intensity of the superposed object fields results with Equ.(2.20)-(2.22):

$$I(\mathbf{r}_i) = 2I_o(\mathbf{r}_i) + I_o(\mathbf{r}_i) \cdot [\exp(-i\delta) + \exp(i\delta)] \cdot C_h(\mathbf{d}_i) \quad (2.23)$$

Equ.(2.23) shows that the contribution of the interference term depends essentially on the *correlation* of the imaged surface roughness C_h:

$$C_h(\mathbf{r}_i) = \int\int_{-\infty}^{\infty} \rho_1(\mathbf{r}_o) \cdot \rho_2^*(\mathbf{r}_o + \mathbf{d}_i) \cdot h(\mathbf{r}_i - M\mathbf{r}_o) \cdot h^*(\mathbf{r}_i + \mathbf{d}_i - M\mathbf{r}_o) d\xi d\eta \quad (2.24)$$

Since C_h defines also the so-called *speckle size*, i.e. the correlation length of the amplitude or intensity in the image (19), the interference fringes are only visible as long as the mutual shift of the speckle patterns is smaller than the speckle size. For non vanishing average interference Equ.(24) requires that the two random phase distributions $\rho_1(r_o)$ and $\rho_2(r_o+d_i)$ are correlated, i.e., $\langle \rho_1(\mathbf{r}_o)\rho_2^*(\mathbf{r}_o + \mathbf{d}_i)\rangle \neq 0$, at least for two mutual positions r_o and (r_o+d_i). This means that the two rough surfaces have to be microscopically identical. Equ.(2.23) can be written as

$$I(\mathbf{r}) = 2I_o(\mathbf{r}) \cdot [1 + C_h(\mathbf{r}) \cdot \cos\delta(\mathbf{r})]. \quad (2.25)$$

The speckles in the image of the coherently illuminated rough surface cause substantial intensity noise. In order to emphasize this influence in accordance with the signal model, see Equ. (2.13), a *multiplicative noise factor* $R_S(r)$ is introduced:

$$I(\mathbf{r}) = 2I_0(\mathbf{r}) \cdot [1 + C_h(\mathbf{r}) \cdot \cos\delta(\mathbf{r})] \cdot (1 + R_S(\mathbf{r})) \quad (2.26)$$

The assumption of multiplicative speckle noise is correct as long as the medium intensity $I_0(r)$ is constant over the speckle correlation length (20). For objects without sharp edges and for sufficiently broad interference fringes this condition is ensured.

Other noise components which have to be considered are mainly additive ones such as the time dependent *electronic noise* $R_E(r,t)$ due to the random signal fluctuations within the electronic components of the photodetector and the image processing equipment (21):

$$I(\mathbf{r},t) = 2I_o(\mathbf{r}) \cdot [1 + C_h(\mathbf{r}) \cdot \cos\delta(\mathbf{r})] \cdot (1 + R_S(\mathbf{r})) + R_E(\mathbf{r},t) \quad (2.27)$$

However, electronic noise usually is only relevant for small signal levels and plays a less important role than speckle noise in coherent optical metrology.

A further quantity of influence that can cause considerable disturbances of the theoretically assumed cosine shaped intensity modulation of the interference fringes is the varying *background intensity* $I_o(r)$. This intensity variation is also called *shading*. It changes the peak to valley values of the intensity due to the Gaussian intensity profile of the laser beam and its imaging onto the surface, the spatially varying surface characteristics of the object under test, and parasitic interferences as e.g. macroscopic diffraction patterns caused by dust particles on lenses and mirrors. The shading is highly worthy of note with respect to the digital processing of fringe patterns, see section 3.

In comparison to the speckle noise and to the intensity shading the distortions due to the digitization and quantization of the continuous signal are not as relevant. In most practical applications a quantization into 256 grey values and a digitization into an array of 1024x1024 pixels is quite sufficient to meet the necessary requirements for a precise phase reconstruction. For phase shifting evaluation (see section 3) the resulting error because of quantization into 8 bits was estimated to be 3.59×10^{-4} wavelengths (22). Fig. 2.5 illustrates the effect of the main components that disturb the signal.

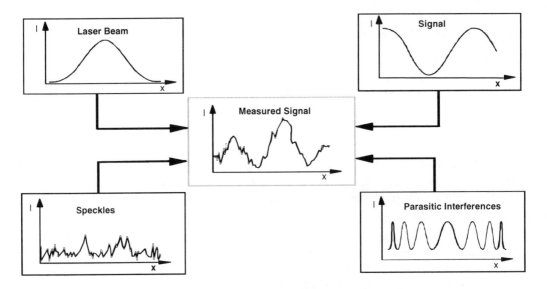

Fig. 2.5: Effect of different disturbances on the resulting signal.

2.3 Computer Simulation of Holographic Interference Patterns

The basis for *computer simulation* of fringe patterns is Equ. (2.27). For simplification of calculation and interpretation the following asumptions are made:

- Nonlinearities of the sensor are neglected. This limitation is according to the behaviour of modern CCD-cameras used in the standard interferometric systems.

- Localization phenomena of the fringes caused by decorrelation of speckle fields (see (19)) will not be taken into account. This assumption is based on the fact that the correlation can be assured by enlarging the speckle size with smaller apertures of the imaging system. This results in an increase of speckle noise, of course.
- Two coherent wavefields are assumed, which differ from each other only by a constant amplitude factor α and the phase difference $\delta(r)$.

The two interfering waves can then be expressed as:

$$A_{01}(\mathbf{r}) = a_1(\mathbf{r}) \cdot \exp[i\varphi_1(\mathbf{r})] \tag{2.28}$$

$$A_{02}(\mathbf{r}) = \alpha \cdot A_{01}(\mathbf{r}) \cdot \exp[i\delta(\mathbf{r})] \tag{2.29}$$

Since the medium intensity within the image plane corresponds with the incoherent image $I_0(r)$ the intensities of the two object waves can be written as

$$I_1(\mathbf{r}) = I_o(\mathbf{r}) \cdot [1 + R_S(\mathbf{r})] \tag{2.30}$$

$$I_2(\mathbf{r}) = \alpha^2 \cdot I_0(\mathbf{r}) \cdot [1 + R_S(\mathbf{r})] \tag{2.31}$$

For the intensity of the resulting fringe pattern follows

$$I(\mathbf{r},t) = I_0(\mathbf{r}) \cdot [1 + \alpha^2 + 2\alpha \cdot \cos\delta(\mathbf{r})] \cdot [1 + R_S(\mathbf{r})] + R_E(\mathbf{r},t) \tag{2.32}$$

With a given *fringe contrast*

$$V = \frac{I_{max} - I_{min}}{I_{max} + I_{min}} = \frac{2\alpha}{1+\alpha^2} \tag{2.33}$$

the modulation term in Equ. (2.32) can be calculated

$$\alpha = 1/V\left[1 - \sqrt{1-V^2}\right] \text{ for } \alpha \le 1. \tag{2.34}$$

Equ. (2.32) can be approximated by

$$\begin{aligned} I(\mathbf{r},t) &\approx I_0(\mathbf{r}) \cdot [1 + V\cos\delta(\mathbf{r})] \cdot [1 + R_S(\mathbf{r})] + R_E(\mathbf{r},t) \\ &\approx [I_0(\mathbf{r}) + I_1(\mathbf{r})\cos\delta(\mathbf{r})] \cdot [1 + R_S(\mathbf{r})] + R_E(\mathbf{r},t) \end{aligned} \tag{2.32a}$$

Consequently for the computer simulation of fringe patterns the following functions have to be derived:
- the phase difference distribution $\delta(r)$

- the incoherent image of the illuminated object $I_O(\mathbf{r})$ and the relation between the two intensities
- the speckle noise $R_S(\mathbf{r})$
- the electronic noise $R_E(\mathbf{r})$.

The calculation of the *phase difference* $\delta(\mathbf{r})$ starts with the modeling of the *displacement field* $\mathbf{d}(\mathbf{r})$ that the surface undergoes between the two states of double exposure (other methods may be calculated in a similar way). For this purpose analytical solutions as e.g. the bending of beams and plates or finite elements calculations can be used. Assuming an interferometer with defined observation and illumination directions $\mathbf{e}_B(\mathbf{r})$ and $\mathbf{e}_Q(\mathbf{r})$, respectively, Fig. 4, the phase difference can be calculated pointwise using the well known *basic equation* of holographic interferometry (23):

$$\delta(\mathbf{r}) = \frac{2\pi}{\lambda}\left[\mathbf{e}_B(\mathbf{r}) + \mathbf{e}_Q(\mathbf{r})\right] \cdot \mathbf{d}(\mathbf{r}) . \qquad (2.35)$$

For the approximation of the incoherent image of the object $I_O(\mathbf{r})$ the projection of the expanded laser beam onto the surface has to be calculated taking into account the geometrical and optical conditions of the illumination system.

Speckle noise can be simulated using a simplified model of the formation process of *subjective speckle patterns* (24). Speckles may be described by adding independent complex amplitudes to yield a circular Gaussian-distributed sum owing to the central limit theorem. A good approach is to follow this definition and to overlay Gaussian-distributed random complex amplitudes $A(m,n)$ at pixel locations (m,n) within the diffraction spot of the imaging system:

$$I_S(k,l) = \left|\sum_{m,n} \mathbf{h}(k-m, l-n) \cdot A(m,n)\right|^2 \qquad (2.36)$$

with

$$A(m,n) = |A(m,n)| \exp[i\varphi(m,n)] \qquad (2.37)$$

The measured intensity is denoted by I_S. The point spread function $\mathbf{h}(i,j)$ can be simulated by a rect-function. Fig. 2.6a shows the result of such a simulation process with the corresponding intensity histogram, Fig. 2.6b. This probability density function is in good agreement with the theoretically expected negative exponential distribution (18).

Digital Processing and Evaluation of Fringe Patterns

305

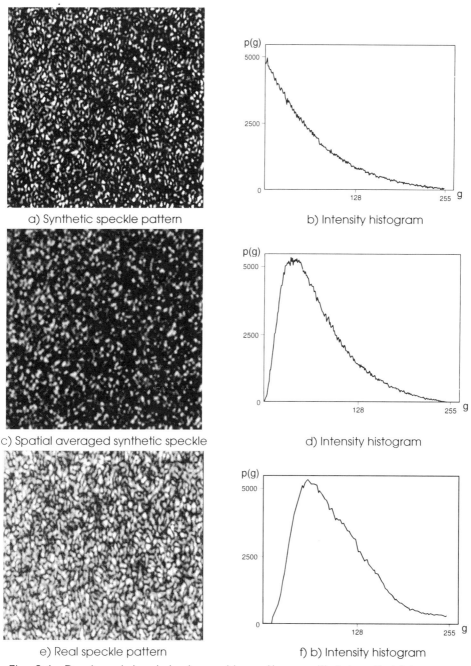

a) Synthetic speckle pattern
b) Intensity histogram
c) Spatial averaged synthetic speckle
d) Intensity histogram
e) Real speckle pattern
f) b) Intensity histogram

Fig. 2.6: Real and simulated speckle patterns with intensity histogram

In many practical situations, speckles are not resolved completely by the sensor. In this case the spatial averaging on the target - which is comparable to low pass filtering - results in a change of the statistics within the pattern. A comparison between Fig. 2.6c and Fig. 2.6e shows that a low pass filtered synthetic speckle pattern corresponds even better with a real speckle pattern captured by a CCD-camera.

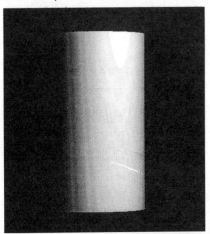

a) background illumination

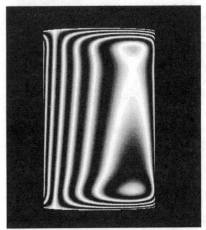

b) interferogram

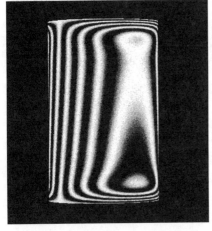

c) interferogram with electronic noise

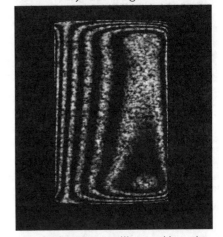

d) interferogram with speckle noise

Fig. 2.7: Simulation of the image formation process in holographic interferometry on exapmle of a cylindrical shell loaded with inner pressure

Electronic noise $R_E(r)$ in photodetectors is a sum of numerous independent random processes obeying different statistical laws, but the central limit theorem of probability theory states that the overall process will be directed by a Gaussian distribution. For the simulation of Gaussian distributed electronic noise the following procedure to transform two uniform distributed pseudo random numbers x_n and x_{n+1} into two independent pseudo random numbers y_n and y_{n+1} of a Gaussian sequence of random numbers can be applied:

$$y_n = \sqrt{-\ln(x_n)} \cdot \cos(2\pi x_{n+1}), \qquad (2.38)$$

$$y_{n+1} = \sqrt{-\ln(x_n)} \cdot \sin(2\pi x_{n+1}). \qquad (2.39)$$

The complete simulation process is demonstrated on example of a cylindrical shell loaded with inner pressure, Fig. 2.7. The displacement field on the surface was calculated with the finite element method.

3. TECHNIQUES FOR DIGITAL PHASE RECONSTRUCTION

Using modern metrological methods the absolute shape as well as the deformation of loaded technical components can be measured in a wide range by fringe evaluation. The quantity of primary interest is the phase of the fringes carrying all the necessary information. During the last 15 years several techniques for the automatic and precise reconstruction of phases from fringe patterns were developed (8), (10), (12-15). In this section attention is only paid for those basic concepts where *digital image processing* is relevant:

– *Fringe Tracking* or *Skeleton Method* (25),
– *Fourier-Transform Method* (26),
– *Carrier-Frequency Method* or *Spatial Heterodyning* (27) and
– *Phase-Sampling* or *Phase-Shifting Method* (28).

All these methods have significant advantages and disadvantages, so the decision for a certain method depends mainly on the special measuring problem and the boundary conditions. For simplification our following discussion of these methods is based on a modification of Equ. (2.27):

$$I(x,y,t) = a(x,y,t) + b(x,y) \cdot \cos[\delta(x,y) + \varphi(x,y,t)] . \qquad (3.1)$$

Here the variables $a(x,y,t)$ and $b(x,y)$ consider the additive and multiplicative disturbances, respectively, and $\varphi(x,y,t)$ is an additionally introduced *reference phase* that categorizes the different phase measuring techniques.

Although phase reconstruction by *fringe tracking* is generally time consuming and suffers from the non-trivial problem of possible ambiguities resulting from the loss of directional information in the fringe formation process, it is sometimes the only alternative for fringe evaluation. Its main advantages are that it works in almost each case and that it requires neither any additional equipment such as phase shifting devices nor additional manipulations in the interference field.

The *Fourier-transform method* (FTE) is applied to the interferogram without any manipulation during the interferometric measurement, too. The digitized intensity distribution is Fourier-transformed leading to a symmetrical frequency distribution in the spatial domain. After an unsymmetrical filtering including the regime around zero the frequency distribution is transformed by the inverse Fourier transformation resulting in a complex valued image. On the basis of this image the phase can be calculated by the arctan-function. The disadvantage of the method is the need of individually adapted filters in the spatial frequency domain.

The most accepted techniques, however, involve calculating the phase $\delta(x,y)$ at each point, either by shifting the fringes through known phase increments (*Phase-Sampling Method*) or by adding a substantial tilt to the wavefront, causing carrier fringes and Fourier-transformation of the resulting pattern (*Carrier-Frequency Method*). Both these types of *phase-measurement interferometry* (PMI) can be distinguished as *phase modulation methods*: *temporal and spatial phase modulation techniques* (10). In the first case a temporal phase modulation is used (28). This can be done by stepping the reference phase with defined phase increments and measuring the intensity in consecutive frames (*temporal phase stepping*) or by integrating the intensity while the phase shift is linearly ramped (*temporal phase shifting*). In the second case the phase is shifted spatially by adding a substantial tilt to the wavefront (*spatial heterodyning*) or by producing several spatial encoded phase shifted fringe patterns (*spatial phase stepping*) by introducing at least three parallel channels into the interferometer, which simultaneously produce separate fringe patterns with the required phase shift (30). In either case, the phase is calculated modulo 2π as the principal value. The result is a so-called *saw-tooth* or *wrapped phase image*, and *phase unwrapping* has to be carried out to remove any 2π-phase discontinuities.

With respect to the successful application of time consuming image processing algorithms to the unprocessed data or to the images that are already improved by some pre-processing, it has been proved to be useful to assess the quality of the data before they are fed to the image processing system. Some parameters that should be evaluated to test the quality of the fringes are:

- the spatial frequency distribution of the fringes over the total frame,
- the fringe contrast,
- the signal-to-noise-ratio (e.g. the speckle index (31) or the image fidelity (32)),
- the fringe continuity (analyzed e.g. with the fringe direction map, see section 3.1.2) and
- the linearity of the intensity scale (saturation effects).

Based on this evaluation the fringe pattern can be accepted for further processing or rejected. In the case of rejection it is usually more effective to prepare the fringe pattern again under improved experimental conditions as to expend disproportionate image processing effort.

The application of digital image processing methods is shown in exemples for the two types of fringe evaluation: with (temporal phase shifting or stepping) and without (fringe tracking) fringe manipulation. Because the improvement of the signal-to-noise-ratio is very essential for all methods, the image processing section starts with a description of useful pre-processing techniques. Afterwards some problems are discussed which are only relevant for the one or other of the methods such as segmentation for fringe tracking,, and phase unwrapping for phase sampling. At the end of that chapter the topical problem of absolute phase measurement is explained.

3.1 Methods for pre-processing of fringe patterns

3.1.1 Smoothing of time dependent electronic noise

The *electronic noise* in photodetectors is recognized as a random fluctuation of the measured voltage or current and is caused by the quantum nature of matter. It is a sum of numerous random processes obeying different statistical laws, but the central limit theorem of probability theory states that the overall process will be directed by a Gaussian distribution. Because electronic noise is a time-dependent process its influence on the intensity distribution can be diminished by averaging over a sequence of frames. With respect to this averaging process it is of interest to consider what happens when several Gaussian distributions are added together. This is particularly important in the area of signal averaging, where this property is used to increase the *signal-to-noise ratio* (SNR). For an intensity signal which consists of a steady signal value plus a random noise contribution with standard deviation σ, the average standard deviation after the averaging of n frames is

$$\sigma_{av} = \frac{\sigma\sqrt{n}}{n} \tag{3.2}$$

The signal-to-noise ratio after averaging of n frames, superscript (n), in contrast to the signal-to-noise ratio without averaging, superscript (1), is as follows:

$$SNR^{(n)} = \sqrt{n} \cdot SNR^{(1)} . \tag{3.3}$$

Thus, if the number of frames is increased from n to (n+m), the SNR is improved by the factor $\sqrt{(n+m)/(n)}$.

In practical applications of holographic interferometry the electronic noise plays a minor role in comparison to the speckle noise. However, in the case of *speckle correlation interferometry* the *correlogram* typically results from the subtraction of two frames. The time dependent noise amplitudes can be observed in real-time and are in the range of the signal amplitude. Without time averaging the speckle dominated intensity distribution is so noisy that sometimes only the skilled eye can observe fringes. Here special care has to be paid to electronic noise. As an example, Fig. 3.1 shows a correlogram of a centrally loaded circular disc without and after averaging of 20 frames.

Between the two images the SNR is increased by about 6 dB ($SRN[dB] = 10\log_{10}\sqrt{n}$).

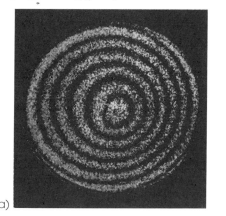 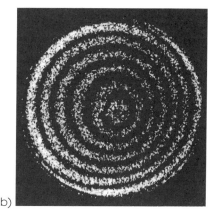

Fig. 3.1: Speckle correlogram of a centrally loaded disc: a) without and b) with averaging of 20 frames

3.1.2 Smoothing of speckle noise

Fringe patterns obtained by coherently illuminated rough surfaces are contaminated with a special kind of noise called *speckle*. Because of its role in the image formation process, see section 2.2, and its function as the carrier of the information to be measured, speckle is unavoidable in coherent optical metrology. However, with respect to the reconstruction of the continuous phase distribution from noisy fringe patterns, it is an impediment that needs to be eliminated to improve the quality of the measured intensity distribution. Since speckle is usually modelled as signal-dependent noise with a negative exponential probability distribution (18) and high contrast, the reduction of *speckle noise* needs more effort than for the suppression of other noise components. There have been two main approaches for reducing speckle noise. One approach is to prevent the creation of speckles by introducing incoherency through the sensor itself. A simple method is random spatial sampling using a rotating diffuser in front of the sensor and averaging a number of frames (33). The second approach, which will be considered here, is to process the speckled fringe patterns to reduce the effects of speckle noise. Different methods for smoothing speckles were proposed in the past. Since the speckles are noise as well as the carrier of the information there is no ideal approach that operates effectively in all cases. Consequently, in general a pragmatic approach has to be chosen.

Following SADJADI (34), three general methods can be distinguished: *temporal*, *spatial* and *geometric filtering*:

- In *temporal filtering*, multiple uncorrelated registered frames of the same scene with randomly changing speckle noise are needed. Using the intensity model described previously (see section 2.2)

$$I(x,y) = \bar{I}(x,y)\cdot\left[1+R_s(x,y)\right] \qquad (3.4)$$

it can be shown that the ensemble average of the n frames of I(x,y) is the maximum likelihood estimate of $\bar{I}(x,y)$, which corresponds to the undegraded image (35). If it is possible to receive multiple uncorrelated frames, this technique is a very effective way of improving the SNR in speckled images, and it is used for instance in high altidude synthetic aperture radar (SAR) processing. However, for optical metrology the registration of numerous uncorrelated frames is practically impossible.

- *Spatial filtering* of speckled images means that the signal processing is limited to a single frame. The standard method of spatial filtering is performeded by averaging the intensity values of adjacent pixels (linear *low-pass filtering*). This process is described by the convolution of the intensity distribution I(x,y) with the impulse response or *convolution kernel* h(x,y) of the filter:

$$I'(x,y) = I(x,y) \otimes h(x,y) = \sum_a \sum_b I(x,y)\cdot h(x-a, y-b) \qquad (3.5)$$

Two types of low-pass filters are commonly used: the *box-kernel* and the *Gaussian-kernel*, respectively:

$$h_{box} = \frac{1}{9}\begin{bmatrix}1 & 1 & 1\\1 & 1 & 1\\1 & 1 & 1\end{bmatrix} \quad , \quad h_{gauss} = \frac{1}{16}\begin{bmatrix}1 & 2 & 1\\2 & 4 & 2\\1 & 2 & 1\end{bmatrix} \quad . \qquad (3.6)$$

If such (3x3)-kernels are applied several times to the same image, filters with larger window dimensions are achieved. A very effective method is the connection of elementary (2x2)- or (3x3)-kernels to a *filter cascade* (8) with $(2^n \times 2^n)$ and $(3^n \times 3^n)$ windows, respectively. The implementation of linear low-pass filters is easy and they run very fast on general purpose processors. But they always result in image blurring, which adversely affects the following segmentation procedure negatively. In the case of fringe patterns with high fringe density a special designed mean filter shows better results (36). This filter compares for all pixels the spatial average M of the grey values in the kernel with the grey value g of the central pixel. The grey value g is only replaced by M if the following rule is fulfilled:

if $|M - g| > T$, then $g := M$, otherwise $g := g$ (3.7)

with T as threshold. This averaging process is usually repeated two or four times. The parameter T ensures that fine structures are preserved while noise is suppressed. Fig. 3.2 shows the rough and mean filtered holographic interferogram of a centrally loaded disc with the grey value profile.

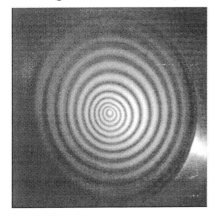

a) Rough interferogram

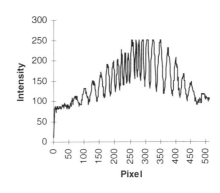

b) Intensity plot

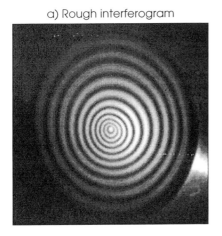

c) Mean filtered interferogram

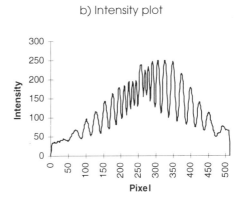

d) Intensity plot

Fig. 3.2: Smoothing of speckle noise: Application of mean filters

Investigations made by GUENTHER et al (37) have shown that *nonlinear filters* such as square root and squaring filters give no improvements over linear spatial average filters. An other nonlinear filter that is well known for reducing the so-called *salt and pepper noise* in images is the *median filter*. This filter belongs to the class of *rank filters* (38) and has been used successfully for speckle suppression. It operates by moving a rectangular window across the degraded image, and by replacing the intensity value of the central pixel with the median of the values of all the pixels in the window. The median filter is effective in removing the speckle noise and does not blur the fringes as much as the spatial average filter. The disadvantage of the median filters is the increased processing time compared with simple low-pass filters. Some more time effective implementations are described in (39), (40).

Methods of image restauration such as *Wiener filtering* (8), (41) can also be applied successfully for noise reduction, provided that some knowlege of the spectral power density of the noise and the signal is available and an addidive noise model is assumed. However, with a logarithmic transformation it is possible to convert the multiplicative model into an additive model. This technique is known as *homomorphic filtering* where

$$\log I(x,y) = \log \bar{I}(x,y) + \log R_s(x,y) \tag{3.8}$$

and has been frequently used for the removal of multiplicative noise (42), (43). Once the noise becomes additive, any standard technique for removing additive noise can be applied. JAIN and CHRISTENSEN (35) have compared low pass, median, spatial average, and Wiener filters in the logarithmic space on a set of test patterns and concluded that the Wiener filter performed slightly better than the rest of the filters. However, the investigations did not show conclusive evidence of superior performances in using homomorphic over conventional filtering.

Directional averaging or median filtering is useful to protect the fringes from blurring while smoothing. Some *directional filters* adapted for fringe patterns were proposed and applied to pre-process holographic interferograms (44), (45). The filters search for local fringe tangent directions and a so-called *fringe direction map* is constructed. This direction map contains the relevant fringe directions within the interference patterns. Based on this map the filter mask of the averaging or median filter is controlled in such a way that the pattern is only filtered in fringe direction. In Fig 3.3 a holographic interferogram with high fringe density in the central region is filtered with and without directional filter control. The fringe blurring of the directional smoothed pattern is not so strong as in the case of conventional filtering.

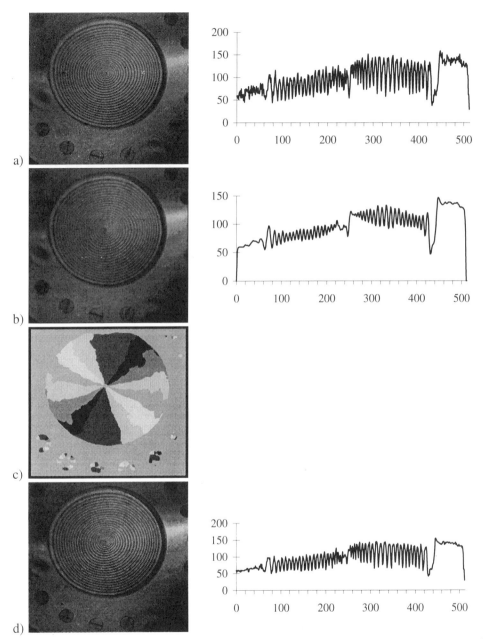

Fig. 3.3: Comparison between isotropic and anisotropic filtering of holographic interferograms a) rough holographic interferogram, b) spatially isotropically averaged interferogram, c) fringe pattern direction map, d) directionally filtered interferogram

A completely different approach for speckle filtering was proposed by CRIMMINS (31), (46). This so-called *geometric filter* is more effective in speckle noise suppression than is frame and spatial averaging, without disturbing the fringe pattern adversely (24). The image is viewed as a three-dimensional surface where speckles appear as narrow, tall towers, but the signal components to be cleaned from noise appear as relatively broad-based, short towers. The filter applies an iterative convex hull algorithm[1] where the narrow and tall towers are cut down more quickly and effectively. The result is a practical elimination of noise while the fringe pattern is preserved. To compare the performance of the geometric filter with the frequently used spatial low-pass filters, Fig. 3.4 shows the result of speckle smoothing on example of a holographic interferogram (Fig. 3.4a, b, c) and a speckle shearogram (Fig. 3.4d, e, f). The different performances of these filter types with respect to blurring effects is visible in the centre of the circular fringes. Here the geometric filter delivers the best result.

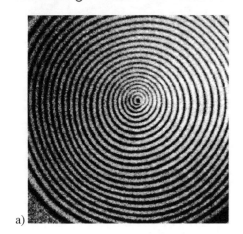

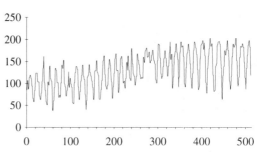

a)

---

[1] The convex hull is the smallest region which contains the object, such that any two points of the region can be connected by a straight line, all points of which belong to the region [47].

Digital Processing and Evaluation of Fringe Patterns 317

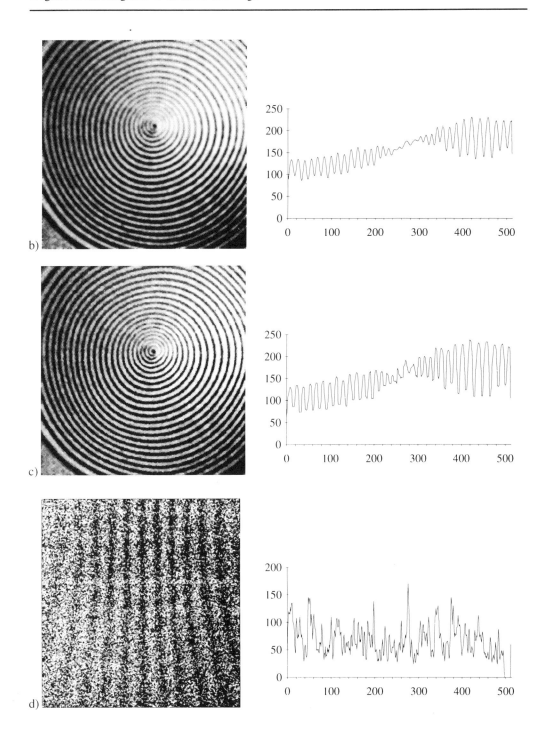

b)

c)

d)

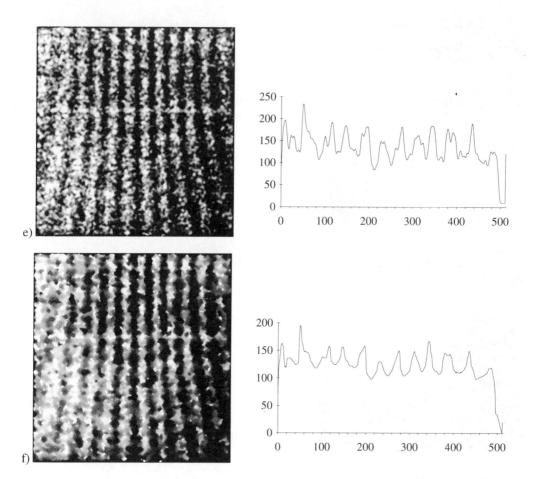

Fig 3.4: The application of low-pass and geometric filtering to a holographic interferogram and a speckle shearogram (the intensity profile is taken from a line through the middle of the image): a) rough holographic interferogram, b) local averaged interferogram, c) geometric filtered interferogram d) speckle shearogram, e) local averaged shearogram, f) geometric filtered shearogram

3.1.3 Shading correction

The *shading correction* of the fringe patterns is an important preprocessing-step in the case of intensity based analysis methods such as *fringe tracking* or *skeletonizing* (see section 3.2.1). Refering to the intensity model, Equ. (2.32a), the fringe pattern is still disturbed by local varying fringe amplitudes $I_1(x,y)$ and background modulation $I_0(x,y)$ after electronic and speckle noise filtering:

$$I(x,y) \approx I_0(x,y) \cdot [1 + V \cdot \cos\delta(x,y)] \approx I_0(x,y) + I_1(x,y) \cdot \cos\delta(x,y) \qquad (2.32b)$$

These disturbances are mainly caused by variations of the illumination due to the intensity distribution of the laser beam, inhomogenities within the intensity distribution of the illumination beam (e.g. diffraction patterns caused by dust particles on lenses or mirrors and distortions due to the projection of the intensity distribution on the object surface), varying reflectivity of the surface under test and speckle decorrelations. Fig. 3.5 conveys an impression of the influence of shading on the fringe pattern for a centered Gaussian intensity distribution.

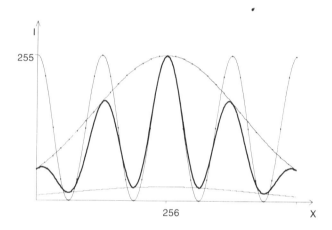

Fig. 3.5: Gaussian modulated intensity distribution with varying fringe amplitude and varying background (bold line: modulated intensity, dashed line: normalized intensity, thin lines: upper and lower hull)

Simple binarisation with fixed thresholds followed by skeletonization will not work in this case, Fig. 3.6. Furthermore it is important to notice that the variation of $I_1(x,y)$ and $I_0(x,y)$ cause a deviation of the fringe peaks from their real positions without shading. In fringe tracking or skeleton techniques where the fringe centerlines are reconstructed first, this systematic phase error has to be taken into account. It makes the skeleton lines deviate from the loci where the phase is an integer multiple of π. A careful investigation for the behaviour of the phase error is carried out in (48) and a general method for removing or reducing the error is proposed. An important conclusion of these investigations is that the systematic phase error caused by variation of $I_1(x,y)$ and $I_0(x,y)$ oscillates from zero in the middle between bright and dark fringes to its extreme values at the fringe centerlines for every fringe, and this error is not accumulated over the whole field. In other words, the systematic phase errors are relevant only to the local fringe configuration, e.g. the local fringe density, the local gradient of the intensity background, and the local fringe amplitude.

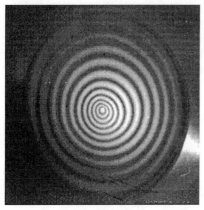

a) Interferogram b) Interferogram after binarization (level 128)

Fig. 3.6: Binarization of a shaded holographic interferogram

Several algorithms have been developed to normalize fringe patterns (8), (40), (48-51) Assuming a multiplicative superposition of the background with the fringe modulation term, the shading correction with background division is the simplest way:

$$I^N(x,y) = \frac{I(x,y)}{I_0(x,y)} \cdot 128 \qquad \text{(256 grey levels)} \qquad (3.9)$$

To reconstruct the background $I_0(x,y)$ the incoherent image, Fig. 3.7, of the object or an approximation such as the low-pass or median filtered fringe pattern can be used (39).

Under the same assumptions as before the method of *homomorphic filtering* also delivers good results. Here the Fourier-transform is applied to the logarithm of the intensity distribution and the low frequency components of the spatial frequency spectrum are filtered out using an appropriate filter function as e.g. the inverted Gaussian low-pass. However, both methods remove the systematic phase error only in the special situation where both the background and the fringe amplitude coincide.

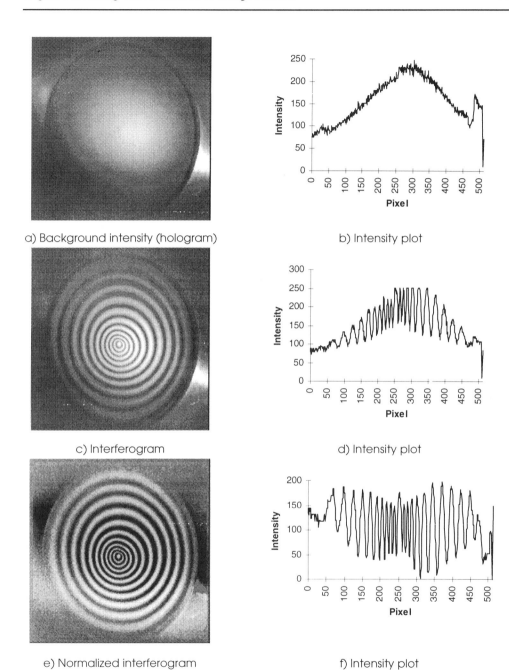

3.7: Shading correction of fringe patterns by background division

A method that approximates the upper and lower hull, see Fig. 3.5, is more effective, in general. In (50), (51) steplike envelopes I_{max} and I_{min} are constructed by connecting the local maxima and minima, respectively, of the fringe pattern.

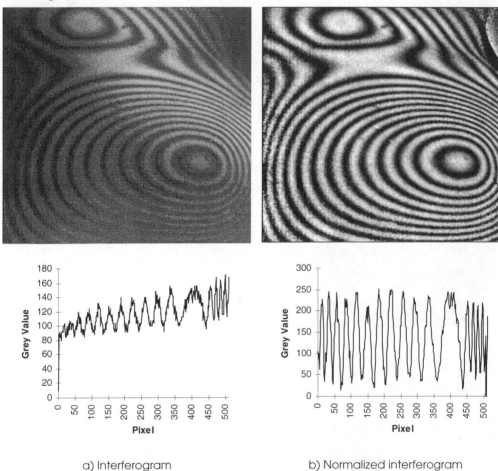

a) Interferogram b) Normalized interferogram

Fig. 3.8: 2D-envelope transformed holographic interferogram with nearly correct envelopes

With these envelopes a fringe pattern $I^N(x,y)$ with constant background I_{0c} and constant amplitude I_{1c} is derived using a *2D-envelope transform*:

$$I_{max}(x,y) = I_0(x,y) + I_1(x,y)$$
$$I_{min}(x,y) = I_0(x,y) - I_1(x,y)$$
$$I^N(x,y) = A\frac{[I(x,y) - I_{min}(x,y)]}{[I_{max}(x,y) - I_{min}(x,y)]} + B \quad\quad (3.10)$$
$$= A/2 + B + A/2 \cos\delta(x,y)$$
$$= I_{0c} + I_{1c}\cos\delta(x,y)$$

A is a constant equal to double the fringe amplitude after the transform, i.e. $A=2I_{1c}$ and B is a constant equal to minimum intensity after the transform, i.e. $B=I_{0c}-I_{1c}$. Yu et al (51) have improved this method by least squares fitting the 2D-envelopes from fringe skeletons. Using these nearly-correct envelopes for the normalization of the fringe pattern, Fig. 3.8, the systematic phase error is almost completely removed (48).

After the fringe pattern is cleaned by removing random noise and the shading correction is performed, various processing methods can be implemented and applied more easily with respect to the reconstruction of the continuous phase distribution.

3.2 Methods for automatic phase measurement

3.2.1 Fringe tracking or skeleton method

The *Fringe Tracking* or *Skeleton method* is based on the assumption that the local extrema of the measured intensity distribution correspond to the maxima and minima of a 2π-periodic function, given by Equ. (3.1). The automatic identification of these intensity extrema and the tracking of fringes is perhaps the most obvious approach to fringe pattern analysis since that method is focussed on reproducing the manual fringe counting process. Different techniques for tracking fringe extrema in two-dimensional implementations are known (25), (36), (52-55). They all are based on several pattern segmentation methods that involve fast Fourier transform, Wavelet transform, adaptive and floating thresholding, and either gradient operators or the piecewise approximation of elementary functions. Some of them are tailored to a specific type of fringe pattern and thus may require substantial modifications when applied to other types of fringe patterns. Others fail on interferograms that have sinusoidal intensity modulation or are inefficient if the shape and the frequency of the fringes vary within the pattern. An efficient segmentation algorithm (36) that extracts very rapidly line structures from interferograms is described in a special section dedicated to the segmentation problem (see sec. 3.3.1). A general processing scheme for digitally recorded and stored fringe patterns consists of the following steps (8):

- Improvement of the signal-to-noise-ratio in the fringe pattern by spatial and temporal filtering, section 3.1.
- Specification of the region of interest to be analyzed.
- Extraction of the raw skeleton by tracking of intensity extrema or pattern segmentation in combination with skeletonization (see section 3.3.1).
- Enhancement of the skeleton by linking interrupted lines, removal of artifacts and adding missing lines (this procedure can be implemented very comfortably by following some simple rules (3), (56)).
- Numbering of the fringes with corresponding order numbers (see section 3.3.2).
- Reconstruction of the continous phase distribution by interpolation between skeleton lines (56).
- Calculation of the quantity to be measured using the phase values.

An example of the fringe tracking method is presented in Fig. 3.9.

3.2.2 Fourier-transform method

The Fourier-transform method is based on fitting a linear combination of harmonic spatial functions to the measured intensity distribution $I(x,y)$ (26,27,57). The admissible spatial frequencies of these harmonic functions are defined by the user via the cutoff frequencies of a bandpass filter in the spatial frequency domain. Neglecting the time dependency and avoiding a reference phase Equ. (3.1) is transformed to

$$I(x,y) = a(x,y) + c(x,y) + c^*(x,y) \qquad (3.11)$$

with the substitution

$$c(x,y) = \frac{1}{2}b(x,y) \cdot \exp[i\delta(x,y)] \qquad (3.12)$$

Here the symbol * denotes the complex conjugation. A 2-dimensional Fourier-transformation of Equ. (3.11) gives

$$I(u,v) = A(u,v) + C(u,v) + C^*(u,v) \qquad (3.13)$$

with (u,v) being the spatial frequencies and A, C and C* the complex Fourier amplitudes.

Digital Processing and Evaluation of Fringe Patterns

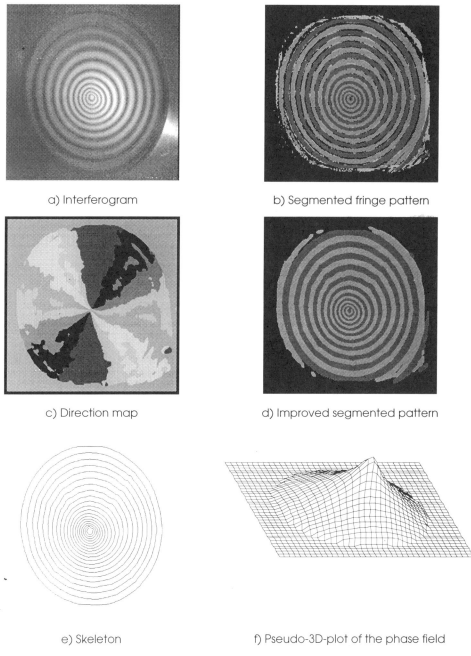

a) Interferogram

b) Segmented fringe pattern

c) Direction map

d) Improved segmented pattern

e) Skeleton

f) Pseudo-3D-plot of the phase field

Fig. 3.9: Phase reconstruction by skeletonization

Since $I(x,y)$ is a real valued function, $I(u,v)$ is a Hermitean distribution in the spatial frequency domain:

$$I(u,v) = I^*(-u,-v) \tag{3.14}$$

The real part of $I(u,v)$ is even and the imaginary part is odd. Consequently the amplitude spectrum $|I(u,v)|$ is symmetric with respect to the dc-term $I(0,0)$. Refering to Equ. (3.13) $A(u,v)$ represents this zero-peak and the low frequency components which originate from background modulation $I_0(x,y)$. $C(u,v)$ and $C^*(u,v)$ carry the same information as is evident from Equ. (3.14). Using an adapted bandpass filter the unwanted additive disturbances $a(x,y)$ can be eliminated together with the mode $C(u,v)$ or $C^*(u,v)$. If for instance only the mode $C(u,v)$ is preserved, the amplitude spectrum is no longer Hermitean and the inverse Fourier transform returns a complex-valued $c(x,y)$. The phase $\delta(x,y)$ can be calculated then with

$$\delta(x,y) = \arctan \frac{\operatorname{Im} c(x,y)}{\operatorname{Re} c(x,y)} \tag{3.15}$$

Taking into account the sign of the numerator and the denominator the principal value of the arctan-function having a continous period of 2π is reconstructed. As a result a mod 2π-*wrapped phase profile* - the so-called *saw-tooth-map* - is received and *phase unwrapping* is necessary, section 3.3.3.

The drawback of this method is that the sign ambiguity of holographic interferometry due to the cosine shaped intensity modulation remains (58):

$$\cos(\delta) = \cos(s \cdot \delta + N \cdot 2\pi) \quad s \in \{-1,1\}, \ N \text{ integer} \tag{3.16}$$

This uncertainty is preserved in Equ. (3.14). In the simple Fourier-transform method the distinction between increasing or decreasing phases requires additional knowledge about the displacement of the object. KREIS (26), (59) has proposed the application of two phase-shifted reconstructions or the twofold evaluation with differently oriented bandpass filters, to overcome the problem.

The general processing scheme for the Fourier transform method consists of the following steps:
− Fourier transformation of the rough fringe pattern
− masking the amplitude spectrum by using a bandpass filter to suppress the zero term and one part of the spectrum
− applying the inverse Fourier Transformation to return a complex-valued function and calculating the phase mod 2π (saw-tooth image)

- unwrapping the saw-tooth image to reconstruct the continuous phase distribution (see sec. 3.3.3)

The complete processing of the centrally loaded plate is shown in Fig. 3.10.

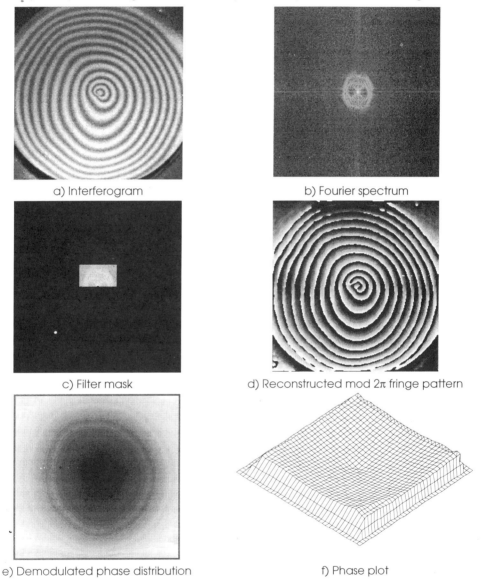

Fig. 3.10: Fourier Transform method on example of a centrally loaded circular plate

3.2.3 Carrier-frequency method

The *Carrier-Frequency method* (27), (60) is based on the introduction of a substantial tilt to the wavefront by tilding the reference wavefront using the reference mirror or by an artificial tilt introduced e.g. in the computer (61). The measurement parameter is encoded as a deviation from straightness in the fringes of the pattern. Without loss of generality it can be assumed that the carrier fringes are parallel to the y axis with a fixed spatial carrier frequency f_0. The recorded intensity distribution is given by

$$I(x,y) = a(x,y) + b(x,y) \cdot \cos[\delta(x,y) + 2\pi f_0 x]$$
$$= a(x,y) + c(x,y)\exp(2\pi i f_0) + c^*(x,y)\exp(-2\pi i f_0 x) \tag{3.17}$$

with $c(x,y)$ defined by Equ. (3.12).

Refering to Equ. (3.1) this method can be classified as a *spatial phase shifting technique*. TAKEDA et al (27) use the FFT algorithm to separate the phase $\delta(x,y)$ from the reference phase $\varphi(x,y) = 2\pi f_0 x$. To this purpose Equ. (3.17) can be rewritten with the substitution (3.12). After a one-dimensional FFT with respect to x and taking into account the Fourier shift theorem, the following holds:

$$I(u,v) = A(u,y) + C(u - f_0, y) + C^*(u + f_0, y) \tag{3.18}$$

Since the spatial variations of $a(x,y)$, $b(x,y)$ and $\delta(x,y)$ are slow compared to the spatial-carrier frequency f_0, the Fourier spectra A, C and C* are well separated by the carrier frequency f_0, as shown on example of a simulated interferogram, Fig. 3.11.

C and C* are placed symmetrically to the dc-term and centered around $u = f_0$ and $u = -f_0$. The following procedure makes use of either of the two spectra on the carrier. By means of digital filtering e.g the sideband $C(u-f_0, y)$ is filtered and translated by f_0 towards the origin of the frequency axis, to remove the carrier frequency. C* and the term $A(u,y)$ are eliminated by bandpass filters. Consequently $C(u,y)$ is obtained, as shown in Fig 3.11d. The inverse FFT of $C(u,y)$ returns $c(x,y)$ and the phase can be calculated mod 2π using Equ. (3.15). Because the phase is wrapped into the range from $-\pi$ to π it has to be corrected by using a phase-unwrapping algorithm (see sect. 3.3.3).

The general processing scheme for the Fourier transform method consists of the following steps:
- Fourier transformation of the rough fringe pattern
- masking the amplitude spectrum by using a bandpass filter to suppress the dc-term and one part of the spectrum
- shifting the relevant part of the spectrum towards the origin of the frequency axis

- applying the inverse Fourier Transformation to return a complex-valued function and calculating the phase mod 2π (saw-tooth image)
- unwrapping the saw-tooth image to reconstruct the continuous phase distribution (see sec. 3.3.3)

The complete processing of a practical interferogram (centrally loaded circular plate) is shown in Fig. 3.12.

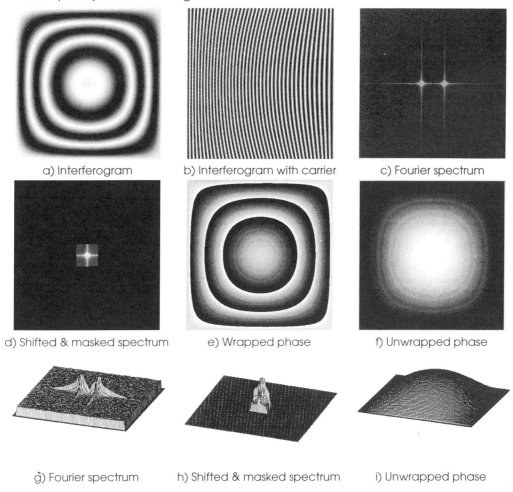

a) Interferogram b) Interferogram with carrier c) Fourier spectrum

d) Shifted & masked spectrum e) Wrapped phase f) Unwrapped phase

g) Fourier spectrum h) Shifted & masked spectrum i) Unwrapped phase

Fig. 3.11: Carrier frequency method applied to a synthetic interferogram

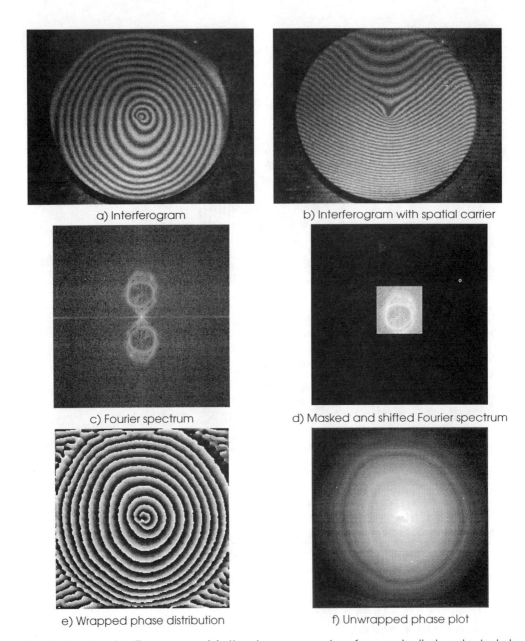

Fig. 3.12: Carrier Frequency Method on example of a centrally loaded plate

A way of processing which is equivalent to the frequency domain processing (*Fourier transform method*) can be performed in the space signal domain.

From Equ. (3.18), the spectrum passed by the filter function $H(u-f_0, y)$ can be written as

$$C(u-f_0,y) = H(u-f_0,y) \cdot I(u,y). \tag{3.19}$$

Taking the inverse Fourier transform of Equ.(3.19), the equivalent processing in space domain is represented by

$$c(x,y)\exp(2\pi if_0 x) = [h(x,y)\exp(2\pi if_0 x)] \otimes I(x,y) \tag{3.20}$$

where $\otimes$ denotes the convolution operation, and $h(x,y)$ is the impulse response defined by the inverse Fourier transform of $H(u,y)$. The sinosoid convolution of Equ. (3.20) was proposed by MERTZ (62), and its relation to the frequency domain analysis was described by WOMACK (61). Sinusoid fitting techniques in which the interferogram is approximated by a pure sinusoid are used by MACY (63) and later improved by RANSOM and KOKAL (64).

3.2.4 Phase-sampling method

Phase Sampling or *Phase Shifting interferometry* is based on the reconstruction of the phase $\delta(x,y)$ by sampling a number of fringe patterns differing from each other by various values of a discrete phase φ. If φ is shifted, for instance temporally in n steps of φ_0, then n intensity values $I_n(x,y)$ are measured for each point in the fringe pattern:

$$I_n(x,y) = a(x,y) + b(x,y) \cdot \cos[\delta(x,y) + \varphi_n] \tag{3.21}$$

with $\quad \varphi_n = (n-1)\varphi_0, \; n = 1,\ldots,m, \; m \geq 3$

and e.g. $\quad \varphi_0 = 2\pi/m$.

In general only three intensity measurements are required to calculate the three unknown components in Equ. (3.1): $a(x,y)$, $b(x,y)$ and $\delta(x,y)$. However, with m>3 a better accuracy can be ensured using a least squares fitting technique (28), (65), (66). If the reference phase values φ_n are equidistant distributed over one or a number of periods, the orthogonality relations of the trigonometric functions provide a useful simplification. Equ. (3.1) is first rewritten in the form

$$I(x,y) = K(x,y) + L(x,y) \cdot \cos\varphi + M(x,y) \cdot \sin\varphi \tag{3.22}$$

where

$$K(x,y) = a(x,y)$$
$$L(x,y) = b(x,y) \cdot \cos\delta(x,y) \tag{3.23}$$
$$M(x,y) = -b(x,y) \cdot \sin\delta(x,y)$$

It can be shown on the basis of a least square fit that L and M satisfy the following equations in analytical form (8):

$$L = \frac{2}{m}\sum_{n=1}^{m} I_n(x,y) \cdot \cos\varphi_n$$

$$M = \frac{2}{m}\sum_{n=1}^{m} I_n(x,y) \cdot \sin\varphi_n$$

(3.24)

A combination of Equ. (3.23) and (3.24) delivers the basic equation for the phase sampling method, where the minus sign is neglected because of the ambiguity of the sign in interferometry,

$$\delta'(x,y) = \arctan\frac{M}{L} = \arctan\frac{\sum_{n=1}^{m} I_n(x,y) \cdot \sin\varphi_n}{\sum_{n=1}^{m} I_n(x,y) \cdot \cos\varphi_n}$$

(3.25)

Equ. (3.25) is sufficient to determine the phase modulo π. In order to compute the phase modulo 2π, the sign of the numerator and denominator must be examined. As result a mod 2π wrapped phase distribution $\delta'(x,y)$ is measured. The *unwrapping* or *demodulation* of this wrapped signal delivers the continuous phase field

$$\delta(x,y) = \delta'(x,y) + N(x,y) \cdot 2\pi$$

(3.26)

with N as the integer fringe number. In the same way expressions for the variables $a(x,y)$ and $b(x,y)$, which give indications for the *background intensity* $I_0(x,y) \approx a(x,y)$ and the *fringe contrast* $V(x,y) \approx b(x,y)/a(x,y)$ can be derived:

$$a(x,y) = \frac{1}{m}\sum_{n=1}^{m} I_n(x,y)$$

$$b(x,y) = \sqrt{L^2 + M^2}$$

(3.27)

m	φ_0	δ'	I_0	V
3	$2\pi/3$	$\delta' = \arctan\sqrt{3}\dfrac{I_3 - I_2}{2I_1 - I_2 - I_3}$	$I_0 \approx \frac{1}{3}\sum_{i=1}^{3} I_i$	$V \approx \dfrac{\sqrt{\frac{1}{3}(2I_1 - I_2 - I_3)^2 - (I_2 - I_4)^2}}{\sqrt{3}I_0}$
4	$\pi/2$	$\delta' = \arctan\dfrac{I - I_4}{I_3 - I_1}$	$I_0 \approx \frac{1}{4}\sum_{i=1}^{4} I_i$	$V \approx \dfrac{\sqrt{(I_1 - I_3)^2 - (I_2 - I_4)^2}}{2I_0}$

Table 1: Solutions of Equ. (64) for the cases m=3 and m=4

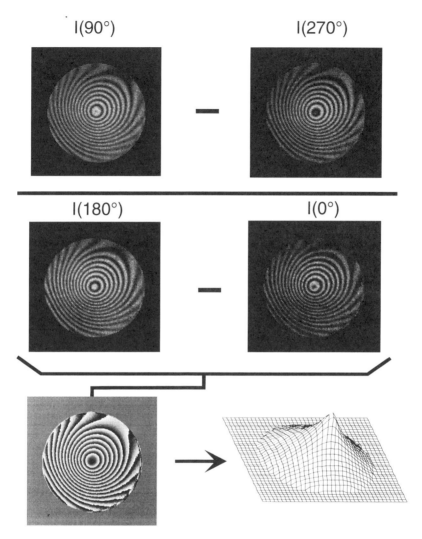

Fig. 3.13: Processing scheme in phase shifting interferometry with 4 phase shifted fringe patterns ($\varphi=90°$)

There are many of solutions for Equ. (3.25) (29), (67). For m=3 and m=4 two well known solutions are given in Tab. 1.

An example with m=4 is shown in Fig. 3.13. Another *three frame technique* for a $\pi/2$ phase shift was published by WYANT (68):

$$m = 3: \quad \varphi_0 = \pi/2, \quad \delta' = \arctan\frac{I_3 - I_2}{I_1 - I_2} \tag{3.28}$$

The general processing scheme consists of the following steps:
- Pre-Processing:
  - Specification of the region of interest to be analyzed and improvement of the signal-to-noise-ratio by spatial filtering
- Sampling
  - Calculation of the mod2π phase image according to the equations given in Tab. 1 or Equ. (3.28), (69), (70)
- Improvement
  - Improving of the saw-tooth image by using special filters and masking the relevant regions to be unwrapped using the inconsistency check (see sec. 3.3.3)
- Unwrapping
  - Unwrapping the improved saw-tooth image to reconstruct the continous phase distribution (see sec. 3.3.3)
- Representation
  - Plotting the continuous phase map as pseudo-3D-plot

The solutions above require the calibration of the phase-shifter device to ensure a defined amount of phase shift. A technique already presented by CARRÉ (69) in 1966 and improved by JÜPTNER et al (70) is independent of the amount of phase shift. The solution is based on the fundamental Equ. (3.21) for the intensity distribution with an unknown value of φ_0. In this case at least four interferograms are needed to solve the equation system for the four unknown quantities. The main variables of interest are $\varphi_0(x,y)$ and $\delta(x,y)$:

$$\varphi_0 = \arccos \frac{I_1 - I_2 + I_3 - I_4}{2[I_2 - I_3]} \qquad (3.29)$$

$$\delta = \arctan \frac{I_1 - 2I_2 + I_3 + [I_1 - I_3]\cos\varphi_0 + 2[I_2 - I_1]\cos^2\varphi_0}{\sqrt{1 - \cos^2\varphi_0}\,[I_1 - I_3 + 2(I_2 - I_1)\cos\varphi_0]} \qquad (3.30)$$

The additional phase shift $\varphi_0(x,y)$ is calculated as a function of the point $P(x,y)$. This allows control over the phase shifter as well as the reliability of the evaluation, which might be disturbed by noise.

A detailed desription of the error sources in PMI taking into account such influences as inaccuracies of the reference phase values, disturbances due to extraneous fringes, coherent noise, and high spatial frequency noise caused by dust particles was made by SCHWIDER et al. (71). Based on this work HARIHARAN et al (72) published another *five-frame technique* which uses $\pi/2$ phase shifts to minimize phase shifter calibration errors:

m = 5: $\varphi = -\pi, -\pi/2, 0, \pi/2, \pi$

$$\delta' = \arctan \frac{2(I_2 - I_4)}{2I_3 - I_5 - I_1} \tag{3.31}$$

Temporal phase shifting is another method of introducing an additional, known phase change: In that case the intensity is integrated while the phase shift is linearly ramped between $(\varphi_n - \Delta\varphi/2)$ and $(\varphi_n + \Delta\varphi/2)$. One frame of integrated recorded intensity data can be written as (73)

$$I_n(x,y) = \frac{1}{\Delta\varphi} \int_{\varphi_n - \Delta\varphi/2}^{\varphi_n + \Delta\varphi/2} \{a(x,y) + b(x,y) \cdot \cos[\delta(x,y) + \varphi_n(t)]\} d\varphi(t) \tag{3.32}$$

where φ_n denotes the average value of the phase shift for the n-th intensity measurement. The evaluation of that integral gives

$$I_n(x,y) = a(x,y) + \text{sinc}(\Delta\varphi/2) \cdot b(x,y) \cdot \cos[\delta(x,y) + \varphi_n] \tag{3.33}$$

Equ. (3.33) shows that the only difference between *temporally phase stepping* and *shifting* is a reduction in the fringe visibility by the factor $\text{sinc}(\Delta\varphi_n/2)$. With respect to the evaluation both methods are equivalent. A general survey of phase shifting methods combined with extensive simulations concerning the influence and compensation of disturbances was given by CREATH (74,75).

3.3 Methods for post-processing of fringe patterns

The objective of *post-processing* of fringe patterns is the reconstruction of the continuous phase distribution using intensities or grey values taken from the pre-processed images. In the following sections those post-processing problems are discussed that are significant for the *skeleton* and the *phase-sampling method*, respectively. That means *segmentation* and *fringe numbering* for the skeleton method and *phase-unwrapping* for the phase-sampling method.

3.3.1 Segmentation of fringe patterns

Segmentation is an important processing step in the skeleton method, sect. 3.2.1. Here the fringe pattern is considered as an array of two-dimensional contour fringes with fixed contour intervals. In cosine-shaped fringe patterns the phase increment between two consecutive contour lines is π. They are represented by the center lines of dark and bright fringes. To extract these lines in the form of a fringe pattern skeleton the image has to be segmented into regions with distinguishable properties - the so-called ridges, slopes and valleys in the grey value „mountain", Fig. 3.14b. A direct segmentation approach is fringe pattern binarisation and then skeletonization of the binary fringes (54). However, for most real fringe patterns it is difficult to determine a global threshold level that delivers good results, see Fig. 3.6. Consequently the intensity distribution must be normalized by shading correction or adaptive and local thresholds (76). Another segmentation method that can be applied successfully in most of cases was developed by Yu et al (77). This so called 2D-derivative-sign-binary fringe method is based on the *fringe direction map* and operates without thresholds.

In the case of fringe patterns with distinct varying background and fringe amplitude an algorithm that considers relevant fringe structures was applied successfully by EICHHORN et al (36). For each pixel (i,j) the grey value difference $\Delta^R(i,j)$

$$\Delta^R(i,j) = [g(i,j) - g(i+R, j+R)], \quad g(i,j): \text{ grey value of the pixel } (i,j) \quad (3.34)$$

between that pixel and the eight neighbouring ones that are centro-symmetrically located around (i,j) at a distance R is calculated, Fig. 3.15. These differences are compared with a pre-defined threshold T. To each such neighbourhood relation, a neighbourhood relation byte is assigned separately to possible ridge and valley candidates using the following rule:

If the $\Delta^R(i,j)$ is positive and greater than T, then the bit-position of the assigned neighbour point in the ridge candidate relation-byte is set to 1; if $\Delta^R(i,j)$ is negative but absolutely greater than T, then the bit-position of the assigned neighbour point in the valley-candidate relation-byte is set to 1. 256 codes of neighbourhood relations can be derived. Owing to the isotropic character of an interferogram, these 256 configurations can be reduced to 36 rotational invariant patterns. Based on experimental investigations 10 significant patterns were selected from these 36 prototypes and proved to be representative for ridge and valley points, respectively. The possibility to choose various distances R between the central point and its neighbouring points allows the use of this method for a relatively wide range of different fringe patterns. To avoid gaps between neighboring pixels on the discrete plane the same procedure is repeated after rotation around a fitting angle α. In the case of interferograms with varying fringe density the results after the application of different distances R can be merged.

Fig. 3.14a shows an interferogram with varying fringe density and background and its pseudo-3D intensity profile (Fig. 3.14b). The segmented image that was improved with fringe direction controled region growing, Fig. 3.14c, and binary filtering is given in Fig. 3.14d. The skeleton (Fig. 3.14e) is derived by line thinning and the continuous phase distribution (Fig. 3.14f) is computed finally by interpolation between the numbered skeleton lines.

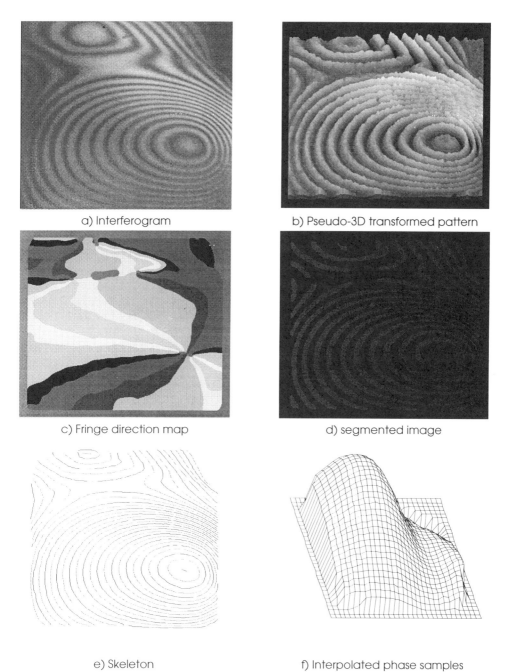

a) Interferogram b) Pseudo-3D transformed pattern
c) Fringe direction map d) segmented image
e) Skeleton f) Interpolated phase samples
Fig. 3.14: Segmented and skeletonized holographic fringe pattern

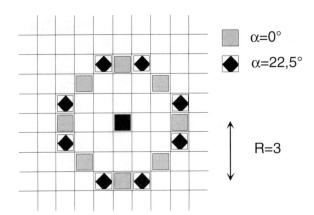

Fig. 3.15: Comparison of the grey values between the cental and the neighbouring pixel, Equ. (3.34)

3.3.2 The numbering of fringe patterns

3.3.2.1 The fringe counting problem

In general the fringes to be analyzed in optical shape and displacement analysis have a sinusoidal variation in intensity which is related to the quantity to be measured such as the phase of the intensity distribution projected on the surface and the phase difference of the interfering wavefronts representing different states of the object under test, respectively. As a good approximation Equ. (3.1) can be used for the description of the intensity distribution $I(x,y)$ in the image as a function of the phase angle $\delta(x,y)$. The objective in quantitative fringe processing is to extract the phase distribution $\delta(x,y)$ from the more or less disturbed intensity distribution $I(x,y)$. Constant values of $\delta(x,y)$ define fringe loci of the object surface and can be extracted with intensity based analysis methods as for instance skeletonization or temporal and spatial phase measurement methods (see section 3.2).

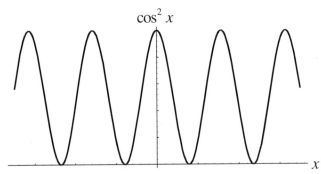

Fig. 3.16: Cosinusoidal intensity modulation function

However, the reconstruction of the phase $\delta(x,y)$ from the intensity distribution $I(x,y)$ by a kind of inversion of Equ.(40) raises the problem that the cosine is not a one-to-one function, but is even and periodic, Fig. 3.16:

$$\cos(\delta) = \cos(s \cdot \delta + N \cdot 2\pi), \quad s \in [-1,1], \quad N \in Z \qquad (3.35)$$

Consequently the phase distribution determined from a single intensity distribution remains indefinite to an additive integer multiple of 2π and to the sign s. Each inversion of expressions like Equ. (3.35) contains an inverse trigonometric function. However, all inverse trigonometric functions are expressed by the arctan-function and this function has its principal value between $-\pi/2$ and $+\pi/2$. If the argument of the arctan-function is assembled by a quotient as in the case of phase measuring interferometry, where the numerator characterizes the sine of the argument δ and the denominator corresponds to the cosine of δ, then the principal value is determined consistently in the interval $(-\pi, +\pi)$. But a modulo 2π-uncertainty as well as the sign ambiguity still remains and this is called the *fringe counting problem*.

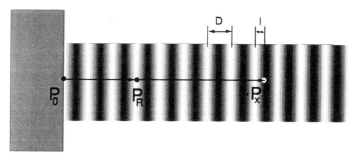

Fig. 3.17: Determination of the absolute phase in the point under investigation

This uncertainty has important practical consequences (78):
When the phase angle δ extends beyond 2π to $N2\pi$ - where N denotes the fringe number of the point P_x to be measured - the absolute phase value $N2\pi$ can only be reconstructed if N can be determined. Using intensity based methods this is performed by counting the bright and dark fringes or their skeleton lines along a path starting from a reference point P_R with known fringe order N_R until the point P_x, Fig. 3.17. The resulting relative fringe order between P_R and P_x consists of an integer part $\tilde{N}$ and a remaining fraction $\hat{N} = l/D$ with the quantity D as the line spacing between two bright fringes. For simplification P_R is often chosen to be placed in the middle of a bright fringe. In that case N_R is an integer and considers the number of fringes between the reference point P_R and a point P_0 with zero fringe order (such a point must not

be given in the image). Consequently for the absolute phase of the point P_x can be written

$$\delta(P_x) = N \cdot 2\pi = [N_R + \tilde{N} + \hat{N}] \cdot 2\pi \qquad (3.36)$$

If spatial or temporal phase measurement methods are used the remaining fraction $\hat{N}$ can be calculated from the phase δ by solving an equation of the form:

$$\hat{N} = \frac{1}{2\pi}\delta'(P_x) \qquad (3.37)$$

where δ' is the result of an evaluation according to Equ. (3.25). However, this method give only a direct approach to the principal value of the phase as well. This principal value $\delta'(P_x)$ corresponds to the fraction of the absolute fringe order $\hat{N}$:

$$\delta'(P_x) = \hat{N} \cdot 2\pi \qquad (3.38)$$

Consequently in practice the fringe counting problem remains since the integer part of N has to be provided by methods known as phase unwrapping (see section 3.3.2.3). ROBINSON (79) has given an excellent definition of this procedure: "Phase unwrapping is the process by which the absolute value of the phase angle of a continous function that extends over a range of more than 2π (relative to a predefined starting point) is recovered. This absolute value is lost when the phase term is wrapped upon itself with a repeat distance of 2π due to the fundamental sinusoidal nature of the wave functios used in the measurement of physical properties." The removal of all 2π-phase discontinuities to obtain the unwrapped result

$$\delta(x,y) = \delta'(x,y) + [N_R(x,y) + \tilde{N}(x,y)] \cdot 2\pi \qquad (3.39)$$

leads to the major difficulties and limitations of the phase sampling procedure. The key to phase unwrapping is the reliable detection of the 2π phase jumps (see section 3.3.2.3). An important condition is that neihgbouring phase samples satisfy the relation

$$-\pi \leq \Delta_i \delta(i,j) < \pi \text{ with } \Delta_i \delta(i,j) = \delta(i,j) - (i-1,j) \qquad (3.40)$$

If the object to be analysed is not simply connected, i.e. isolated objects or shaded regions occur, then the mod 2π unwrapping procedure fails because of violated neighbourhood conditions and unknown phase jumps greater than π. These difficulties can be explained clearly on example of the projected fringe technique using a light source with a transparency to project a fringe pattern on the surface of the object and a CCD-detector to observe the varying intensity across the object, Fig. 3.18. A relative simple shaped object with a rectangular solid on a flat body is taken to show the problems which are typical for the evaluation of more complex components (e.g. shaded regions, unresolved fringes, phase jumps greater than π). The regions I and V on the flat body as well as the higher area III are normally illuminated and can be observed without problems. However, region I and III are separated by a shaded area II. On the other side of the rectangular solid the regions III and V are simply connected. However, due by the steep edges, the fringes projected on the suface IV are so narrow that they cannot be resolved by the detector. The observed fringe map is shown in Fig. 3.18b. Consequently, the only advantages of PMI-methods in comparison with intensity based methods are the easier way to automate the processing scheme, the higher accuracy of the results, and the on-line recognition of points where the monotony of the phase angle is changing. Both methods, however, suffer from the fringe counting problem.

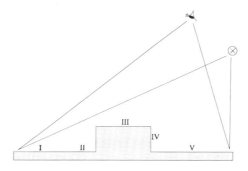

a) Optical setup in fringe projection

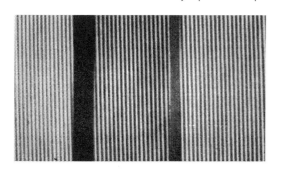 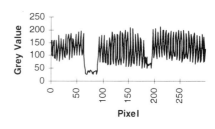

b) Observed fringe pattern   b) Intensity profile along a line

Fig. 3.18: Outline of the optical set-up and classification of different regions for the evaluation of projected fringes

3.3.2.2 Methods of fringe numbering

The phase reconstruction from its *contour lines* such as skeletons is a non-trivial problem due to the ambiguities discussed above and resulting in the loss of directional information. Thus it is impossible to determine the slope of the phase at a given point without additional knowledge. The information that can be derived from a single interference pattern may be explained by the comparison with contour lines in a geographical map: the skeleton lines can be considered as contour lines of a phase surface, with the bright fringes corresponding to integer multiples of 2π ($N \cdot 2\pi$) and the dark ones to intermediate values ($(2N+1) \cdot \pi$). By analyzing the properties of such contour lines, some important features can be pointed out:
- two fringes never cross each other,
- two adjacent lines of same gray level represent the same phase value,
- two adjacent lines of different gray level represent a phase difference of $\pm\pi$,
- a line can nowhere end except at the border of the area,
- the numbering along closed pathes delivers the same value at the starting and final point.

However, these properties are insufficient to ensure a unique numbering, especially concerning the problem of non-monotonic changes. Due to the periodicity and the evenness of the cosine function, Equ. (3.1), it is impossible to determine whether the phase difference between two adjacent fringes is $+\pi$ or $-\pi$. The *interference phase* δ is only modulo 2π definite. Consequently a priori knowledge is necessary.

Additional information concerning the sign of the phase gradient (increasing/decreasing) can be provided by some real time manipulations in the fringe pattern. An artificial phase shift results for instance in a movement of the fringes. If the moving direction of each fringe is considered the fringe numbering can be performed automatically (80). In some circumstances, this problem can be bypassed by introducing a substantial tilt to the interferogram (81). The result is that the *fringe number N* increases monotonically from one fringe to the next and the fringe numbering can be carried out automatically. However, in general the deformation or the shape of the object to be investigated is more complex and changes in the sign of the phase gradient have to be considered, Fig. 3.14a.

The common way to reconstruct the phase using intensity-based methods is to start from the skeletonized fringe pattern and to number the fringes manually with their relative fringe order. This way of numbering can be performed on a simple basis. The operator draws a line across several fringes and the system assigns stepwise an increasing or decreasing number to each fringe according to the presumed sign of the phase gradient. By repeating this operation over several clusters of fringes in the image, a consistent numbering using the operators's knowledge of the monotony.

Another *graph-based method* that computes automatically a consistent numbering of the skeleton lines without a-priori knowledge is proposed in (82). To model the problem, a graph of adjacency is built from the skeleton. A *graph* must be understood here as a structure made of nodes linked by arcs. An arc represents a relationship between two nodes. In the case of fringe patterns, each fringe in the image is associated to a node in the graph, and two nodes are linked together by an arc if and only if both corresponding lines are adjacent. The efficiency of the graph based method is demonstrated by the evaluation of an interferogram with a saddle shown in Fig. 3.19.

The original interferogram, Fig. 3.19a, has been skeletonized, Fig. 3.19b, and the graph was calculated automatically, Fig. 3.19c. By means of this graph the phase distribution could be reconstructed, Fig. 3.19d. To build the graph, a region-filling algorithm is used. It seems easier and more robust than trying to automatize the manual line drawing method. However, the main drawback of using region-filling is that perfectly separated regions are necessary, i.e. closed fringes or fringes ending only at the border of the image.

Real interferograms do not always ensure this. Two kinds of problems have to be considered: discontinuities in the skeleton and border effects (that prevent the fringes from reaching the border of the picture). The first problem can be solved by improving the skeleton (drawing a line or an arc) to make the fringes continuous. The second problem can be managed either by prolonging the fringes to the border, or taking a region of interest that excludes the bordering zone of the skeleton. Once the graph is built, it must be analyzed. Only a limited number of graph configurations are possible. The numbering itself is achieved by enforcing a few rules:

- Two adjacent nodes of same gray level will have the same number
- Two adjacent nodes of differing gray level will have a phase difference $\pm\pi$
- Any non ramified path, i.e a path without branches, made of alternately bright and dark nodes will have a monotonous numbering.
- Two adjacent nodes of same gray level on a non ramified path indicate a change sign of phase gradient.

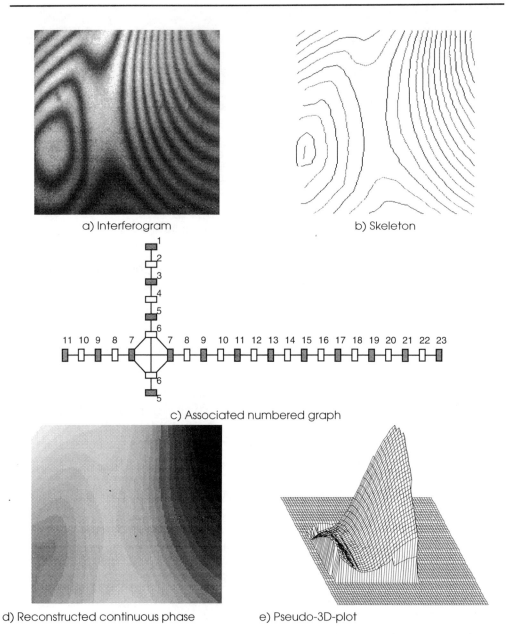

Fig. 3.19: Example of graph based fringe numbering

These rules can easily be applied on a non ramified path. The only extra parameters that are needed to completely determine the solution are the *absolute phase* and the counting direction at one starting point. The problem

gets more difficult when several branches in the graph occur. Then the absolute phase at one node of the graph, and the sign of the phase gradient at one node per branch is necessary. The algorithm does not take into account any external parameter and will assign an arbitrary sign of phase gradient to each branch. The whole numbering process includes the following steps:
- Select a starting node (if possible, choose a node at the end of a branch)
- Choose a starting number (e.g, N=0), and a starting sign of phase gradient.
- Browse the whole graph (for each node, treat all the connected untreated nodes first), and enforce the previously given rules.

Two ways of numbering are possible: the algorithm may either proceed automatically to a consistent but arbitrary result or the algorithm stops at ambiguous points and the operator can decide on the correct sign of phase gradient. In both cases a consistent numbering is delivered.

3.3.3 Unwrapping of mod2π-phase disributions

The key to phase unwrapping is the reliable detection of the 2π phase jumps. For noise-free wrapped phase distributions a simple algorithm as illustrated in Fig. 3.20 can be implemented:
- scan line by line through the image
- detect the pixels where the phase jumps and consider the direction of the jump
- integrate the phase by adding or subtracting 2π at these pixels.

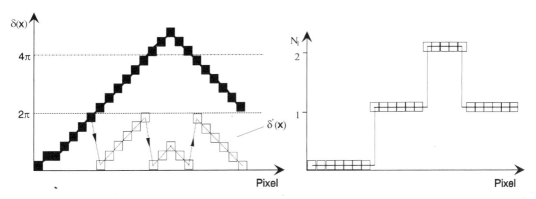

Fig. 3.20: 1D-phase unwrapping scheme

An analysis of the 2-D unwrapping algorithm was provided by GHIGLIA et al (83). According to that the processing scheme for unwrapping the sequence of principal values consists of a cascade of three operations: *differencing, thresholding* and *integrating*. On condition that neighbouring phase samples satisfy the relations

$$-\pi \leq \Delta_i \delta(i,j) < \pi \quad \text{with} \quad \Delta_i \delta(i,j) = \delta(i,j) - (i-1,j) \tag{3.41}$$

$$-\pi \leq \Delta_j \delta(i,j) < \pi \quad \text{with} \quad \Delta_j \delta(i,j) = \delta(i,j) - \delta(i,j-1) \tag{3.42}$$

over the 2-D array both ways of unwrapping along colums or lines and any other combination, as for instance the unwrapping along a spiral starting from a certain point within the pixel matrix, yield identical results. Thus, the process of unwrapping is path independent. Otherwise inconsistent values exist in the wrapped phase field. The checking of wrapped phase distributions for such possible *inconsistencies* is an approved mean for the identification of „pathological" areas and the selection of suitable unwrapping pathes. Unfortunately, in practice the conditions (3.41) and (3.42) are not always satisfied due to the presence of noise, the violation of the sampling theorem and the influence of the object shape as well as the effect of its deformation. These phenomena are discussed briefly now.

For a physically correct unwrapping it is necessary to distinguish between true mod 2π-discontinuities and apparent ones caused by noise or aliasing which should be corrected, or by the object as for instance gaps, boundaries, shadows and non-continuous deformations (e.g. cracks) which should be evaluated. Various strategies are proposed to avoid unwrapping errors in the phase map but until now there has been no general approach to avoid all types of error without user interaction - especially if objects with complex shape undergo non-continuous deformations. The main error sources in phase unwrapping can be classified into four classes: *noise, undersampling, object discontinuities* and *non-continuous phase fields*. Some simulations of the influence of these types of error are given in (84). *Electronic* and *speckle noise* are the most familiar error sources in automatic evaluation of interference patterns. In distinction to other evaluation techniques which operate with only one interferogram (e.g. fringe tracking) the stability of the intensity distribution during the reconstruction is essential for phase sampling methods. That means, the noise configuration must be identical for all considered frames. Consequently, a real-time environment with interferometric stability, in contrast to photographic stability for conventional double-exposure techniques, has to be guaranteed during the reconstruction of the m ($m \geq 3$) interferograms. In most practical situations, this condition is not fulfilled because of the time-dependent character of electronic noise, speckle displacements and so on. Hence, computed principal phases values are corrupted by noise to such an extent that locally inconsistent regions are caused. Following the statistics of speckle fields it is obvious that dark speckles dominate in the image. At a point where the intensity is zero, the wave-front contains a singularity, Fig. 3.21, and the phase is indeterminate. Around the singularity the phase forms a corkscrew (85). In consequence such pixels cause phase dislocations and inconsistencies (86).

A simple method to detect inconsistent regions in the saw-tooth image is given by computing the sum of the wrapped phase differences for all 2x2-pixel windows along the path depicted in Fig. 3.22 (87). In the phase map black pixels represent $-\pi$ and bright π. The difference between two adjacent pixels, (i,j) and $(i,j+1)$ can be calculated as

$$\Delta(i,j) = \{[\delta(i,j) - \delta(i,j+1)]/2\pi\} \qquad (3.43)$$

where { } denotes rounding to the nearest integer. Possible values of $\Delta(i,j)$ are -1, 0 or 1. The sum of all differences along the closed path in the 2x2-pixel window

$$S = \sum_{i=1}^{4} \Delta(i,j) \qquad (3.44)$$

gives an indication for the consistency of the wrapped phase in that region. All points within the window are consistent if the sum is equal zero. Otherwise the points are labeled as inconsistent.

 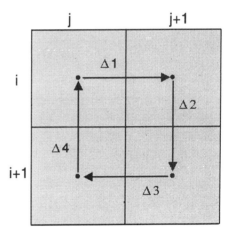

Fig. 3.21: Phase map of a speckle field with singularities

Fig. 3.22: Inconsistency check in a 2x2-window

HUNTLEY (87) introduced the name „dipoles" for those points where the path integral remains nonzero. These dipoles indicate errors in the derived phase map, either a 2π edge detection where there is no edge or a failure in edge detection where there is an edge. Data in regions well away from these points are influenced by the errors if they are not considered. Beginning from the inconsistent regions, obvious streaks are drawn through the image.

The basic assumption for the validity of the phase unwrapping scheme described above is that the phase between any two adjacent pixels does not change by more than π (88). This limitation in the measurement range results from the fact that sampled imaging systems with a limited resolving power are used. There must be at least two pixels per fringe - a condition that limits the maximum spatial frequency of the fringes to half of the *sampling frequency* (Nyquist frequency) of the sensor recording the interferogram. Fringe frequencies above the Nyquist frequency are aliased to a lower spatial frequency. In such cases the unwrapping algorithm is unable to reconstruct the modified data. If the fringe frequency is higher than the Nyquist frequency the unwrapping algorithm fails. The simplest solution is to improve the sampling using sensors with higher resolution or to process magnified regions of the interferogram. Discontinuities in the object, as for instance gaps and shaded regions as well as irregular boundaries, give also reasons for an incorrect unwrapping because the intensity does not change in such regions according to the phase sampling principle. Consequently, abrupt phase changes in the range of π can appear in the wrapped phase. To avoid such errors corresponding regions should be excluded from the unwrapping process. Noncontinuous deformations, as for instance cracks due to the material behaviour under load give rise to dislocations in the phase distribution. In the surroundings of the crack the structure of the fringe pattern is influenced in such a way that the fringes are cut and displaced from each other. The non-continuous behaviour of the phase surrounding the discontinuity results in inconsistent areas and consequently in an incorrect unwrapping if straightforward procedures are used.

In practice, however, these factors appear in common. Fig. 3.23a shows an interferogram of a vibrating industrial component recorded with a double pulse laser technique. The fringe pattern is corrupted with speckle noise, electronic noise and specular reflections which contribute to a distortion and fusion of the fringes. In some sections the signal-to-noise-ratio and contrast are very low. Other influences are object structure and irregular boundaries. A conventional unwrapping procedure starting with the column at the right border and following the lines will meet a lot of inconsistent areas (see Fig. 3.23c) and consequently generate streaks with incorrect phase offset over the whole phase map (see Fig. 3.23f).

Digital Processing and Evaluation of Fringe Patterns

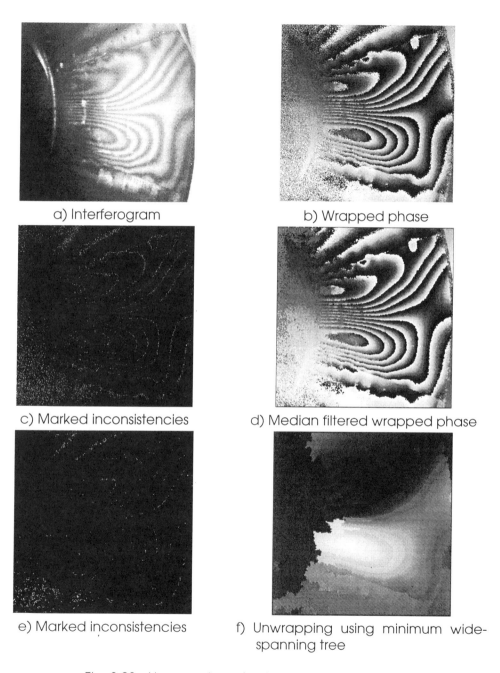

a) Interferogram

b) Wrapped phase

c) Marked inconsistencies

d) Median filtered wrapped phase

e) Marked inconsistencies

f) Unwrapping using minimum wide-spanning tree

Fig. 3.23: Unwrapping of noisy interferograms

It is obvious that alternative strategies for a correct unwrapping, including preprocessing of the fringe patterns and smoothing the wrapped phase map, are necessary. A lot of work has been done already in this field but there is no ideal recipe that works satisfactory in all cases (84-97). To preprocess the phase shifted interferograms the methods described in section 3.1 can be applied. An approved and simple mean for noise reduction is the smoothing of the intermediate results of the numerator and denominator of Equ. (3.45). The special case with m=4 (see Tab. 1) gives for instance:

$$I_4(x,y) - I_2(x,y) = 2b(x,y) \cdot \sin[\delta(x,y)]$$
$$I_3(x,y) - I_1(x,y) = 2b(x,y) \cdot \cos[\delta(x,y)]$$
(3.45)

Noisy wrapped phase images can be smoothed by *median filtering* (see Fig. 3.23d) to preserve true phase discontinuities and to decrease the number of inconsistencies. However, directional averaging is more convenient in this case as shown for example in Fig. 3.24. A modified rank order filter designed for ESPI phase data was proposed and compared to some other procedures by OWNER-PETERSEN (98).

a) Wrapped phase   b) Smoothed image

Fig. 3.24: Directional averaging of wrapped phase images

An useful segmentation step consists in the global partitioning of the interferograms into areas with and without relevant data. According to the phase sampling principle described above more information than the phase distribution can be reconstructed from the phase shifted interferograms. Areas with insufficient contrast, disturbing spots and irregular boundaries as shown in

Fig. 3.23a give rise to inconsistencies and should be excluded from further processing by the generation of masks.

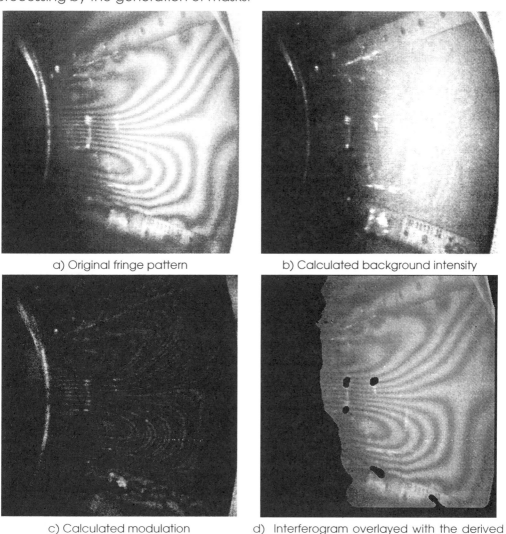

a) Original fringe pattern

b) Calculated background intensity

c) Calculated modulation

d) Interferogram overlayed with the derived mask

Fig. 3.25: Masking of phase shifted interferograms with the objective of separating areas with relevant and irrelevant data

The generation of these masks for labeling irrelevant areas can be done with help of the computed *background intensity* and *modulation* (see Tab 1 The derived background intensity and modulation distribution, Fig. 3.25b and

3.25c, are thresholded to receive binary masks. Both masks can be combined, Fig. 3.25d, to frame the relevant area for further processing.

Various *noise-immune algorithms* have been proposed to cope with the inconsistent points (79), (83-102). Typical examples of these are the *region oriented method* by GIERLOFF (99), the *cut-line method* by HUNTLEY (87), the wide *spanning tree method* by JUDGE et al. (90, 96), the *pixel ordering technique* by ETTEMEYER et al. (100), the *line detection method* by ANDRÄ et al. (90), and LIN et al. (101), and the *distributed processing method* using cellular automata by GHIGLIA et al. (83) or using a neural network by TAKEDA et al (92) and KREIS et al. (102). Common to all these techniques is that the most essential part of the principle is solving combinatorial optimization problems (97) such as minimization of the overall length of the cut lines using simulated annealing (87), minimization of the overall phase changes between neighbouring pixels by PRIM´s algorithm (91) and minimization of the overall smoothness of the phase distribution through the dynamic state changes of cellular automata or neurons (83, 92, 102).

A new an prospective method was shortly proposed by MARROQUIN et al (103-105). Here the unwrapping considers the reconstruction of phase fields from intensity measurements as an indirect problem and solves the problem by regularization. Therefore a short introduction to the term *indirect problem* is given (103).

Direct and indirect problems

Processes with a well-defined causality such as the process of image formation are called *direct problems*. Direct problems need information about all quantities which influence the unknown effect. Moreover, the internal structure of causality, all initial and boundary conditions and all geometrical details have to be formulated mathematically (104). To them belong wellknown initial and boundary value problems which are usually expressed by ordinary and partial differential equations. Such direct problems have some excellent properties which make them so attractive for physicists: If reality and mathematical description fit sufficiently well, the direct problem is expected to be uniquely solvable. Furtheron it is in general stable. That means small changes of the initial or boundary conditions cause also small effects only (in contrast to chaotic processes). Unfortunately numerous problems in physics and engineering deal with unknown but non-observable values. If the causal connections are investigated backwards we come to the concept of *inverse problems*. Based on indirect measurements, i.e. the observation of effects caused by the quantity we are looking for, one can try to identify the missing parameters. Such identification problems are well known in optical metrology. For instance the recognition and interpretation of subsurface flaws using HNDT, see sec. 5., and the reconstruction of phase distributions from the observed intensity values as discussed here are quite common. However, inverse problems have usually some undesirable properties: they are in general ill-posed, ambiguous and unstable.

The concept of *well-posedness* was introduced by HADAMARD (105) into the mathematical literature. He defined a CAUCHY problem of partial differential equations as well-posed, if and only if, for all CAUCHY data there is a uniquely determined solution depending continuously on the data; otherwise the problem is ill-posed. In mathematical notation an operator equation

$$F(x) = y \qquad (3.46)$$

is defined as *well-posed* with a linear operator $F \in \pounds(X,Y)$ in Banach spaces X and Y if the following three Hadamard conditions are satisfied (108):

1. $F(x)=y$ has a solution $x \in X$ for all $y \in Y$ (*existence*),
2. This solution x is determined uniquely (*uniqueness*),
3. The solution x depends continuously on the data y, i.e., the convergence $||y_n-y|| \to 0$ of a sequence $\{y_n\}=\{F(x_n)\}$ implies the convergence $||x_n-x|| \to 0$ of corresponding solutions (*stability*).

If at least one of the above conditions is violated, then the operator equation is called ill-posed. Simply spoken ill-posedness means that we have not enough information to solve the problem uniquely.

For the solution of inverse/ill-posed problems it is important to apply a maximum amount of *a-priori knowledge* or *predictions* about the physical quantities to be determined and we always have to answer the question if the measured data contain enough information to determine the unknown quantity uniquely. In case where the data y result from the integration of unknown components, thus this results in smoothing. Consequently the information about any single component gets lost and very different causes may give almost the same effect after integration. A well known example for specialists in HNDT is the ambiguous relation between an observed fringe pattern (the effect) on the surface and its corresponding cause (one or several subsurface flaws) under the surface. The response of the flaw on the applied load is smoothed since only the displacement on the surface gives rise to the observed fringe pattern. These fringe patterns are very noisy and their topology is strongly limited (see sec. 5). Therefore the conclusion from the observed pattern to the cause behind is ambiguous in case of simple and straightforward inspection procedures. Later we will come back to this example. This instability effect of ill-posedness is difficult to handle and to overcome in the numerical solution process of inverse problems.

In order to overcome the disadvantages of ill-posedness in the process of finding an approximate solution to an inverse problem, different techniques of *regularization* are used. Regularizing an inverse problem means that instead of the ill-posed original problem a well-posed neighboring problem has to be formulated. The key decision of regularization is to find out an admissible compromise between stability and approximation (107). The formulation of a sufficiently stable auxiliary problem means that the original problem has to be changed accordingly radically. As a consequence one cannot expect that the properties of the solution of the auxiliary problem coincide with the properties of the original problem. But convergence between the regularized and the original solution should be guaranteed if the stochastic character of the experimental data is decreasing. In case of noisy data the identification of unknown quantities can be considered as an estimation problem. Depending on the linearity or non-linearity of the operator F, we than have linear and non-linear regression models, respectively. Consequently, *least-square methods* play an important role in the solution of inverse problems:

$$\|F(x) - y_\varepsilon\|^2 \to \min \qquad (3.47)$$

with $y_\varepsilon = y + \varepsilon$.

Marroquin et al (103-105) applied for the reconstruction of the continous phase field from noisy mod2π data the THIKHONOV regularization theory (109-110) to find solutions that correspond to minimizers of positive-definite quadratic cost functionals. This method can be considered as a generalization of the above discussed classical least-squares solution by introducing a so-called regularization parameter and a stabilizing functional.

Another way to categorise the various methods was given by ROBINSON (79). Robinson discriminates between *path-dependent* and *path-independent methods*. To the first class where the success of the unwrapping process is dependent on the path taken through the wrapped phase distribution belong such methods as sequential linear scanning (line by line), pixel ordering or random walk methods. Variations on this method use two or more routes through the data and compare the results to increase the probability of analyzing the data correctly. Path independent methods such as graph-based and cellular automata integrate the phase along all possible paths between a starting pixel (or reference point) and all pixels to be processed.

A detailed introduction to the theory, algorithms and software of two-dimensional phase unwrapping is given in the book of GHIGLIA and PRITT (111).

3.3.4 Absolute phase measurement

In the previous sections it was pointed out that all phase measurement techniques deliver the phase value mod 2π only and consequently phase unwrapping has to be carried out. However, even a correct phase unwrapping results in a relative phase map with an unknown bias given by the integer N_R in Equ. (3.36); but it is generally necessary to measure the *absolute phase* of each measuring point instead of its relative phase. Problems arising from this fact are for example

- incorrect numbering of the fringes and derived quantities if no stationary region of the object can be identified to locate the zero-order fringe;
- incorrect evaluation if multiple interferograms with differing illumination beams have to be analysed with respect to vectoral deformations (in general there is no way to know how the phase at a point in one interferogram relates to the phase at the same point in another) (112);
- if the object to be analysed is not simply connected, i.e. isolated objects or shaded regions occur, then the mod 2π-unwrapping procedure fails because of violated neighbourhood conditions and unknown phase jumps greater than π.

The last-mentioned difficulties was already discussed in section 3.3.2.1 on example of Fig. 3.20.

Different approaches that partly solve the problems of *absolute phase measurement* are known (113):

- the application of a rubber band drawn between the object and a fixed part within the measured scene to determine the fringe number (114);
- the manipulation of components of the optical set-up that influence the fringe pattern but not the measuring quantity (e.g. the recording and evaluation of two fringe patterns with two different wavelengths (115) or two different illumination directions (116));
- the use of sensitivity vector variations to provide absolute displacement analysis (112);
- the application of coded light techniques (117, 118) for time-space encoding of the projected fringe patterns with respect to the fringe number;
- the combination of two fringe patterns with different spatial frequencies (119) or different orientations (120).

The last-mentioned approach has been proven to be suitable for a wide range of practical applications. To avoid ambiguity the general principle of these methods consists in the generation of a synthetic wavelengths by tuning one of the flexible system parameters (wavelength, angle separation of the light sources, spatial frequency or orientation of the projected fringe pattern) in such a way that only one fringe period covers the object. But with only one fringe a sufficient accuracy cannot be ensured (121). Therefore in practice absolute phase measurement is carried out in a hierarchical way. At first the synthetic wavelength is chosen to receive only one fringe. Then the flexible parameter is changed step by step to cover the measuring field with an increasing number of fringes to get a better accuracy.

An important advantage of this method is the avoidance of the conventional unwrapping process that is based on the evaluation of neighbourhood relations to compare the relative phases of adjacent pixels. However, neighbourhood-based procedures are very sensitive against definite signal and object properties such as speckle noise, edges, isolated regions as discussed above. Therefore several neighbourhood independent algorithms were developed in the middle of the 90th. Such procedures are known under the names *synthetic wavelength interferometry, absolute interferometry* and *wavelength scanning* (115), (122-127), respectively, *hierarchical demodulation* (113, 121) and *temporal phase unwrapping* (128-129) bekannt. The objective of such algorithms is it to increase the limited range of unambiguity (mod2π) by the generation of adapted synthetic wavelength and to enable a pixel-based absolute evaluation of the mod2π-phase ditribution. Since accuracy of evaluation is limited by using of only one synthetic wavelength adapted sequences of different synthetic wavelength are used.

In the following a short description of the working principle of the hierarchical type of *absolute phase measurement* is given. The description is oriented on the variation of the mutual fringe orientation of two fringe patterns to be superimposed with respect to generate a synthetic wavelength by moiré-effect (121). The procedure can be generalized for other methods of synthetic wavelength generation as shown in Table 2 (113). The derived sequence of fringe patterns is strongly oriented on the available accuracy for the primary measured quantity - the phase of the fringe pattern - and ensures a more robust measurement. or angle orientation.

If two fringe patterns are projected and superimposed on the surface of the object the relevant part of the observed moiré intensity that represents the difference between the fundamental pattern masking up the two basic patterns can be described with Equ.(3.1):

$$I(x, y) = a(x, y) + b(x, y) \cdot \cos[\delta_1(x, y) - \delta_2(x, y)] \tag{3.48}$$

where (see Fig. 3.26)

$$\delta_i(x,y) = N_i(x,y) \cdot 2\pi = \left[N_{Ri} + \hat{N}_i + \tilde{N}_i \right] \cdot 2\pi \tag{3.49}$$

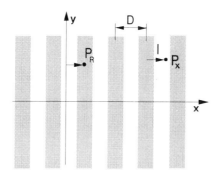

Fig. 3.26: Geometric interpretation of the phase values in the projected fringe technique $\left(N_R = N(P_R),\ \hat{N} = 1/D \right)$

Assuming that the two patterns are oriented with angles γ_1 and γ_2, respectively, between them and the x axis of the coordinate system, the two phase functions $\delta_1(x,y)$ and $\delta_2(x,y)$ can be written as

$$\delta_1(x,y) = \frac{2\pi}{D_1} \cdot (x \cdot \sin\gamma_1 + y \cdot \cos\gamma_1)$$
$$\delta_2(x,y) = \frac{2\pi}{D_2} \cdot (x \cdot \sin\gamma_2 + y \cdot \cos\gamma_2) \tag{3.50}$$

If the two patterns are oriented symmetrically to the x axis $(\gamma_1 = \gamma;\ \gamma_2 = -\gamma)$, the phase difference in Equ. (3.48) is

$$\Delta\delta(x,y) = \delta_1(x,y) - \delta_2(x,y) = \frac{4\pi x \cdot \sin\gamma}{\overline{D}} + \frac{2\pi y \cdot \cos\gamma}{\Lambda} = (N_1 - N_2)2\pi \tag{3.51}$$

with the average line spacing

$$\overline{D} = \frac{D_1 + D_2}{2} \tag{3.52}$$

and the *synthetic* or *beat wavelength* of the two patterns

$$\Lambda = \frac{D_1 D_2}{D_1 - D_2} \tag{3.53}$$

Two separate cases are relevant here. When $D_1 = D_2 = D$, the second term of Equ.(90) is zero and the equation can be rewritten as

$$\Delta\delta(x,y) = \frac{4\pi x \cdot \sin\gamma}{D} = \Delta N \cdot 2\pi \tag{3.54}$$

The other simple case occurs when the patterns are parallel to each other with $\gamma=0°$. This makes the first term of Equ.(3.51) zero and the moiré pattern will then be described by

$$\Delta\delta(x,y) = \frac{2\pi y}{\Lambda} = \Delta N \cdot 2\pi \qquad (3.55)$$

The resulting moiré does not change the quantity to be measured but only the fringe pattern. In the first case with the two symmetrically oriented patterns the equation for the x coordinate can be written

$$x = \frac{\Delta N \cdot D}{2 \cdot \sin\gamma} \qquad (3.56)$$

This equation shows some interesting properties:
- The primary quantity to be measured is no longer the absolute fringe order N but the difference ΔN that can be calculated point-by-point.
- For every limited phase field it is always possible to find a sufficiently small angle γ that exactly one integer ΔN fulfils Equ.(3.56). This essential property is illustrated in Fig. 3.27a showing two symmetrical orientations of a projection grid and the resulting moiré fringe along the x-axis.

Consequently the phase can be measured along the x-axis without ambiguity if γ is controlled in such a way that exactly $\Delta N=1$ fits for the whole measuring field. However, with only one fringe a sufficient accuracy cannot be guaranteed (121), Fig. 3.28. Therefore, in practice, absolute phase measurement is carried out in a hierarchical way. At first the phase is tuned for $\Delta N=1$ to ensure the unambiguity. Then the angle γ is increased step-by-step to cover the measuring field with an increasing number of fringes to ensure the accuracy.

To ensure the lack of ambiguity with respect to the moiré fringe number ΔN small angles γ are necessary. Concerning the accuracy of x, the coefficient $1/\sin\gamma$ in Equ. (3.56) acts as a weight factor. Consequently small angles γ will lead to a drastic decrease in accuracy. The task consists of the construction of a sequence of pattern orientations and angles γ, respectively, which contributes to a more robust measurement, i.e. in the elimination of the influence of the weight factor $1/\sin\gamma$. This sequence of angles $\gamma^{(k)}$ requires the following properties:
- The determination of the sequence $\gamma^{(k)}$ is oriented on the phase inaccuracy ε of the applied phase measuring technique.
- In every successive step (k+1) it must be ensured that the corresponding period length $\Lambda^{(k+1)}$ of the still ambiguous solution (including its tolerance at

the borders) is greater than the remaining tolerance region $\{\Lambda^{(k)} \cdot 4\varepsilon\}$ of the previous step (k) (compare Fig. 3.27b):

$$\Lambda^{(k+1)} \cdot (1 - 4\varepsilon) \geq \Lambda^{(k)} \cdot 4\varepsilon \qquad (3.57)$$

This requirement implies an upper limit on every successive $\gamma^{(k)}$.
- The tolerance region has to be reduced to the limit $\{\Lambda \cdot 2\varepsilon\}$ for $\gamma^{(k)}=90°$ in as few steps as possible.

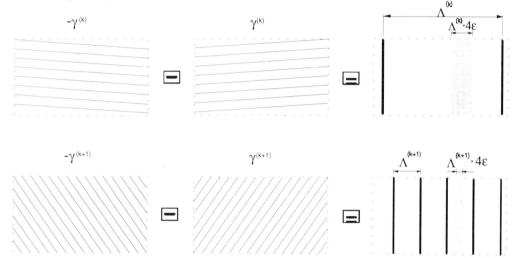

Fig. 3.27: Schematic illustration of the generation of synthetic wavelengths by the superposition of two symmmetrically oriented fringe patterns: a) two symmetrically oriented grids for the k-th measurement with $\pm \gamma^{(k)}$ and the resulting moiré with the period $\Lambda^{(k)}$ and the tolerance ε; b) following orientations for $\pm \gamma^{(k+1)}$ with the condition that the resulting period $\Lambda^{(k+1)}$ including its tolerance is larger than the tolerance of the measurement with $\Lambda^{(k)}$

Such an algorithm is described and experimentally verified in (121). Using this algorithm and assuming a phase measuring system with an ε of 1/25 an adapted sequence of angles can be derived: $\gamma^{(k)}=(0,12°, 0,63°, 3,31°, 17,64°, 90,0°)$. With respect to the test object of Fig. 3.18b, for the first orientation, $\gamma=0,12°$, the absolute phase distribution is calculated with the algorithm. The results are shown in Fig. 3.29.

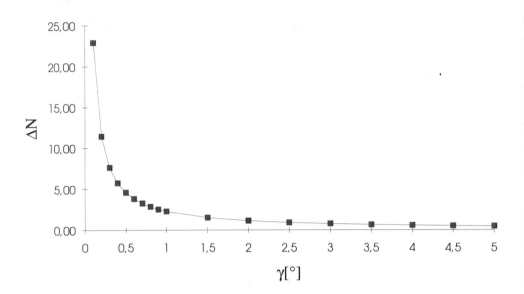

Fig. 3.28: Dependency of the phase error $\Delta N=\Delta\xi/l$ on the angle γ ($\varepsilon = 1/25$)

The data are absolute and do not have to be unwrapped, since the wrapped input data containing the principal values of the phase $\hat{\delta}_1^{(k)} = \hat{N}_1 \cdot 2\pi$ and $\hat{\delta}_2^{(k)} = \hat{N}_2 \cdot 2\pi$ are used point-by-point for the calculation of the absolute phase values. Consequently, *phase unwrapping* becomes unnecessary. The only disadvantage is the higher noise level of several fringe orders. Within the shaded region II it is impossible to measure the phase. However, there is no resolution problems in the region IV, since the fringes of the two primary patterns to be subtracted are almost perpendicular to the usual fringe orientation if the conventional procedure is used. Now the interim results for the next two orientations {0.63°, 3.31°} are skipped here, and the result for the orientation $\gamma = \pm 17.64°$ is given, Fig. 3.29d. The noise components are almost completely compensated. In spite of the discontinuity between region I and III, the boundary regions I and V fit together perfectly. The maximum accuracy is finally achived using the standard orientation $\gamma = \pm 90°$. Because of the relationship between the direction of the fringes, the orientation of the surface and the observation direction the resolution of the CCD-sensor is too small to resolve the fringes within the region IV and consequently the phase values of this region are lost, Fig. 3.29e. The combination of the phase values resulting from the different evaluation steps leads to an optimum compromise between accuracy and completeness.

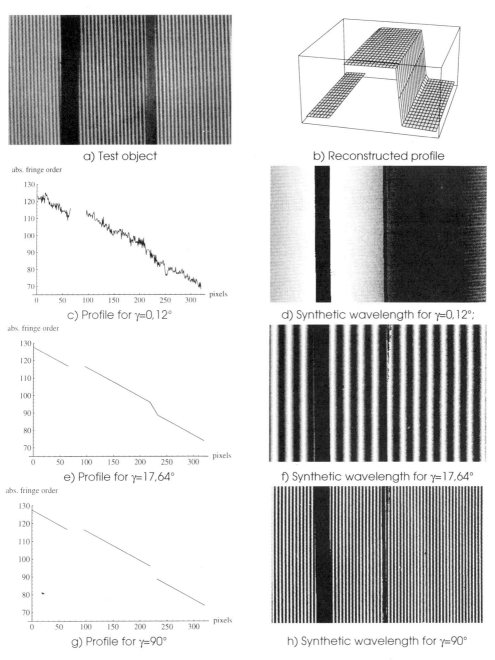

Fig. 3.29: Absolute fringe order after the evaluation along a sensor line

Fig. 3.29b shows a pseudo-3d plot of the final phase distribution, where a certain correspondance to the height distribution was made by subtraction of the linear phase component.

The results exemplary derived for the case of the combination of two symmetrically rotated grids can be generalized for other applications in fringe projection and interferometry where synthetic wavelengths are generated for absolute phase measurement. The basis for the calculation of the sequence of synthetic wavelengths is Equ.(3.57). If the error of the measured principal value is $\delta(\hat{N}) = \pm\varepsilon$ the synthetic wavelength $\Lambda^{(k)}$ has to be controled according Equ.(3.57). The method for the generation of $\Lambda^{(k)}$ is free. Tabele 2 shows some examples.

Method	Parameter	Synthetic Wavelength
Symmetrical Rotation of two Grids	Rotation Angle γ	$\Lambda^{(k)} = \dfrac{D}{2\sin\gamma^{(k)}}$
Variation of Spatial Frequencies of two Grids	Difference of Spatial Frequencies $\Delta D = D_1 - D_2$	$\Lambda^{(k)} = \dfrac{D^2}{\Delta D}$
Variation of the Wavelength of the Illumination Beam	Difference of the Wavelengths $\Delta\lambda = \lambda_1 - \lambda_2$	$\Lambda^{(k)} = \dfrac{\lambda^2}{\Delta\lambda^{(k)}}$
Variation of the Angle Separation of the Illumination Sources	Angle separation α	$\Lambda^{(k)} = \dfrac{\lambda}{2\sin(\alpha^{(k)}/2)}$

Table 2: Methods for the generation of synthetic wavelengths

HUNTLEY und SALDNER (128), (129) prefer a pixelbased absolute evaluation along the time axis - the so called *temporal phase unwrapping*. Here the phase is tracked during the load is varied in time. The demodulated phase results for every pixel from an evaluation along the time axix by summation of the phase differences of succeeding displacement intervals. These intervals have to be controlled in such a way that a temporal sampling rate of at least two samples per -π...π-phasen cycle is ensured.

4. MEASUREMENT OF 3D-DISPLACEMENT FIELDS

The main effort in laser metrology during the last fifteen years has been focussed on the development of high precision phase measurement techniques and their digital implementation, since the phase is the primary quantity for metrological testing. However, the phase distribution $\delta(x,y)$ gives only a first impression of the deformation or the shape of the surface. However, in practice the *three-dimensional displacement field* is required if the mechanical behaviour of the object under load is to be investigated. To calculate displacement components some further quantities are necessary, e.g. the three coordinates of the object points. Although the contour measurement can also be reduced to a phase measurement problem, the determination of three-dimensional displacements is more complex than a high precision phase evaluation. From the practical point of view five main tasks have to be performed (130):

- the planning of the experiment including the choice of the measurement principle, the loading conditions, the sensitivity of the optical set-up with respect to the measuring quantity and the laser source;
- the design of the interferometer with respect to its sensitivity and its robustness concerning the propagation of input data errors in the computed results;
- the acquisition of the primary phase data:
  - the evaluation of the interferogram with respect to the reconstruction of interference phase distribution $\delta(x,y)$
  - the evaluation of the projected fringe pattern concerning the determination of the cartesian coordinates and *sensitivity vectors* of the measuring points;
- the combination of all data sets acquired with different measuring systems (e.g. the transformation of the point coordinates measured in a topometric set-up onto the object surface in the holographic interferometer) with respect to the calculation of the displacement field and
- the evaluation of the data with respect to their accuracy and the graphic presentation of the results.

In the following sections these tasks are discussed.

4.1 Planning of the experiment
4.1.1 Basic relations

The the basic relation of holographic interferometry (23) gives the connection between the measured phase difference $\delta(x,y)$ and the displacement vector $\mathbf{d}=\mathbf{d}(u,v,w)$ of a point $P=P(x,y,z)$:

$$\delta(P) = \frac{2\pi}{\lambda} \cdot \left[\mathbf{e}_B(P) + \mathbf{e}_Q(P)\right] \cdot \mathbf{d}(P) = \mathbf{S}(P) \cdot \mathbf{d}(P) \quad \text{with} \quad \delta(P) = N(P) \cdot 2\pi, \qquad (4.1)$$

with N as the fringe number, λ as the wavelength, $\mathbf{e}_B$ as the unit vector in observation direction, $\mathbf{e}_Q$ as the unit vector in illumination direction and $\mathbf{S}$ as the sensitivity vector. For the measurement of the three displacement components $\mathbf{d}(u,v,w)$ at least 3 independent equations are necessary. Usually 3 or more observation and illumination directions, respectively, are chosen. To avoid perspective distortions it is convenient to use independent illumination directions as illustrated in Fig. 4.1. In this case the following equation system can be derived:

$$N_i(P) \cdot \lambda = \left[\mathbf{e}_B(P) + \mathbf{e}_{Qi}(P)\right] \cdot \mathbf{d}(P), \qquad i = 1\ldots n, n \geq 3 \qquad (4.2)$$

To write in simplified terms a matrix notation of Eq. (4.2) is used:

$$\mathbf{N} \cdot \lambda = \mathbf{G} \cdot \mathbf{d}, \qquad (4.3)$$

with the column vector $\mathbf{N}$ which contains the n fringe numbers N_i and the (n,3)-matrix $\mathbf{G}$ which contains in the i-th line the 3 components of the i-th sensitivity vector $\mathbf{S}_i$. Because of measuring errors the solution $\mathbf{d}=\mathbf{d}(u,v,w)$ of the equation system (4.3) differs from the true value $\mathbf{d}_0=\mathbf{d}_0(u_0,v_0,w_0)$ by $\Delta\mathbf{d}$. Sources of experimental error can be divided into two categories: errors $\Delta\mathbf{G}$ of the geometry matrix $\mathbf{G}$ due to the limited accuracy of the measuring tools for coordinate measurement and errors $\Delta\mathbf{N}$ of the fringe number vector $\mathbf{N}$ due to the limited accuracy of the applied phase measurement technique. All these errors cause an error $\Delta\mathbf{d}$ of the displacement vector. Consequently we have to write:

$$\lambda \cdot (\mathbf{N}_0 + \Delta\mathbf{N}) = (\mathbf{G}_0 + \Delta\mathbf{G}) \cdot (\mathbf{d}_0 + \Delta\mathbf{d}), \qquad (4.4)$$

$$\mathbf{N}_0 \cdot \lambda = \mathbf{G}_0 \cdot \mathbf{d}_0, \qquad (4.5)$$

$$\mathbf{G} = \mathbf{G}_0 + \Delta\mathbf{G}, \qquad (4.6)$$

$$\mathbf{N} = \mathbf{N}_0 + \Delta\mathbf{N}, \qquad (4.7)$$

$$\mathbf{d} = \mathbf{d}_0 + \Delta\mathbf{d}. \qquad (4.8)$$

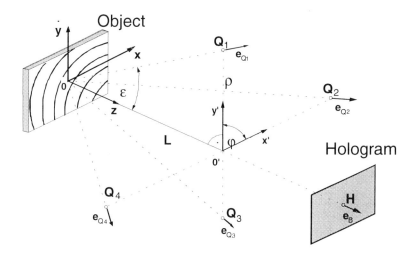

Fig. 4.1: Scheme of a holographic interferometer with 4 illumination directions and 1 observation direction (The declaration of the parameters ε, ρ, φ and L is given in section 4.1.3.)

To improve the accuracy of the displacement components overdetermined systems (n>3) are often used. In our case more than 3 illumination directions e_{Qi} (i=1...n, n>3) have to be provided then, see Fig. 4.1 with n=4. Such equation systems can be solved with the least squares error method:

$$grad_\mathbf{d}\left[(\mathbf{N}\cdot\lambda-\mathbf{G}\cdot\mathbf{d})^T\cdot(\mathbf{N}\cdot\lambda-\mathbf{G}\cdot\mathbf{d})\right]=\mathbf{0} \quad (4.9)$$

which results in the so called normal equation system

$$\mathbf{G}^T\cdot\mathbf{N}\cdot\lambda=\mathbf{G}^T\mathbf{G}\cdot\mathbf{d} \quad (4.10)$$

The matrix

$$\mathbf{F}=\mathbf{G}^T\cdot\mathbf{G} \quad (4.11)$$

is denoted as normal matrix with $\mathbf{G}^T$ as the transposed one of $\mathbf{G}$. Its certain importance for the construction of the interferometer and the error analysis is explained in section 4.3. and 4.4.

4.1.2. Valuable a-priori knowledge

For an inspection problem it is convenient to start with a certain amount of a-priori knowledge concerning the object under test and its loading conditions. This information gives indications for the construction of the interferometer with respect to its sensitivity and for the choice of the loading principle as well as the loading strength. For the most simple case if the displacement direction is known coincident illumination and observation directions can be used and the sensitivity of the interferometer can be controlled in such a way that the amount of the displacement is measured directly:

$$d = N \cdot \lambda / 2 \tag{4.12}$$

Consequently some serious problems that will be discussed in the next sections are irrelevant then. But in the most inspection tasks a lot of knowledge about the object is available: e.g. shape, structure, orientation and material; the working load, probable deformations and possible positions as well as types of weak points and faults. Using this knowledge the interferometer can be controlled with respect to its sensitivity and robustness and adapted loading techniques can be applied. Furtheron model based simulations as a combination of finite element modeling of surface displacement fields and interferometric pattern computation using an analytical approach to the image formation can deliver good orientations for the practical inspection process (131). A just as simple as powerful tool for the optimal arrangement of objects within the holographic setup was developed by Abramson already in the first decade of holographic interferometry (132). Abramson has demonstrated the advantage of the so called Holo-Diagram on an example of a more than 2m long steel beam using a 60 mW He-Ne laser with a coherence length of only 30 cm (133). Concerning the loading technique BIRNBAUM and VEST (134) distinguish in their review on holographic non-destructive evaluation between five basic principles for the solution of different inspection problems:

- direct mechanical stressing
- pressure or vacuum stressing
- thermal stressing
- impulse loading and
- vibrational excitation.

The choice of a specific loading technique depends mainly on the following criteria:

- the load shall simulate the working conditions of the object under test,
- the load shall generate surface displacements within the measuring range (rule of the thumb: 100 nm ... 50 µm) and
- the load shall cause irregular fringe patterns if surface or subsurface flaws occur.

Further boundary conditions for the experiment are determined by the required accuracy of data, the choice of the laser source and the distribution of the measuring points across the object. The accuracy mainly depends on the objective of the experiment. If related mechanical quantities as strains and stresses should be calculated by numerical differentiation of the displacements a high accuracy phase measurement has to be guaranteed on a sufficient narrow mesh (135). This affects the necessary effort for the phase reconstruction considerably. The choice of the laser source depends on the loading conditions, the quantity to be measured and the environmental influences. Pulse lasers are better suited for industrial inspection problems than cw-lasers because of their high energy and short pulse duration. The distribution of the measuring points is mainly conditioned by the special inspection problem. In connection with FEM-simulations the topology of the mesh is strictly conditioned. Otherwise expected stress concentrations and weak points can give orientations.

General inspection problems make it necessary to measure 3D-displacements on the surface of 3D-objects. For this purpose at least three independent phase measurements for each measuring point as well as the 3 coordinates of each point within a predefined coordinate system have to be measured. Additionally, some geometrical parameters of the interferometer must be known such as the central point of the entrance pupil of the observation system to define the observation direction and the locus of the focal point of the illumination system to define the illumination direction. All this has to be done in a concrete interferometer setup. Therefore in the next section the physical and mathematical background for a suitable interferometer design is discussed.

4.1.3 Interferometer design

For the calculation of the 3 displacement components d(u,v,w) at least 3 phase measurements are required to get an equation system of the type (4.3). According to the theory of linear equation systems these equations must be linear independent. That means independent input data sets have to be measured. The best way to acquire linear independent data is the choice of an interferometer geometry with 3 orthogonal sensitivity vectors. Such an orthogonal three-leg ensures a minimum propagation of input data errors on the computed results as well as a best possible sensitivity for all displacement components. But from the practical point of view this kind of geometry is highly complex. Thus the task consists in finding a compromise between effort and advantage with respect to the error sensitivity of the interferometer. Such a so called *optimized interferometer* is described in this chapter.

A proven method to find a geometry that is characterized by minimum propagation of the experimental errors $\Delta \mathbf{N}$ and $\Delta \mathbf{G}$ on the displacement vector is an optimization with help of the perturbation theory of linear equation systems. The condition number κ of the matrix $\mathbf{G}$ is a numerical measure for the propagation of input data errors on the computed results (136):

$$\frac{|\Delta d|}{|d|} \leq \kappa(G) \cdot \left[\frac{\|\Delta G\|}{\|G\|} + \frac{\|\Delta N\|}{\|N\|} \right], \qquad (4.13)$$

where $\|G\|$ denotes the norm of the matrix G. Using the spectral norm the condition number for $n \geq 3$ can be calculated by (136)

$$\kappa(G) = \|G\| \cdot \|G^{-1}\| = \sqrt{\frac{\chi_{max}(F)}{\chi_{min}(F)}}, \quad \kappa \geq 1 \qquad (4.14)$$

with χ_{max} and χ_{min} as the maximum and minimum eigenvalues of the normal matrix F. Since G consists of the n sensitivity vectors S_i the number $\kappa(G)$ characterizes also the influence of the mutual arrangement of these vectors on the error propagation. Consequently the investigation of the influence of numerous geometrical parameters on the accuracy of the displacement vector can be simply reduced to the analysis of only one quality number. In our sense the objective of the optimization consists in the minimization of κ by variation of the vector orientation considering the demand for a low complexity of the optical setup (137).

In comparison with an orthogonal three-leg set-up the complexity of the interferometer is considerably reduced if we use an interferometer geometry where the n illumination sources and observation points, respectively, are placed in one selected plane as shown in Fig. 37. This plane is chosen to be parallel to the (x,y)-plane of the object coordinate system. Such a set-up is very often used because of its good conditions for the data acquisition. Fig 38 shows an interferometer with 90°-symmetry applied for the investigation of microcomponents (138).

Now it is shown that this simplification can result also in a robust error behaviour if some rules are considered. For the case of n illumination points the greatest possible separation of the sensitivity vectors can be guaranteed with the following simple conditions:

- All illumination points Q_i are arranged within one plane parallel to the (x,y)-plane and placed on a circle with the greatest possible radius ρ. (In the case of n observation points all the points H_i where the unit vectors in observation direction e_{Bi} break through the hologram are arranged within one plane parallel to the (x,y)-plane, i.e the hologram plane(s) is/are parallel to the (x,y)-plane)
- The centre of this circle is the point H where the unit vector in observation direction e_B breaks through the hologram. (In the case of n observation directions the centre of the circle is defined by the point Q where the unit vector

in illumination direction breaks through the plane spread by the n hologram points H_i)
- The hologram point H is arranged in a plane parallel to the (x,y)-plane of the object coordinate system and in an orthogonal distance L.
- The angle φ between two adjacent lines connecting the source points Q_i with the hologram point H is

$$\varphi = 360'' / n .\tag{4.15}$$

Fig. 4.2: Optimized interferometer set-up with 90°-symmetry in the orientation of the illumination directions

On these conditions the norm of all sensitivity vectors is equal - a necessary criterion of well conditioned systems. Additional to the angle φ only one further parameter is necessary to describe the interferometer - the effective aperture ξ:

$$\xi = \tan \varepsilon = \frac{\rho}{L} .\tag{4.16}$$

An additional result of this geometry is the simplification of the error analysis. For the object point $P_0(0,0,0)$ placed in the origin of the coordinate system the normal matrix **F** has now only non-zero elements in the main diagonal:

$$\mathbf{F} = n \cdot \frac{\xi^2}{\xi^2+1} \begin{bmatrix} \frac{1}{2} & 0 & 0 \\ 0 & \frac{1}{2} & 0 \\ 0 & 0 & \left[\frac{1+\sqrt{\xi^2+1}}{\xi}\right]^2 \end{bmatrix} \qquad (4.17)$$

Eq. (4.17) shows that the condition number of the interferometer optimized in the way described above is independent on the number of equations. This is essential if least squares error methods are used. Taking into account that the expression

$$\left[\left(1+\sqrt{\xi^2+1}\right)/\xi\right]^2 \geq 1 \quad \forall \xi \qquad (4.18)$$

is valid we can find for the condition number with Eq. (4.14):

$$\kappa(\mathbf{G}) = \sqrt{2}\,\frac{1+\sqrt{\xi^2+1}}{\xi} \qquad (4.19)$$

The minimum of κ can be derived for $\xi \to \infty$:

$$\kappa^{\min}(\mathbf{G}) = \sqrt{2} \,. \qquad (4.20)$$

In practice the apertures ξ can be enlarged by increasing the distance between the source points Q_i and the z-axis and/or by decreasing the distance between the object and the hologram. For both cases the separation of the sensitivity vectors is improved. Fig. 4.3 illustrates the dependence of the condition number κ on the distance L for three different values of the radius ρ. Inconvenient set-ups can result in very large condition numbers and consequently in a high sensitivity with respect to phase and coordinate measuring errors.

Eq. (4.19) gives an orientation for the quality of the interferometer with regard to the error propagation. But the results are only valid in the origin $P_0(0,0,0)$ of the object coordinate system. In practice generally the measurement of a point field of extended objects is necessary. Here the optimization should be performed using generalized condition numbers applied to the point field. For instance it is useful to optimize the aperture ξ by variation of the distance L with fixed ρ using the medium or maximum condition number of a point field as the function to be minimized:

$$\overline{\kappa} = \frac{1}{m}\sum_{i=1}^{m}\kappa(P_i) \quad , \text{m - number of object points,} \qquad (4.21)$$

$$\hat{\kappa} = \max[\kappa(P_i)] \quad . \quad . \tag{4.22}$$

Fig. 4.4 shows the distribution of the condition number over a 510mm×330mm area after the optimization of ξ using $\bar{\kappa}$ (ρ=86,5 mm, n=3, φ=120°).

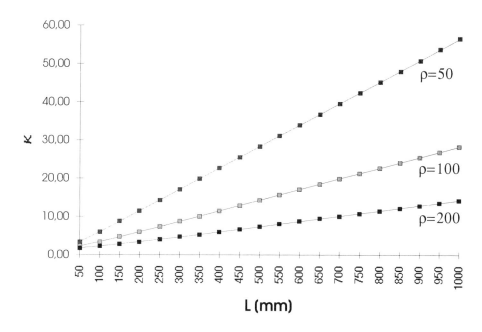

Fig. 4.3: Dependence of the condition number κ of an optimized interferometer with n=3 (φ=120°) on the distance L for three different radius ρ (mm) calculated with Equ. (4.19)

The same results can be derived for n observation directions and one illumination direction because of the symmetry of Equ. (4.1) concerning the observation and illumination direction. An optimized interferometer for the fringe counting method ensuring the best possible condition κ=1 is based on the same geometric assumptions (137, 139).

4.1.4 The influence of sensitivity

Equ. (4.1) shows that the measured phase difference δ(x,y) is an equivalent for the component of the displacement vector d_S in direction of the sensitivity vector **S**. However, this direction depends on the location P(x,y,z) where the phase δ(x,y) is measured since the observation and illumination direction vary with the corresponding object point. Often this dependence is neglected by assuming a constant sensitivity across the surface of the object. This way the measurement is simplified considerably because the geometry of the object is of no importance. Consequently the shape of the object doesn't have to be measured and the displacement measurement reduces to the reconstruction of the interference phase for every object point. As shown in the following experiments this simplified technique causes serious errors.

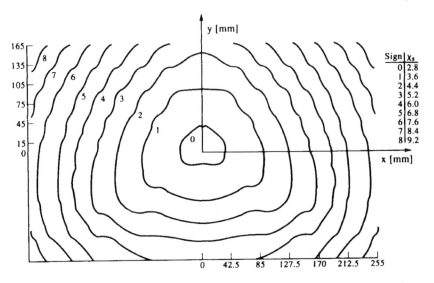

Fig. 4.4: Distribution of the condition number κ across a 510mmx330mm surface after the optimization of ξ using $\bar{\kappa}$

The experiment is performed with a closed cylinder of aluminium (diameter 100 mm, length 200mm) that is fixed at the bottom, Fig. 4.5. A load is applied by changing the inner pressure of a gas. The coordinates, measured with the fringe projection technique (see chapter 4.2), are used for the dermination of all point dependent sensitivity vectors. For comparison purposes a constant sensitivity vector is assigned to all object points, too.

For every illumination direction the double-exposed hologram is made using only one plate and mutually incoherently adjusted reference beam paths. Fig. 4.6 shows the corresponding interferograms with a pressure difference of 1.2 bar. The phase distribution $\delta(x,y)$ was reconstructed with the phase shifting technique, chapter 3.2.4.

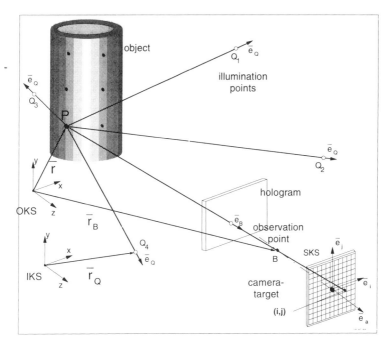

Fig. 4.5: Experimental set-up for the investigation of the displacement field of an aluminum cylinder loaded by inner pressure using different illumination directions $e_{Q1}, ..., e_{Q4}$

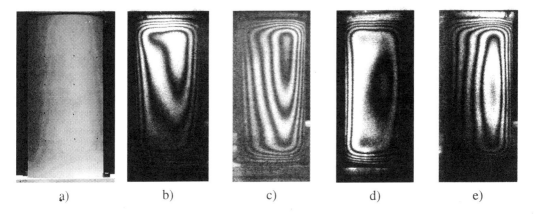

a)   b)   c)   d)   e)

Fig. 4.6: Cylinder and the corresponding interferograms for the illumination direction $e_{Q1}, ..., e_{Q4}$

Based on the measured geometry and phase the equation system (4.2) is solved for every point P(x,y,z). The result is a data field with every point P and its corresponding displacement d(u,v,w), Fig 4.7.

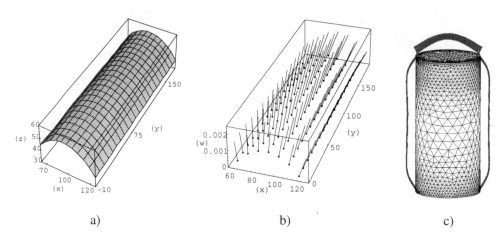

Fig. 4.7: a) measured geometry data of the cylinder, b) displacement field considering the locally variable sensitivity across the surface of the cylinder ($|d_s| \approx 1,5 \mu m$) in mm), c) deformation of the cylinder calculated by FEM simulation (all measures in mm)

The comparison between the calculation with and without consideration of a locally dependent sensitivity is shown in Fig. 4.8. The evaluation for the radial component along the cross section of the cylinder provides similar results in both cases. However, the tangential component shows clear deviations such as its high amount and its changing sign - both in conflict with the expected rotational symmetry. Also the axial component along the longitudinal axis differs increasingly.

The reason for this is that holographic interferometry is mainly sensitive for the out-of plane component. In ususal set-ups the sensitivity vector shows in this direction. Consequently the measurement of the in-plane components such as the tangential and axial one in case of the cylinder are measured less sensitive and less robust. Measuring errors of the primary quantities, the interference phase and the coordinates, affect the in-plane components more than the out-of-plane component.

The conclusion we can draw from this experiment is that the consideration of the geomery and consequently the local variation of the sensitivity vector can be recommended if in-plane components have to be measured for extended objects.

The effect of this global and systematic source of errors can also be studied by FEM calculations.

Digital Processing and Evaluation of Fringe Patterns

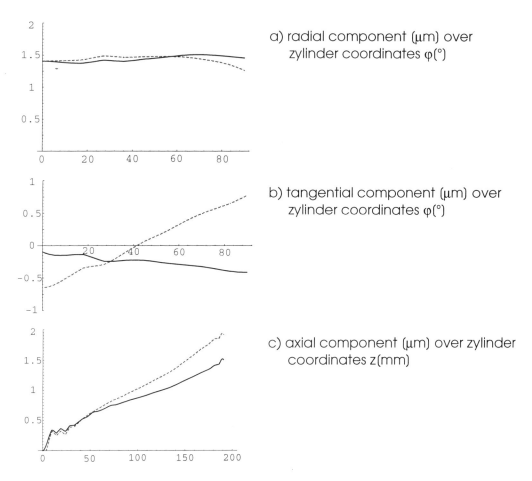

Fig. 4.8: Comparison between the calculation of the displacement components with and without the consideration of a locally dependent sensitivity (--- constant sensitivity, ——— variable sensitivity)

As boundary condition for modeling only the clamp at the lower border (no displacement) and the pressure load are used. Fig. 4.7c shows the deformed shell due to the inner pressure as a cut through the FEM-net. The simulation returns a radial displacement that is constant because of the rotational symmetry along the cross section, Fig 4.9a. As expected, the tangential component is zero, Fig. 4.9b. The axial displacemet increases almost linearly along the longitudinal axis.

In comparison with the simulation the experimentally measured displacement components (taking into account a locally variable sensitivity), Fig. 4.9, show marginal deviations of the constant course and a negative slightly decreasing tangential component.

Reasons for the appearing differences can be: rigid body displacements of the cylinder caused by variations of the inner pressure, possible phase offsets due to inaccurate location of the zero order fringe during the evaluation of the fringe pattern and simplifications concerning the geometry and the boundary conditions for the simulation. Nevertheless it can be stated that the consideration of the locally variable sensitivity for the in-plane components contributes to a cosiderable improvement of the measured displacements.

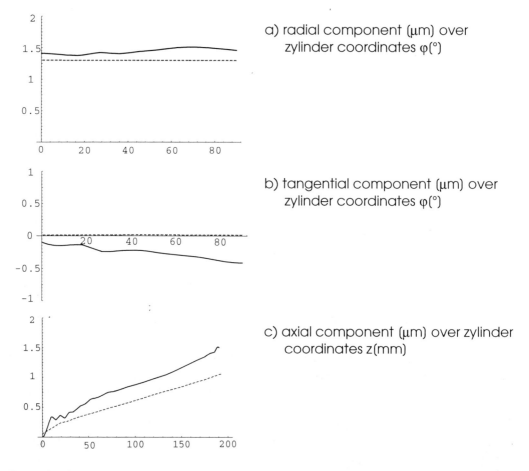

Fig. 4.9: Computed (---) and measured (——) displacement components

a) radial component (µm) over zylinder coordinates φ(°)

b) tangential component (µm) over zylinder coordinates φ(°)

c) axial component (µm) over zylinder coordinates z(mm)

4.2 Acquisition of the data

The objective of the measurement is the determination of vector displacements of a point field. To this purpose the three Cartesian coordinates of all relevant object points and some other geometrical parameters have to be known in a predefined coordinate system to determine the sensitivity vectors. In practice the measurement of these coordinates is performed more and more with special topometric methods using structured illumination by projected fringes and evaluation with the triangulation principle (122), (140). Consequently the conventional time consuming and inaccurate process using e.g. a measuring tape is replaced by high precision phase measuring techniques. In this sense interferometric displacement measurement including optical shape measurement gets an uniform basis in optical phase measurement, see section 3. To present the complete process, the next chapter gives an introduction in modern optical methods for shape measurement.

4.2.1 Shape measurement by projected fringe technique

This technique is based on the structuring of the illumination of the object by the projection of 2d-transparencies (e.g. fringe patterns) onto its surface. Dependent from the topology of the surface the fringes are distorted. A schematic set-up of the projected fringe technique with the two main components - the projector and the camera - and the relevant geometric parameters are shown in Fig. 4.10. The process of measurement can be divided into two steps:

- To each image point of the object a phase or phase difference value is assigned as the primary measuring quantity
- Based on a geometric model of the image formation process the 3D-coordinates are determined using these phase values and certain system parameters of the measurement unit have to be identified in advance.

For the first step the same methods as described in section 3. can be used. To avoid the unwrapping problems discussed above absolute techniques should be applied. Two promising techniques are known: the coded light or Gray Code technique (118) and the combination of at least two projected patterns with different spatial frequencies or rotation positions to increase the range of uniqueness by the active generation of adapted synthetic wavelengths Λ, see section 3.3.4 (120), (121). Sign ambiguities and problems with edges or shadows are irrelevant in this case.

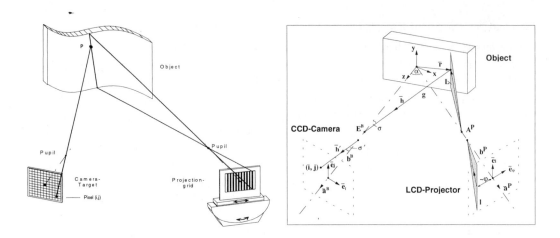

a) Schematic set-up   b) Relevant geometric parameters

Fig. 4.10: Schematic set-up of the fringe projection method for coordinate measurement

The Gray Code technique (118) for time space encoding of the fringe pattern with respect to the fringe number is a form of sequential distance numbering, often used in position encoders, arranged so that here is a change on one bit only in each measurement step. Illumination of the whole scene is via a programmable liquid crystal light valve projector (LCLV), which can project successive patterns of stripes, the scene is viewed at an angle by a camera, in an arrangement that enables triangulation, Fig. 4.11. The first projected pattern is half black, half white; the second has (offset) alternate stripes one quarter the width of the field, the next has ones one-eight width, and so on. The images derived from successive fields are stored in separate binary framestores (one bit per pixel). By examining the bit pattern in the successive stores relating to any one pixel, its „z" position can be calculated unumbiguously. Thus it takes only ten images capture operations to obtain a complete x,y,z map for every pixel in the entire field.

However, the binary nature of the Gray Code results in a limited accuracy. Threrefor a combination of Gray Code and phase shifting (141) ensures as well uniqueness as high resolution of the results.

The basis of the evaluation is the *triangulation principle*: A light point projected onto the surface is observed under the so-called triangulation angle θ, Fig. 4.12a. Using an optical system this point is imagined on a light sensitive sensor such as a CCD-line. Consequently the measurement of the hight Δz is reduced to the measurement of the lateral position Δx on the CCD-Chip. For the calculation of Δz the imaging geometry and the triangulation angle is needed. The derivation of a relation between the measured quantities and the distance Δz taking into account the imaging of the scattered light can be found in (142).

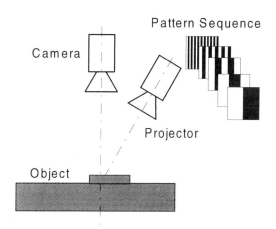

Fig. 4.11: Triangulation set-up and Gray Code pattern sequence

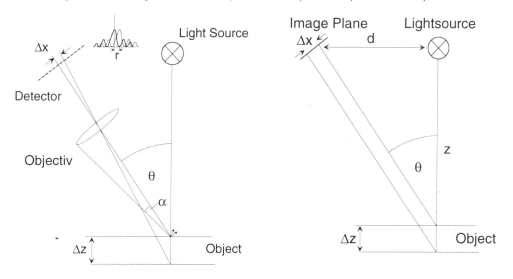

a) Triangulation principle

b) Simplified model of a triangulation sensor

Fig. 4.12: Triangulation principle

However a simplified model as given in Fig. 4.12b clarifies already essential dependences. For the position change of the spot on the sensor is valid

$$\Delta z = \frac{\Delta x}{\sin \theta} \quad \text{with} \quad \theta = \arctan\frac{d}{z} \quad . \tag{4.23}$$

Equ. (4.23) shows that the sensitivity of the range sensor with respect to Δz increases with increasing triangulation angle θ.

In complex optical set-up's as used in optical shape measurement the 3d-coordinates of the object are calculated on the basis of a geometric model (143). The model depicted in Fig. 4.10b and the related equations between the input and output data are based on the idea that object coordinates can be determined from the intersection of principal observation rays and projector generated surfaces, that means surfaces of constant phase value φ. In the case of fringe projection these are the light sections L resulting from the projection of a particular grating line into the object space. The constant phase value φ assigned to each light section can be geometrically interfpreted as the distance of the corresponding grating line to some fixed point on the grating, for instance the crossing point of the optical axes. On the observation side the known image coordinates (i,j) of each object point determine the direction of the principal ray emerging from that object point. To realize such a modelling approach the first step is to find a suitable geometric description of principal rays. From the well-known relationships of geometrical optics a vector formula can be derived, connecting the directions of the input ray $\bar{h}$, the the optical axes $\bar{a}$ and the output ray $\bar{h}'$:

$$\bar{h}' = (\kappa^B - 1)(\bar{a}^B \cdot \bar{h})\bar{a}^B + \bar{h} \tag{4.24}$$

with

$$\kappa^B = \tan\sigma / \tan\sigma' \quad . \tag{4.25}$$

The parameter κ considers the basic properties of the optical system. On the basis of the principal ray formula (119) the equation of the straight line g of the observation ray emerging from the object point with the coordinate vector $\bar{r}$ and passing through the entrance pupil $\bar{r}_{E''}$ to the camera target point with pixel indices i,j can be derived

$$\bar{r} = \bar{r}_{E''} + \lambda\left[\bar{c}_0 + \bar{c}_i(i - i_0) + \bar{c}_j(j - j_0)\right] \tag{4.26}$$

The vector coefficients $\bar{c}$ are functions of the system constants $\{\bar{a}^B, \bar{e}_i, \bar{e}_j, \kappa^B, b^B, ...\}$ described in Fig. 41b and λ is a distance parameter.

In a similar manner the plane equation describing the light section with the constant phase value φ can be written in the form

$$\bar{r} = \bar{r}_{A^P} + v\left[\bar{d}_0 + \bar{d}_\varphi(\varphi - \varphi_0) + \bar{d}_I \mu\right] . \tag{4.27}$$

The vector coefficients $\bar{d}$ are specified as functions of the system constants $\{\bar{a}^P, \bar{e}_\varphi, \bar{e}_I, \kappa^P, b^P, ...\}$. The vector relations (4.26) and (4.27) represent an equation system with the unknown parameters $\{\lambda, \mu, v\}$. Its solution for λ inserted into Equ. (4.26) yields the basic formula of the geometric model which relates the intersecting 3D-coordinates $\bar{r}(x,y,z)$ to the primary measuring quantities i,j and j(i,j) by a set of geometric-optic parameters $\bar{c}$ and k which characterize the overall measuring set-u:

$$\bar{r} = \bar{r}_{E^{II}} + \left[\bar{c}_0 + \bar{c}_i(i-i_0) + \bar{c}_j(j-j_0)\right] \frac{\tilde{k}_0 + \tilde{k}_\varphi(\varphi - \varphi_0)}{k_0 + k_i(i-i_0) + k_j(j-j_0) + \left[k_\varphi + k_{i\varphi}(i-i_0) + k_{j\varphi}(j-j_0)\right](\varphi - \varphi_0)} \tag{4.28}$$

As an alternative way to the necessity of a high accurate measurement and adjustment of the optical setup using external measuring tools a procedure for the identification of the coefficients $\bar{c}$ and k of the geometric model (Equ.123) based on the solution of the inverse problem for a known calibration object is given. This can be realized in a two step calibration procedure with the help of a precise calibration plane and adequate calibration points:

1. A linear camera calibration technique (144) yields the external orientation of the calibration plane (Fig. 4.13), i.e. the parameters $\bar{r}_{E^B}, \bar{c}_0, \bar{c}_i$ und $\bar{c}_j$ of the observation model, Equ.(4.26). To ensure a unique solution of the identification problem, at least two different orientations of the calibration plane should be used, Fig 4.13.

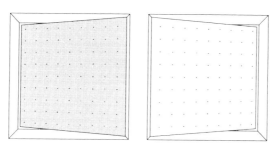

Fig. 4.13: Calibration plane with calibration points

To estimate the relative orientations of the camera with respect to the calibration planes the observation model Equ. (4.26) is transformed to

$$(i_n - i_0)\bar{e}_j \cdot (\bar{r}_n - \bar{r}_{E^{II}}) - (j_n - j_0)\bar{e}_i \cdot (\bar{r}_n - \bar{r}_{E^{II}}) = 0 , \quad (n=1,...,N; N \geq 5) \tag{4.29}$$

using the orthonormal property of the basis sytem $\{\bar{a}^B, \bar{e}_i, \bar{e}_j\}$, i.e. $|\bar{a}^B| = |\bar{e}_i| = |\bar{e}_j| = 1$ and $\bar{a}^B \cdot \bar{e}_i = \bar{a}^B \cdot \bar{e}_j = \bar{e}_i \cdot \bar{e}_j = 0$ with the unknown parameters $\{\bar{e}_i, \bar{e}_j, \bar{r}_{E^B}\}$. This relationship is linear with respect to the external orientation parameters $\{\bar{e}_{i1}, \bar{e}_{i2}, \bar{e}_{j1}, \bar{e}_i \cdot \bar{r}_{E^B}, \bar{e}_j \cdot \bar{r}_{E^B}\}$ if the internal camera parameters are known (144). Using this solution the vector parameters $\{\bar{e}_i, \bar{e}_j, \bar{a}, \bar{r}_{E^B}\}$ can be composed and consequently the original vector coefficients $\bar{r}_{E^B}, \bar{c}_0, \bar{c}_i$ and $\bar{c}_j$ of the model(4.28) can derived. In practice it is important to use a sufficient number of calibration points and their corresponding image coordinates with subpixel accuracy to estimate the system parameters in an optimal manner.

2. Using these parameters $\bar{r}_{E^B}, \bar{c}_0, \bar{c}_i$ and $\bar{c}_j$ for a certain number of calibration points $\bar{r}(i,j)$ the distance parameter λ can be determined:

$$\lambda(i,j) = \frac{\bar{r}(i,j) - \bar{r}_E^B}{\bar{c}_0 + \bar{c}_i(i - i_0) + \bar{c}_j(j - j_0)} \qquad (4.30)$$

Taking into account that the system parameters in the basic formula of the geometric model (4.28) must be determined only up to a common factor and the phase offset φ_0 (provided that it is constant for all measuring points) can be integrated into these parameters, the following linear identification task can be solved with respect to the modified parameters $\tilde{k}'_\varphi, k'_0, k'_i, k'_j, k'_\varphi, k'_{i\varphi}$ and $k'_{j\varphi}$:

$$\lambda(i_n, j_n)\{k'_0 + k'_i(i_n - i_0) + k'_j(j_n - j_0) + [k'_\varphi + k'_{i\varphi}(i_n - i_0) + k'_{j\varphi}(j_n - j_0)]\varphi_n\} - \tilde{k}'_\varphi \varphi_n = 1$$

$$(n = 1, \ldots, N; \ N \geq 7) \qquad (4.31)$$

The index n is valid for the total number of selected calibration points ($N \approx 200$). Because the phase values φ_n are inaccurate just as the previously determined camera parameters the coefficients of the projection unit $\tilde{k}'_\varphi, k'_0, k'_i, k'_j, k'_\varphi, k'_{i\varphi}$ and $k'_{j\varphi}$ can be identified with limited accuracy only. Fig. 4.14 shows the distribution of the systematic residual error with respect to the height coordinate z for a relative phase error of 0.05 fringe periods. Fig 4.15 shows exemplary the residual systematic distortions of the object plane $z = 0$. Although the derived error distributions are mainly caused by the random distribution of the input errors their scale will be reproducible within the shown amount.

Fig. 4.16 shows a technical component and ist reconstructed shape using the projected fringe technique.

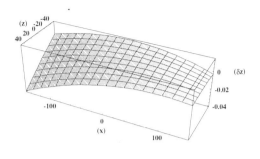

Fig. 4.14: Distribution of the residual distortions of the height coordinate within the plane y = 0 (mm)

Fig. 4.15: Systematic distortions of the object plane z = 0 (mm)

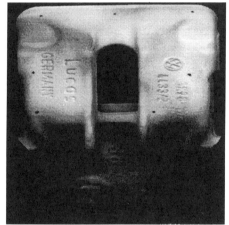

a) Object (car brake saddle)

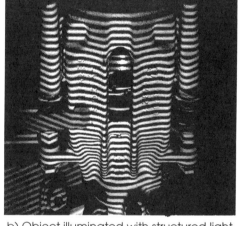

b) Object illuminated with structured light

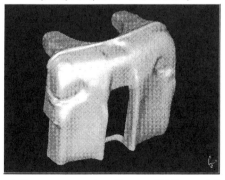

c) Shape reconstruction as CAD-model

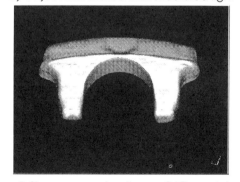

d) Shape reconstruction as CAD-model

Fig. 4.16: Shape measurement on example of a car brake

4.2.2 Shape measurement by holography

Interferometry is the way to transform phase differences $\delta(x,y)$ of wavefronts into observable intensity fluctuations - called interference fringes. In general a change of the interferometer components between two states/exposures as for instance a displacement $\mathbf{d}(u,v,w)$ of the object $P(x,y,z)$, a change of the illumination wavelength $\Delta\lambda = \lambda_1 - \lambda_2$ or the illumination direction $\mathbf{e}_Q(P)$ generates phase differences and consequently fringes with the fringe order $N(P)$. The connection between the observed fringes and the quantities to be measured gives the generalized basic equation of holographic interferometry (23), (145), (146):

$$\delta(x,y) = \frac{2\pi}{\lambda_C} \{ [\mathbf{e}_B(P) + \mathbf{e}_Q(P)] \cdot \mathbf{d}(P) - \mathbf{e}_Q(P) \cdot \mathbf{q} + \mathbf{e}_R \cdot \mathbf{r} + [\mathbf{e}_B(P) + \mathbf{e}_R] \cdot \mathbf{h}$$
$$+ [r_Q(P) + r_H(P) - r_R] \cdot \frac{\Delta\lambda}{\lambda_1 \cdot \lambda_2} \cdot \lambda_C + [r_Q(P) + r_H(P) - r_R] \cdot \Delta n \}$$

(4.27)

where λ_C is the wavelength of the reconstruction wave, $\mathbf{e}_B$ is the unit vector in observation direction; $\Delta n = n_1 - n_2$ denotes the difference of the refractive index between two exposures; $\mathbf{q}$, $\mathbf{r}$ and $\mathbf{h}$ are displacements of the illumination source Q, the reference source R and the hologram H, respectively; r_Q, r_H and r_R are the distances between the source and the object point, the hologram and the object point and the hologram and the reference source, respectively, Fig. 4.17.

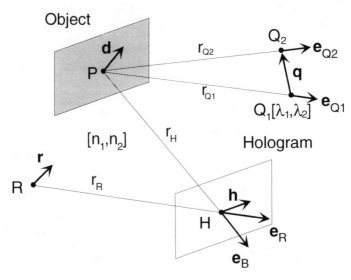

Fig. 4.17: Holographic interferometer with relevant parameters

Holographic techniques can be also applied advantageously for the measurement of coordinates and surface contours of complex 3D objects. Three main techniques can be distinguished which are based on the systematic variation of some interferometer parameters: 2-wavelength-method, 2-refractive-index-method, two-source-method.

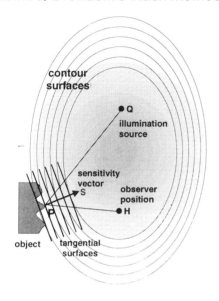

Fig. 4.18: System of rotationally symmetric ellipsoids

Fig. 4.19: Sculpture with contour lines

The *2-wavelength-method* is based on a double exposure of the hologram with two different wavelength λ_1 und λ_2 illuminated in succession by one source Q, Fig. 4.17. In the result of the simultaneous reconstruction and superposition of both wavefields a system of rotationally symmetric ellipsoids is generated in space, Fig. 4.18. The intersections of this system with the object can be interpreted as contour lines. Refering to Equ. (4.27) one can write for the phase difference:

$$\delta(x,y) = \frac{2\pi}{\Lambda}\left(r_Q + r_{II} - r_R\right) \text{ with } \Lambda = \frac{\lambda^2}{\Delta\lambda} \text{ and } \lambda^2 \approx \lambda_1 \cdot \lambda_2 \tag{4.28}$$

The term Λ is the synthetic wavelength as introduced in section 3.3.4. Using a telecentric systems it can be shown that the corresponding height difference Δz of two succeding contour lines is

$$\Delta z \approx \frac{\lambda^2}{2 \cdot \Delta\lambda} \tag{4.29}$$

With the two wavelength λ_1=488nm and λ_2=514,5nm of the argon laser one gets for instance $\Delta z \approx 4,5\ \mu m$. To receive larger synthetic wavelength and height differences, respectively, coherent wavefronts with smaller wavelength difference generated by dye lasers or *external-cavity*-diode lasers (147) are used. Fig. 50 shows a sculpture covered with such contour lines.

The *2-refractive-index method* uses a change of the refractive index between the double exposure. Since this corresponds to a wavelength change a system of rotationally symmetric ellipsoids is generated. The corresponding height difference Δz of two succeeding contour lines is

$$\Delta z \approx \lambda / (2 \cdot \Delta n) \ . \tag{4.30}$$

In case of the 2-source-method two different illumination directions are used. Both directions can be produced by a displacement $\mathbf{d}_Q$ of the source point Q_1 between the second exposure. Similar to the superposition of 2 spherical waves a system rotationally symmetric hyperboloids is generated in space. The intersections of this system with the object can be also interpreted as contour lines. Refering to Equ. (35a) one can write for the phase difference:

$$\delta(x,y) = \frac{2\pi}{\lambda} \cdot \mathbf{e}_{Q1} \cdot \mathbf{d}_Q \ . \tag{4.31}$$

In a sufficient large distance from the source the surfaces of equal phase are approximately plane an parallel to the medium illumination direction. The corresponding height difference Δz of two succeding contour lines is:

$$\Delta z = \frac{\lambda}{2 \cdot \sin(\alpha / 2)} \tag{4.32}$$

with α as angle separation of the two sources. It should be mentioned that the sensitivity of both contouring techniques is complementary.

4.2.3 Shape measurement by digital holography

In conventional holography the analysis of the fringes and their transformation into an absolute interference order is a complicated and imperfect procedure. To increase the reliability and accuracy several phase reconstruction methods have been developed, see section 3. A direct approach to the phase distribution of the interference pattern is delivered by *Digital Holography* (148), (106). Digital Holography (DH) overcomes several drawbacks of conventional holography by storing the hologram directly in the computer using a CCD matrix as the optical sensor. The reconstruction of the object wave is performed directly in the computer. As an important consequence DH allows direct access to both the phase and intensity of the object wave. Fig. 4.20 shows a typical setup using digital holography for deformation analysis of diffusely reflecting objects.

It should be stated that the restricted resolution of the CCD Chip (about 30...50 times lower than conventional holographic photo plates) limits the maximum spatial frequency resolvable by the target. This means a small observation area (volume in which the object to be analyzed can be positioned). It has the shape of a cone with an opening angle of about $\alpha=3°$, Fig. 4.21. However, the limitation of the aperture recommends this method for the investigation of small objects. Larger objects can be recorded with the use of additional lenses. But then the lateral resolution decreases in the same way.

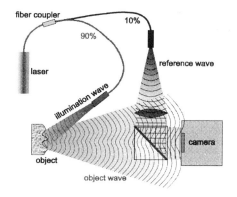
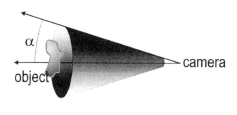

Fig. 4.20: Principle setup for Digital Holography using diffusely reflecting objects

Fig. 4.21: Maximum detectable area using holography. The resolvable spatial frequency on the CCD chip determines the spatial angle α.

Once the holograms of the two (or more) states of the object are stored in the computer, they can be reconstructed numerically. This is done by calculating the convolution of the hologram with the point-spread-function (impulse response of the free space). In a reconstruction distance equal to the original distance between object and CCD there will be the sharp picture of the object combined with its virtual picture and the 0th order. The picture contains the complex amplitude of the object wave on the surface of the object from which one can calculate intensity and phase. Doing digital holographic interferometry (DHI) means to subtract the two reconstructed phases. Consequently a mod2π phase picture can be derived which has to be unwrapped only to get the continuous interference phase $\delta(x,y)$ order. No additional manipulation as e.g. phase shifting is necessary, Fig.4.22.

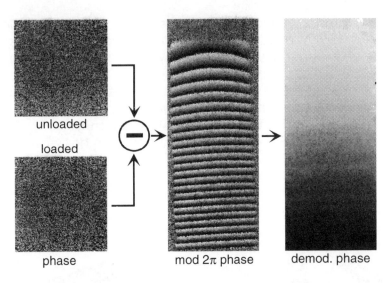

Fig. 4.22: The principle of digital holographic interferometry demonstrated on example of a loaded small beam. The interference phase is the result of the subtraction of the two phases which correspond to the digital holograms of the both object states

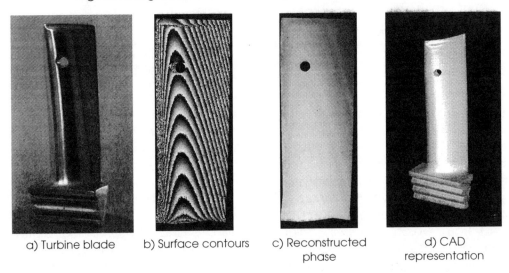

a) Turbine blade   b) Surface contours   c) Reconstructed phase   d) CAD representation

Fig. 4.23: Contour measurement of a turbine blade (height 10cm) using 2-wavelength DHI and transformation of the approximated surface into a CAD system

Besides the electronic processing and the direct phase access a further advantage of digital holography is its flexibility. The availability of the individual reconstructed phase distributions offers some new metrologic applications. Thus one may record a series of digital holograms with increased load amplitude. In the evaluation stage the convenient states can be compared interferometrically (149). All known holographic techniques can easily be implemented. Fig. 4.23 shows the application of the digital holographic 2-wavelength-contouring on a turbine blade.

Further methods determine the sensitivity vectors from known rigid body motions or rotations of the object (150), displacements of the hologram or the illumination source (151) Using Eq.1 for the simultaneous determination of the coordinates and the displacements a non-linear equation system with a corresponding number of equations has to be solved (152). The necessary high accuracy of the phase data can be achieved by modern phase reconstruction techniques but until now it is not clear how to orientate the sensitivity vectors with respect to minimum error propagation if non-linear equation systems are applied. The computational effort can be reduced if a linearized system is used (153).

4.3 Evaluation of the data

If all input data are measured the displacement components can be computed by solving the equation system (4.2). In this section our interest is focussed on the evaluation of these components with respect to their accuracy. With the methods of perturbation theory of linear algebraic systems described in section 3 we got a convenient procedure for the optimization of the interferometer. For the evaluation of the accuracy of data these methods are unsuitable because they give only an orientation for the maximum possible error.

More powerful results can be derived using the statistical error analysis. The basic results were published by NOBIS and VEST (154). The authors showed that the influence of faulty data on the accuracy of the displacement vector is controled by the geometry of the interferometer. For the derivation of the basic error relation a separate consideration of the two types of error (faulty fringe orders and faulty geometric data) was used. This method is possible because the effects of these two types of input error are additive to first order. In the following a general equation is given taking into account the both errors simultaneously and applied to the optimized interferometer (155).

The derivation is based on four basic assumptions which correspond to the physical conditions of the measuring process:

1. The measurement of the vectors **N** and **G** is repeated a large number of times. So we get Δ**N** as a random vector and Δ**G** as a random matrix.
2. All errors ΔN_i are uncorrelated, normally distributed random variables with zero mean and variance σ_{Ni}^2. Therefore

$$E(\Delta \mathbf{N}) = 0 \tag{4.33}$$

with $E(x)$ as the expectation value of the random variable x.

3. The errors of the measured Cartesian coordinates Δx_i, Δy_i and Δz_i are as well uncorrelated, normally distributed random variables with zero mean. It does not follow from this assumption that the components Δg_{ij} ($i=1...n$, $j=1,2,3$) of Δ**G** have the same properties, and we assume only:

$$E(\Delta \mathbf{G}) = 0. \tag{4.34}$$

4. Δ**N** and Δ**G** are uncorrelated:

$$E[\Delta \mathbf{N} \cdot \Delta \mathbf{G}] = E[\Delta \mathbf{N}] \cdot E[\Delta \mathbf{G}] = 0, \tag{4.35}$$

$$E[\mathbf{G}^{-1} \cdot \Delta \mathbf{N}] = 0. \tag{4.36}$$

A simple transformation of Eq. (4.4) gives a relation for the error Δ**d**:

$$\Delta \mathbf{d} = \lambda \cdot \mathbf{G}^{-1} \left[\Delta \mathbf{N} - \Delta \mathbf{G} \cdot (\mathbf{G}_0)^{-1} \cdot \mathbf{N}_0 \right]. \tag{4.37}$$

If we apply the approximation

$$\mathbf{G}^{-1} \cdot \Delta \mathbf{G} \approx \mathbf{G}_0^{-1} \cdot \Delta \mathbf{G}, \tag{4.38}$$

and Eq. (4.36) to Eq. (4.37) we get

$$E(\Delta \mathbf{d}) = 0. \tag{4.39}$$

The objective of the analysis is to find a statistical estimate of the expected error Δ**d** in terms of Δ**N** and Δ**G**. Following NOBIS and VEST we use to this purpose the covariance matrix Γ of **d** (154)

$$\Gamma = E\left[(\mathbf{d} - E[\mathbf{d}]) \cdot (\mathbf{d} - E[\mathbf{d}])^T \right]. \tag{4.40}$$

Considering Eq. (4.39) this relation can be simplified to

$$\Gamma = E[\Delta \mathbf{d} \cdot \Delta \mathbf{d}^T] \tag{4.41}$$

and a general relation, taking into account both types of error, can be derived:

$$\Gamma = \lambda^2 \cdot E\left[\mathbf{G}^{-1} \Delta \mathbf{N} \cdot \Delta \mathbf{N}^T (\mathbf{G}^{-1})^T \right] + E\left[\mathbf{G}^{-1} \Delta \mathbf{G} \cdot \mathbf{d}_0 \cdot \mathbf{d}_0^T \Delta \mathbf{G}^T (\mathbf{G}^{-1})^T \right]. \tag{4.42}$$

Based on Eq. (4.42) two special cases are discussed now.

a)   *Errors of fringe-orders ΔN*

In this case only phase measuring errors are assumed: **ΔG**=0. Consequently Eq. (4.42) is reduced to

$$\Gamma_N = \lambda^2 \cdot \mathbf{G}^{-1} E[\Delta \mathbf{N} \cdot \Delta \mathbf{N}^T] (\mathbf{G}^{-1})^T . \tag{4.43}$$

The expression $\mu = E(\Delta \mathbf{N} \cdot \Delta \mathbf{N}^T)$ is the covariance matrix of the fringe number vector **N**. If we assume that the standard deviations σ_{Ni} are identical for all measured fringe numbers N_i we have

$$\mu = \sigma_N^2 \cdot \mathbf{I} \tag{4.43}$$

with **I** as the identity matrix and finally

$$\Gamma_N = \lambda^2 \cdot \sigma_N^2 \cdot \mathbf{F}^{-1} . \tag{4.44}$$

This result correponds with that derived already by EK and BIEDERMANN (156)

b)   *Errors of geometric data ΔG*

Here it is supposed that **ΔN**=0. With this assumption and taking into account the approximation (4.38) Eq. (4.42) is reduced to

$$\Gamma_G \approx \mathbf{G}_0^{-1} E[\Delta \mathbf{G} \mathbf{d}_0 \cdot \mathbf{d}_0^T \Delta \mathbf{G}^T] (\mathbf{G}_0^{-1})^T . \tag{4.45}$$

If the requirements are met that all elements Δg_{ij} of **ΔG** are uncorrelated and that the standard deviations σ_{Sx}, σ_{Sy} and σ_{Sz} of the three sensitivity vector components are identical for all sensitivity vectors we can use the approximation (8)

$$E[\Delta \mathbf{G} \mathbf{d}_0 \cdot \mathbf{d}_0^T \Delta \mathbf{G}^T] \approx [\sigma_{Sx}^2 \cdot u_0^2 + \sigma_{Sy}^2 \cdot v_0^2 + \sigma_{Sz}^2 \cdot w_0^2] \cdot \mathbf{I} \tag{4.46}$$

and Eq. (4.45) can be rewritten as follows

$$\Gamma_G \approx [\sigma_{Sx}^2 \cdot u_0^2 + \sigma_{Sy}^2 \cdot v_0^2 + \sigma_{Sz}^2 \cdot w_0^2] \cdot \mathbf{F}^{-1} . \tag{4.47}$$

c)   *Errors of fringe-orders and geometric data*

On the condition that both types of error are additive to first order we get a formula to calculate the errors of the displacement components dependent on the phase measuring and coordinate measuring errors:

$$\Gamma \approx \Gamma_N + \Gamma_G \approx [\lambda^2 \cdot \sigma_N^2 + \sigma_{Sx}^2 \cdot u_0^2 + \sigma_{Sy}^2 \cdot v_0^2 + \sigma_{Sz}^2 \cdot w_0^2] \cdot \mathbf{F}^{-1} = C_{NG} \cdot \mathbf{F}^{-1} . \tag{4.48}$$

The main diagonal elements of the covariance matrix Γ are the variances σ_u^2, σ_v^2 and σ_w^2 of the three displacement components. Therefore we describe the square roots of the three main diagonal elements $\bar{f}_{ij}$ (i=j=1,2,3) of the inverse normal matrix $\mathbf{F}^{-1}$ as *error factors* since they control as weight factors the propagation of the input errors (considered in C_{NG}) to the standard deviations of the displacement components:

$$f_u = \sqrt{\overline{f_{11}}} \approx \frac{\sigma_u}{\sqrt{C_{NG}}}$$

$$f_v = \sqrt{\overline{f_{22}}} \approx \frac{\sigma_v}{\sqrt{C_{NG}}} \quad (4.49)$$

$$f_w = \sqrt{\overline{f_{33}}} \approx \frac{\sigma_w}{\sqrt{C_{NG}}} .$$

Because the $\overline{f}_{ij}$ are controlled by the geometrical parameters of the set-up the error factors f_u, f_v and f_w represent also the influence of the geometry on the measuring accuracy. Consequently we have an additional criterion for the assessment of the interferometer geometry. But in contrast to the condition number κ this criterion is more relevant because it gives an approach to the separate evaluation of the error influence for each of the three dislplacement components. For the estimation of the standard deviations of the displacement components follows:

$$\begin{pmatrix} \sigma_u \\ \sigma_v \\ \sigma_w \end{pmatrix} = \sqrt{C_{NG}} \cdot \begin{pmatrix} f_u \\ f_v \\ f_w \end{pmatrix} . \quad (4.50)$$

This result can be applied to the optimized interferometer described in section 4.1.3 In the case of a 360°/n-symmetry the normal matrix **F** has only elements in the main diagonal for the object point $P_0(0,0,0)$ (see Eq. (4.17)). That provides very simple formulas for the calculation of the error factors:

$$f_u(P_0) = \sqrt{\frac{2(\xi^2+1)}{n \cdot \xi^2}} , \quad (4.51)$$

$$f_v(P_0) = \sqrt{\frac{2(\xi^2+1)}{\cdot n \cdot \xi^2}} , \quad (4.52)$$

$$f_w(P_0) = \frac{\sqrt{\frac{\xi^2+1}{n}}}{1+\sqrt{\xi^2+1}} . \quad (4.53)$$

Using these equations it is possible to compare the different behaviour of the in-plane components u, v and the out-off plane component w with respect to the error propagation dependent on the both parameters ρ and L which control the aperture of the interferometer. Such a comparison is shown in

Fig. 4.24. The better sensitivity of the zero-order fringe method for the out-off plane component represented by an out-off plane sensitivity vector (see Eq. (35a)) leads to a lower error sensitivity too. With increasing distance L this quality is changing slowly compared to the increasing error sensitivity for the in-plane components. The consideration of that behaviour is very important for the investigation of large objects where long distances between the object and the hologram are necessary.

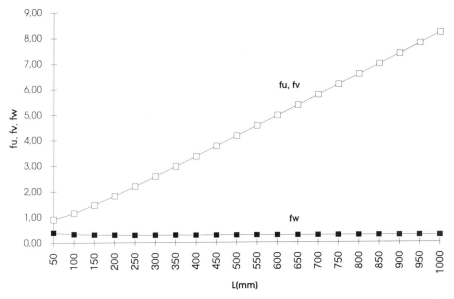

Fig. 4.24: Dependence of the error factors of the in-plane components f_u, f_v and the out-off-plane component f_w on the orthogonal distance L for the object point Po(0,0,0) and a radius $\rho = 100$ mm

Using Eq. (4.51)-(4.53), the dependence of the measuring accuracy on the number n of equations can be investigated. If (n+m) instead of n equations and observation or illumination directions, respectively, are used for all error factors Eq. (4.54) is valid:

$$f_{u,v,w} \propto \frac{1}{\sqrt{n}}$$

$$f_{u,v,w}(n+m) = f_{u,v,w}(n) \cdot \sqrt{\frac{n}{n+m}}$$

(4.54)

Fig. 4.25 shows the decreasing trend of the in-plane error factors with increasing number m of additional measurements. But it is obvious that the more equations are used the lower is the additional gain in accuracy. The percentage increment in accuracy $Z_a(\%)$ for (n+m) instead of n equations can be estimated by Eq. (154):

$$Z_a[\%] = 100 \cdot \left\{ 1 - \sqrt{\frac{n}{n+m}} \right\} \quad . \tag{4.55}$$

In accordance with DHIR and SIKORA (157), the largest increment is already obtained if only one additional equation is used (n=3, m=1). This characteristic behaviour of Z_a is illustrated in Fig. 4.26.

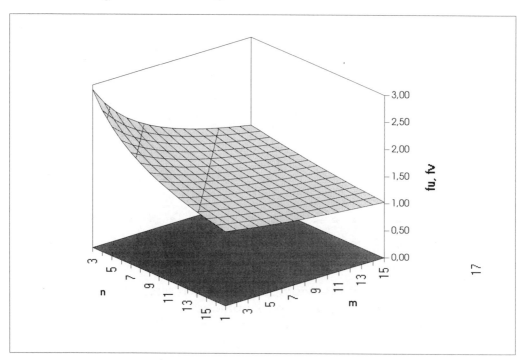

Fig. 4.25: Dependence of the error factors f_u and f_v for in-plane components on the number n of equations

Using these results the experiment can be better planned and controlled with respect to the accuracy of the measured displacement components.

Digital Processing and Evaluation of Fringe Patterns

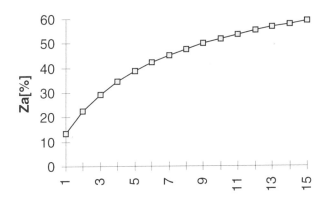

Fig. 4.26: Percentage increment in accuracy $Z_a(\%)$ for (n+m) instead of n equations (n=3)

a) Interferogram of the loaded car brake

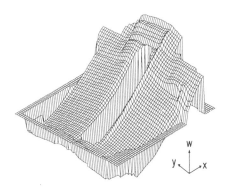

b) Pseudo-3D-displacement plot

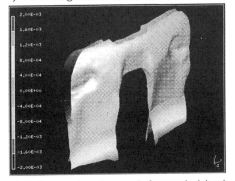

c) Deformed and non-deformed object

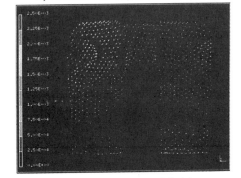

d) 3D-Displacement field

Fig. 4.27: Results of holographic-interferometric displacement analysis

4.5 Displacement calculation and data presentation

If all input data are measured the displacement components can be computed by solving the equation system (4.2). It should be considered, however, that the data acquired with holographic interferometry and optical shape measurement have to be transformed into a joint coordinate system. That means for both data sets a registration problem has to be solved (158), (159). Finally the computed displacement data can be mapped onto a FEM-net to present them within the CAD-system (158), Fig. 4.27, and to provide valuable boundary conditions for the FEM-calculation.

5. TECHNIQUES FOR THE QUALITATIVE EVALUATION OF FRINGE PATTERNS

Holographic and speckle interferometry can be applied not only for high precision quantitative analysis of displacement and strain fields but also for *nondestructive inspection* of single objects and complex structures with regard to surface and internal *flaws*. These flaws are recognized by the evaluation of the resulting fringe patterns with respect to characteristic pattern irregularities as for instance "bull eye"-fringes, distorted fringes, locally compressed fringes, cut and displaced fringes. Based on practical experience and knowledge about the material behaviour as well as on the boundary conditions of the experiment the flaw can be classified as e.g. *void, debond, delamination, weak area, crack* or similar defect. The methods have been applied successfully for industrially relevant testing problems in different fields: quality control of circuit boards and electronic modules, inspection of satellite fuel tanks and pressure vessels, tire testing, glass and carbon fibre reinforced material testing, investigation of turbine blades and motor car inspection, e.g. (160). However, the automatic evaluation of the interferograms failed for a long time, since the recognition of fault indicating changes in a complex interferogram is more difficult than the phase reconstruction for deformation measurements: The task consists not only in the automatic recognition of complex patterns within noisy interferograms but also in the decisive evaluation of these patterns with regard to the variety of flaw-induced deviations. First results were reported by GLÜNDER et al. (15) in 1982. In this work an opto-electronic hybrid processor was used for fast and very effective data reduction. A complete digital system for the analysis of misbrazing in brazed cooling panels was proposed by ROBINSON (52) in 1983. When the plate is slowly pressurized misbrazing is observed as a closed-ring fringe pattern before the standard deformation pattern appears. However, the published evaluation procedure is very time consuming. Another approach (161) was proposed that applies parallel hardware. In less than one minute the skeleton of the fringe pattern is derived and two line features (density and curvature) are determined to recognize flaw induced patterns. The most obvious deficit in all published procedures is the insufficient capability for *flaw classification*. Two modern approaches have good potential to overcome this unsatisfactory situation. The first one applies *knowledge-based systems* or *neural networks* to "learn" different kinds of flaws from simulated or practical examples (162), (163), (164). The second approach combines theoretical *simulation methods* and practical measurements (165), (167).

A short description of the two most promising approaches is given in the following: neural networks and knowledge-based systems. Both approaches have common qualities as e.g. the preprocessing of noisy interferograms and the selection of representative *features*, but also important differences in the recognition architecture used. These aspects are discussed using examples of simple, so-called *basic fringe patterns* (162), (168). For these pattern types, spot checks are generated using mathemetical simulation and practical preparation of loaded samples. Furthermore the choice of robust *features* discriminating different basic patterns (classes), and the proposal for a special system architecture, are discussed.

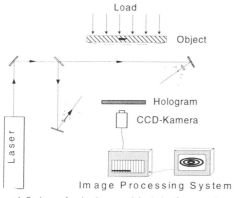

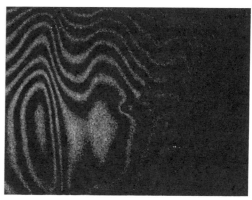

a) Set-up for holographic interferometry

b) Example of an interferogram wiht a flaw indicating pattern

Fig. 5.1: Holographic nondestructive testing

5.1 The technology of HNDT

The basic idea in *holographic non-destructive testing* (HNDT)[2] is that a component with a fault will react in a different manner compared to a sound structure, since the stiffness or the heat conductivity will be modified locally or globally (169). This results in changed fringe pattern when the component is

---

[2] Holographic shall be representative for the different coherent-optical methods in this chapter

tested by holographic interferometry, Fig. 5.1a. The fringe pattern - an example is given in Fig. 5.1b - can be evaluated to determine the displacement and the deformation. However, in HNDT the question is not to evaluate deformations but to determine unacceptable deformations identified by inhomogeneities in the fringe system. Furthermore, in industrial applications the main question is only to detect fault-indicating fringe irregularities and to classify them; but not all inhomogeneities indicate a defect. The idea of determining the full field of deformations and to compare the strains to allowed values is too complex and time consuming and will not match industrial needs, in general.

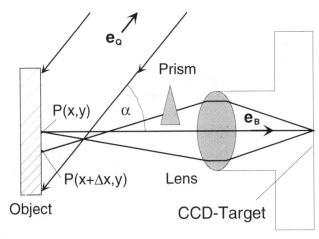

Fig. 5.2: Experimental set-up for shearography

Speckle shearography for instance is a very promising technology for industrial inspection. It delivers a direct approach to derivations of the displacement components by shearing of two wavefronts coming from the same object, Fig 5.2. Consequently one point in the image plane receives contributions from two different points P(x,y) and P´(x+Δx,y) on the object (here a lateral shear Δx was used exemplary). Because of a deformation both points are displaced relatively which results in a relative phase difference δ(P,P´) (171):

$$\delta(P,P') = 2\pi/\lambda \{[\partial w/\partial x] \cdot (1+\cos\theta) + [\partial u/\partial x] \cdot \sin\theta\} \cdot \Delta x \qquad (5.1)$$

with θ as the angle between the z-axis and the observation direction. Shearography is a very robust technique since the both sheared wavefronts play the role of a mutual reference. Consequently the disturbing influence of rigid body motions in HNDT is removed. As example of non-destructive evaluation of a large industrial component Fig. 5.3 shows a shearogram of an aircraft sandwich panel that was loaded by a light flash to show some inner faults.

For some applications as for instance the investigation of cultural objects a higher resolution is necessary as speckle techniques can bring. In this case phase shifting holographic interferometry is of advantage, Fig. 5.4.

 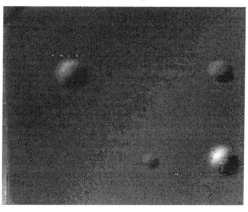

a) Shearogram   b) Demodulated phase image

Fig. 5.3: Shearogram of a aircraft sandwich plate loaded by a light flash

Statistical methods have been applied to the fringe system, since the beginning of automatic evaluation, in order to detect inhomogeneities (170). The idea was to detect anomalously higher and lower fringe densities in either the spatial or frequency domain. This works quite well for simple geometries. However, when the geometry of the component is complex, the fringe system becomes complex and will often contain inhomogeneities even in the absence of defects. So these methods fail for widespread application of HNDT. The fact, that skilled operators can detect fault-indicating fringe pattern changes even in complex structures leads to the approach od deusing processing techniques which apply apply knowledge to the evaluation of the fringes.

5.2 Evaluation with neural networks

An artifical neural network is the attempt of a computer model to simulate the functionality of a biological brain in a fundamental manner (172). The unit comparable to the *neuron* is refered to as a processing element. Every neuron - e.g. neuron j - has a number n of *input paths* x_i, i=1,...,n, representing the dendrites.

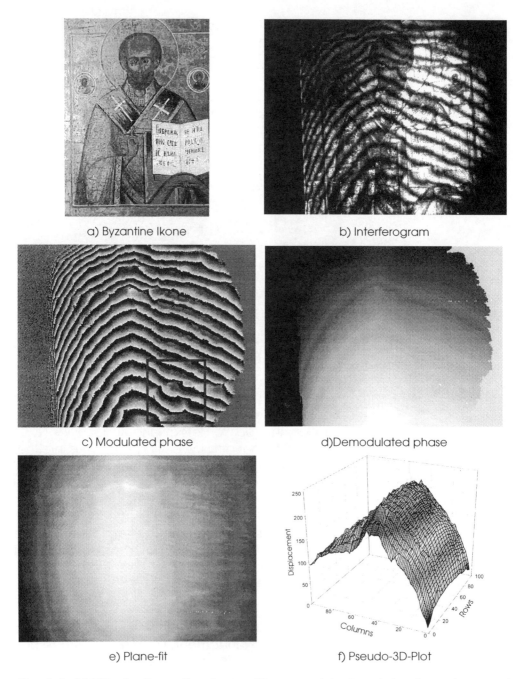

Fig. 5.4: HNDT of a Byzantine icon with respect to the detection of separated layers

Each input is multiplied by a *weight* w_{ij} representing the synaptic strength. The weighted values are summed to an internal activity level I_j, which is modified by a transfer function f. Thus the output y_j of the processing element j is given by

$$y_j = f(I_j) = f\left(\sum_i w_{ij} x_i\right) \qquad (5.1)$$

The transfer function f can be a threshold function to pass information only if the combined activity reaches a certain level, or it can be a continuous function of the internal activity level. The standard for *back-propagation learning* (used in this work) is the *sigmoid function* $f(z)=(1+\exp(z))^{-1}$.

The processing elements are in general organized in groups called layers: A typical network consists of a layer of source elements for the data input, followed by one, two or (sometimes) more so-called hidden layers, and an output layer with the response of the network, Fig. 5.5. The neurons of two successive layers may be fully connected or only special connections may be selected (13).

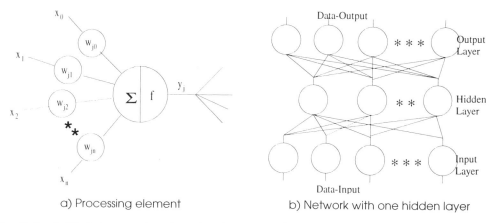

a) Processing element   b) Network with one hidden layer

Fig. 5.5: Artificial neural network: a) processing element; b) network with one hidden layer

Training the neural network means to input a feature vector to the source neurons and judging the output to be right or wrong. The structure of the network is created by changing the weights of the processing elements as long as the output differs from the desired one. The training needs about ten examples for each feature, i.e. each component of the feature vector. This has to be taken into account when the feature vector shall be defined.

The design of a neural network for the evaluation of HNDT fringe patterns with the outputs "Fault" or "No fault" is demonstrated by the following successful system (163), (164). The feature vector is based on the determination of the slope values of the intensity variation in sub-areas of the fringe pattern: The image is divided into squared areas of a certain number of points, e.g. 8x8 pixels. In each of the areas the maximum slope is taken as a feature, where the slope at a given pixel is defined by fitting a two-dimensional plane taking into account the 8 neighbours. The slopes are valued in relation to those of neighbouring areas by a *Laplace-filtering*. The four highest Laplace values of the regarded areas are given to the input neurons of the network; one value to one input neuron. This procedure weights the relative slopes indepently from where they appear and allows the decision about the existence of an inhomogeneity anywhere in the pattern. However, one size of the areas can only be an optimum for a certain size of the defect. Therefore the choice for the running network was to divide the image of 512x512 pixels into areas of different sizes: 8x8, 16x16, 32x32 and 64x64 pixels.

The structure of the neural network starts with the input layer. According to the before-mentioned description of feature selection one has to have 16 input neurons: four neurons for any of the four area sizes. The second layer is a hidden layer. The number of neurons in this layer must be optimized experimentally, i.e. networks with different numbers of neurons in this layer are built-up and tested in terms of learning speed and accuracy. A number of only four neurons are found to be the best. Additionally, the connections between the neurons of the first layer and the second one have to be defined. The best choice made by experimental experience is to connect the maximum (with concern to the Laplace values) input neuron of each area size to the first neuron in the hidden layer, the second largest input neuron of each area size to the second on, and so on. To install all possible connections resulted in a prolongued learning phase with worse results.

Experimentally a neural network with a second hidden layer was found to learn better and faster than a network with only one hidden layer. The optimum number of neurons in this layer was found to be eight. This means there are more neurons in the second than in the first hidden layer. All of the neurons of the second hidden layer were connected to all neurons of the first hidden layer and the network weighted by itself. Then there are two output neurons: "Fault" and "No Fault". This structure has some advantages since it helps to survey the momentary performance of the network.

Learning of the neural network requires samples of interferograms of components with and without defects, produced under varying conditions of load and set-up. Heuristically, a training set needs approximately ten times the number of samples compared to the number of weights for the network to generalize well. For the given task a number of several hundreds were calculated to be necessary. The interferograms are mainly simulated and complemented by some experimentally ones. As test ensemble 1000 interferograms, 500 with and 500 without defect, for a wide-spread variety of conditions are simulated. All parameters are chosen statistically. The whole ensemble of interference patterns is fed to the network randomly again. With the experimental interferograms both the network and the test ensemble can be tested to work sufficiently.

The test ensemble of simulated interferograms contained a wide-spread number of different fringe patterns. Nevertheless the network with the described network structure is able to learn after about 5 - 6.000 learning steps. This work is performed by the neural network programm by itself. The time for learning these 6.000 samples is about 15 min. After the first 6.000 steps the error curve remained at "0" indicating that every decision was correct from this moment. Even really noisy interferograms could be detected after an image preprocessing in a straight on forward manner, Fig. 5.6a. The ability of the network to detect small defects was tested with a simulated interferogram, Fig. 5.6b. The network indicated the fault-indicating fringe pattern inhomogeneity correctly - although it is hard to be found by a test person. Up to now the network only indicates the appearance of a defect anywhere in the component. It is not able to show the location or classify the severity of the defect. For this purpose knowledge-based systems are developed.

a) Interference pattern   b)) Smallest inhomogeneity

Fig. 5.6: Example of an experimental interference pattern and the smallest inhomogeneity that can be detected by neural network

5.3 Evaluation with knowledge based systems

A *knowledge based system* is the alternative to neural networks to include knowledge into the evaluation of fault indicating fringe patterns. The experience e.g. with the HNDT of glass-fibre reinforced plastic (GRP) tubes showed that a limited number of different fault types can be stated for a given component including the material. Furthermore, these faults result in typical forms of the fringe pattern inhomogeneity which can be listed and can serve to qualify the type of the fault. This leads to the hypothesis that any HNDT problem can be described by a limited number of basic fault patterns and a first approach is to list the possible fault-induced fringe distortions, to search directly for these fringe pattern properties, and to classify the faults afterwards by this knowledge base (162), Table 3. An always remaining task is to look for possible additions to the list. The procedure of processing the input fringe patterns is divided into three steps:

Step 1:   Preprocessing
Basis: rough interferograms
Objective: reduction, vectorization and list transformation of skeletons
Result: compressed data list containing relevant metrical and topological information about skeletons

Step 2:   Model based pattern recognition (1st classification step)
Basis: data list and binary pixel image
Objective: feature extraction within the data list aimed at the preclassification of fault indicating patterns
Result: regions within the image pre-classified for flaw presence

Step 3:   Knowledge based flaw classification (2nd classification step)
Basis: roughly classified regions with flaw presence
Objective: knowledge assisted fine classification taking into account additional object knowledge
Result: hypothesis about the flaw type

In each of the steps knowledge is used as well of data as of rules of evaluation.

The test ensemble for the evaluation of the knowledge base was produced experimentally as well as by simulation (168). Five examples are shown which represent the 5 pattern classes: *compression, bend, groove, displacement, eye*, Table 3. Each of the classes of flaw indicating inhomogeneities has special features such as line density, direction curvature, symmetry and shape.

Symptom	Fringe Pattern	Interferogram	Expected Flaw
Compressinon: local spatial frequency change			material separation, weak area
Bend: non-continous direction change			local separation, void
Groove: systematic and directed fringe deformation			extended separation crack under the surface
Displacement: fringe distortion			Crack on the Surface
Eye: circular/elliptic fringe pattern			void, inclusion local separation

Table 3: Classes of characteristic fault-indicating fringe patterns

Table 3 enables a decision if defects are in the component. However, to classify the quality of a defect, more knowledge about the influences of the test conditions are necessary as for instance the kind of load (thermal, mechanical, pressure, vibration, impulse), the material, the shape of the object and its fixing. Together with that process knowledge for every special application a *rule based system* of the kind

„*IF* pattern class X *AND* condition A *AND* condition B *AND* .... *THEN* flaw Y *ELSE* ..."

can be implemented. It is obvious that in practice the variety of influencing parameters in one given test task is limited since nobody would change the set-up, the fixing, the loadig in a remarkable degree.

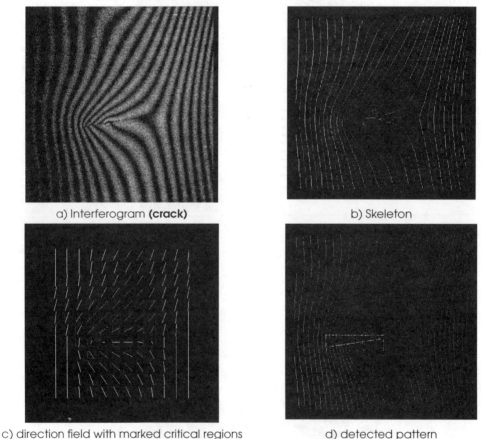

a) Interferogram **(crack)**   b) Skeleton

c) direction field with marked critical regions   d) detected pattern

Fig. 5.7: Detection of flaw indicating patterns on example of a crack

An important step is the *preprocessing* of the original fringe pattern. This step leads to a skeleton of the pattern, Fig 5.7b . By using all the approaches known from image processing research it is meanwhile possible to nearly reach the ability of a human eye in fringe detection. However, some problems remain with very low fringe densities (below 5 fringes in the pattern) and with very high numbers of fringes when the pattern is disturbed by coherent noise. The preclassification - that means the recognition of the fault indicating patterns according to Table 3 - is running automatically. The recognition ability is demonstrated for an example of the class displacement, Fig. 5.7a. The highlighted square indicates where the computer has found a fault pattern based, Fig 5.7d, on the evaluation of such features as the direction and the density of the fringes, Fig. 5.7c.

5.4 Material fault recognition in HNDT using recognition by synthesis

The diversity and ambiguity of fringe patterns in HNDE show that fixed recognition strategies based on known flaw - fringe pattern relations are successful only than if the boundary conditions of the test procedure are limited in an inadmissible way. Since such conditions cannot be guaranteed generally more flexible recognition strategies have to be developed. A new approach is based on a flexible testing strategy that combines the knowledge about the pattern formation process with the possibilities of modern CAE-tools (173). The basic idea of this method is shown in Fig. 5.8. At the beginning an interferometric measurement (holographic interferometry, EPSI or shearography) creates an interferogram containing information about the object under test. A first skeletonization and feature-extraction allows to make a hypothesis about the object and its fault. This hypothesis serves as a basis for a finite-element model of the natural object including the fault and experimental conditions. With the known circumstances of the experimental set-up a synthetic interferogram of the object is generated, followed by a second skeletonization and feature-extraction. All properties (skeleton and features) of both interferograms are compared to determine if the supposed hypothesis is correct. If the difference between synthesized and natural pattern is too big, an iterative process is started which modifies the hypothesis combined with the manipulation of the load or other known boundary conditions to improve the coincidence. As the result of this optimizing feedback loop an improved hypothesis containing the desired information about the tested object such as the type of the fault, its dimension and location is derived, Fig. 5.8.

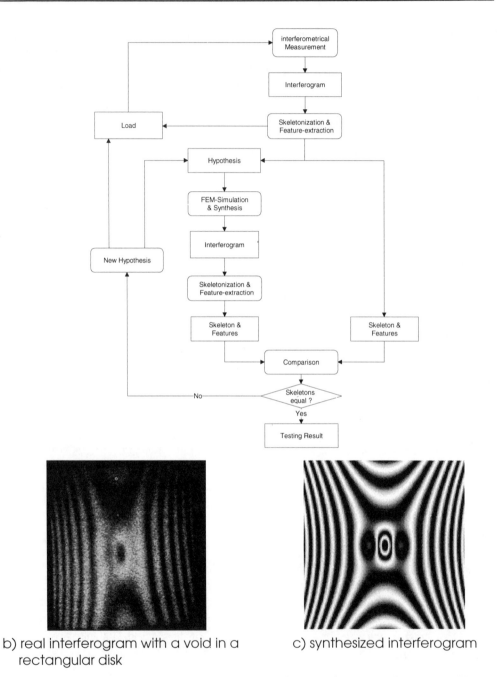

b) real interferogram with a void in a rectangular disk

c) synthesized interferogram

Fig. 5.8: Algorithm for recognition by synthesis: a) recognition algorithm

6. MODERN SOFTWARE SYSTEMS FOR DIGITAL PROCESSING OF FRINGE PATTERNS

For several years various commercial image processing systems with different efficiency and price level have been developed and specialized systems dedicated to the solution of complex fringe analysis problems are commercially available now (1), (2), (3). Advanced hardware and software technologies enable the use of desktop systems with several image memories and special video processors nearby the optical set-up. This on-line connection between digital image processors and optical test equipment opens completely new approaches for optical metrology and non-destructive testing as real time techniques with high industrial relevance. In this presentation as example the BIAS FRINGEPROCESSOR™ is decribed (3), (174).

The BIAS FRINGE PROCESSOR™ is the result of an almost 20 years research work done by many scientists of the Bremen Institute of Applied Beam Technology (BIAS) and the former Berlin Institute of Cybernetics and Information Processing (ZKI) in the field of digital fringe pattern analysis for optical metrology.

The FRINGE PROCESSOR™ is a WINDOWS™ based software system with 32 bit performance that requires the same hardware basis as WINDOWS™ itself. No special hardware such as frame grabbers or video processors are necessary. The complete processing is performed by the general purpose processor of the PC. However, a memory extension of at least 8 MByte should be guaranteed, 16 MByte are recommended. It is a main feature of the FRINGE PROCESSOR™ that the exchange of results between the system and the real world works only off-line via a standard software interface (image and data files with defined format). That makes it more easy to use the system in every hardware environment without laborious adaptations.

With respect to its processing capability the FRINGE PROCESSOR™ offers three ways of processing fringe patterns:

- SIMULATION enables to simulate the complete fringe formation process including the simulation of different noise components such as *speckle* noise and electronic noise as well as the computation of different phase distributions as result of loading defined mechanical objects such as plates and shells.
- PROCESSING provides the most important fringe processing procedures such as *phase shifting, fringe tracking, spatial heterodyning* and *Fourier transformation*.
- TOOL BOX delivers typical processing tools separately.

6.1 The simulation mode

The SIMULATION mode is based on the assumption that fringe patterns generated by holographic interferometry, structured illumination and other metrological methods are nothing but the image of the object under test after a characteristic redistribution of the intensity: the image of the object to be investigated is modified by muliplicative fringe patterns, additive random noise $R_E(\mathbf{r},t)$, and other influences. The relation between the observed intensity distribution and the quantity of interest, for example the *interference phase* $\delta(\mathbf{r})$ in holographic interferometry, is described by an *intensity model* $I(\mathbf{r},t)$. This model is based on the optical laws of interference and imaging. Additionally, the model includes the various disturbances such as background illumination $I_0(\mathbf{r})$ and multiplicative speckle noise $R_S(\mathbf{r})$ which influence both the accuracy of the reconstructed phase distribution and the way of fringe processing:

$$I(\mathbf{r},t) = 2I_o(\mathbf{r}) \cdot \{1 + V(\mathbf{r}) \cdot \cos[\delta(\mathbf{r}) + \varphi(\mathbf{r},t)]\} \cdot R_S(\mathbf{r}) + R_E(\mathbf{r},t) \qquad (6.1)$$

The variable $V(\mathbf{r})$ considers the fringe visibility and $\varphi(\mathbf{r},t)$ is an additionally introduced reference phase that discriminates the different phase measuring techniques. The complete simulation process is demonstrated on example of a cylindrical shell loaded with inner pressure (Fig. 2.7).

Although it is quite impossible to separate the effects of different kinds of noise in a real image, a physical model describing the image formation process with all its influential contributions gives the opportunity to compute synthetic images which approximate step by step the complex structure of a real image. By means of this artificial test environment the performance of algorithms and image processing tools can be studied effectively. Both error detection and the selection of adapted parameters become more convenient.

6.2 The processing mode

The PROCESSING mode is the central part of the system because it contains the most known fringe evaluation procedures with their sequences of processing steps. The quantity of primary interest in optical metrology is the phase of the fringes carrying all the necessary information about the shape and the displacement of the object. During the last 15 years several techniques for the automatic and precise reconstruction of phases from fringe patterns were developed. The FRINGE PROCESSOR™ offers those basic concepts where digital image processing is relevant (12):

– Fringe Tracking or Skeleton Method,
– Phase-Sampling or Phase-Shifting Method,
– Fourier-Transform Method and

– Carrier-Frequency Method or Spatial Heterodyning.

All these methods have significant advantages and disadvantages, so the decision for a certain method depends mainly on the special measuring problem and the boundary conditions (see section 3).

6.3 The tool box

The TOOL BOX contains many useful procedures such as the Fast Fourier Transformation (FFT), several finite impulse response filters (FIR filters) and presentation means which can be dynamically linked during the run-time of the program. In any case the operator can activate the cursor and display the intensity profile of any line cut or can zoom the image to different resolution levels. This concept of Dynamic Link Libraries (DLL) ensures that only those tools are active which are of interest in the moment.

6.4 The FRINGE PROCESSOR™ shell

The frame of the FRINGE PROCESSOR™ is the so-called FRINGE PROCESSOR *Shell*. The *Shell* itself has no image processing functions like filters or other routines, it is mainly a connection of two interfaces. On the one hand there is the visible user interface known for most users, on the other hand there is the programming interface for third party DLLs. The *Shell* and its programming interface are based on a hierarchy of C++ classes representing the scientific image types including some integer formats, a floating-point and a complex format as well as standard computer image formats like Windows™ BMP (bit map). Writing your own application for the *Shell* using the special libraries delivered with the FRINGE PROCESSOR™ you are not faced with the question how to write a Windows™-Program or to display an image in a window. Moreover you can concentrate on your actual purpose - the realization of any image processing algorithm. Most of this functionality is also accessible to DLLs via simple function calls like *ShowPicture()*, *SavePicture()* and so on. However, this structure and the API (Application Programming Interface) of the Shell are hidden under the surface, so that these users who don't want to write a program will only see one form of the Shell, the FRINGE PROCESSOR™.

7. LITERATURE

(1) Steinbichler Optotechnik GmbH, Am Bauhof 4, D-83115, Germany, „FRAMES - FRinge Analysis and MEasuring System"
(2) Warsaw Univ. of Technol., Inst. of Micromechanics and Photonics, Chodkiewicza Street 8, 02-525 Warsaw, Poland, „Fringe Laboratory"
(3) BIAS GmbH, Klagenfurter Str. 2, D-28359 Bremen, Germany, „FRINGE PROCESSOR™ - Computer aided fringe analysis under MS Windows"
(4) Halliwell, N. et al.(Eds.), Fringe Analysis´92, Proc. of the FASIG Conference, Leeds, 1992
(5) Osten, W., Pryputniewicz, R. J., Reid, G. T., and Rottenkolber, H., (Eds.), Fringe´89, Proc. 1st Intern. Workshop on Automatic Processing of Fringe Patterns, Berlin, Akademie Verlag, 1989
(6) Reid, G. T., (Ed.), Fringe Pattern Analysis, Proc. SPIE Vol. 1163, Bellingham, 1989
(7) Jüptner, W., and Osten, W.,(Eds.), Fringe´93 & Fringe´97, Proc. 2nd & 3rd Intern. Workshop on Automatic Processing of Fringe Patterns, Berlin, Akademie Verlag, 1993 & 1997
(8) Osten, W., Digital Processing and Evaluation of Interference Images, (in German), Berlin, Akademie Verlag, 1991
(9) Osten, W., and Höfling, R., "Digital Holographic and Speckle Interferometry," In: Frankowski, G., Abramson, N., and Füzessy, Z., (Eds.), Application of Metrological Laser Methods in Machines and Systems, Berlin, Akademie Verlag, 1991, pp. 265-298
(10) Robinson, D. W., and Reid, G. T., (Eds.), Interferogram Analysis, Bristol and Philadelphia, IOP Publishing Ltd, 1993
(11) Kaufmann, G. H., "Automatic Fringe Analysis Procedures in Speckle Metrology," In: Sirohi, R. J. (Ed.), Speckle Metrology, New York, Basel, Hong Kong, Marcel Dekker Inc., 1993, pp. 427-472
(12) Osten, W.; Jüptner, W.: Digital Processing of Fringe Patterns. In: Rastogi, P.K. (Ed.): Handbook of Optical Metrology. Artech House Publishers, Boston and London 1997
(13) Kreis, Th, "Computer-Aided Evaluation of Holographic Interferograms," In: Rastogi, P. K. , (Ed.), Holographic Interferometry, Berlin, Springer Verlag, 1994
(14) Schwider, J., "Advanced Evaluation Techniques in Interferometry," In: Wolf, E., (Ed.), Progress in Optics, Vol. XXVIII, Amsterdam, Elsevier Science Pupl. B.V., 1990
(15) Glünder, H., and Lenz, R.: „Fault detection in nondestructive testing by an opto-electronic hybrid processor," Proc. SPIE Vol. 370(1982), pp. 157-162

(16) Stetson, K.-A.: A rigorous theory of the fringes of hologram interferometry. Optik 29(1969), 386-400
(17) Lowenthal, S., and Arsenault, H., Image formation for coherent diffuse objects: statistical properties, J. Opt. Soc. Am. 60(1979)11, 1478-1483
(18) Goodman, J.W., "Statistical properties of laser speckle patterns," In: Dainty, J.C. (Ed.), Laser speckle and related phenomena, Berlin, Springer Verlag, 1983, pp. 9-75
(19) Dändliker, R., "Heterodyne holographic interferometry," In: Wolf, E., (Ed.), Progress in Optics, Vol. XVII, Amsterdam, Elsevier Science Pupl. B.V., 1980
(20) Tur, M., Chin, C., and Goodman, J.W., "When is speckle noise multiplicative," Appl. Opt. 21(1982), pp. 1157-1159
(21) Jenkins, T.E., Optical Sensing Techniques and Signal Processing, Englewood Cliffs, Prentice/Hall International, 1987
(22) Creath, K., „Phase measurement interferometry: Beware these errors," Proc. SPIE Vol. 1553, 1991, pp. 213-220
(23) Sollid, J.E., "Holographic interferometry applied to measurements of small static displacements of diffusely reflecting surfaces," Appl. Opt. 8(1969), pp. 1587-1595
(24) Höfling, R., and Osten, W., „Displacement measurement by image-processed speckle patterns," J. Mod. Opt. 34(1987)5, 607-617
(25) Nakadate, S., Magome, N., Honda, T, and Tsujiuchi, J., "Hybrid holographic interferometer for measuring three-dimensional deformations," Opt. Eng. 20(1981)2, 246-252
(26) Kreis, T., "Digital holographic interference-phase measurement using the Fourier-transform method," J.Opt.Soc.Am. 3(1986)6, 847-855
(27) Takeda, M., Ina, H., and Kobayaschi, S., "Fourier-transform method of fringe pattern analysis for computer based topography and interferometry," J.Opt.Soc.Am. 72(1982)1, 156-160
(28) Bruning, J.H., Herriott, D. R., Gallagher, J. E., Rosenfeld, D. P., White, A. D., and Brangaccio, D. J., "Digital wavefront measuring interferometer for testing optical surfaces and lenses," Appl.Opt. 13(1974)11, 2693-2703
(29) Creath, K., „Temporal phase measurement methods," In (10), pp. 94-140
(30) Kujawinska, M., „Spatial phase measurement methods," In (10), pp.141-193
(31) Crimmins, T.R., „Geometric filter for speckle reduction," Appl. Opt. 24(1985)10, pp. 1434-1443
(32) Davila, A., Kerr, D., and Kaufmann, G.H., „Digital processing of electronic speckle pattern interferometry addition fringes," Appl. Opt. 33(1994)25, pp. 5964-5968
(33) Yu, F.T.S., and Wang, E.Y, „Speckle reduction in holography by means of random spatial sampling," Appl.Opt. 12(1973), pp. 1656-1659

(34) Sadjadi, F.A., „Perspective on techniques for enhancing speckled imagery," Opt. Eng. 29(1990)1, pp. 25-30
(35) Jain, A.K., and Christensen, C.R., „Digital processing of images in speckle noise," In: Carter, W.H. (Ed.), Applications of speckle phenomena, Proc. SPIE Vol. 243(1980), pp. 46-50
(36) Eichhorn, N, and Osten, W., „An algorithm for the fast derivation of line structures from interferograms," J.Mod.Opt. 35(1988)10, 1717-1725
(37) Guenther, D.B., Christensen, C.R., and Jain, A., „Digital processing of speckle images," In: Proc. IEEE Conf. on Pattern Recognition and Image Processing, 1978, pp. 85-90
(38) Hodgson, R.M., Bailey, D.G., Nailor. M.J., Ng, A.L., and McNeil, S.J., „Properties, implementation and application of rank filters," Image and Vision Computing 3(1985)1, pp. 3-14
(39) Ruttimann, U.E., and Webber, R.L., „Fast computing median filters on general purpose image processing systems," Opt. Eng. 25(1986)9, pp. 1064-1067
(40) Stanke, G., and Osten, W., „Application of different types of local filters for the shading correction of interferograms," In (5), pp. 140-144
(41) Bieber, E., and Osten, W., „Improvement of speckled fringe patterns by Wiener filtering," Proc. SPIE Vol. 1121(1989), pp. 393-399
(42) Lim, J.S., and Nawab, H., „Techniques for speckle noise removal," Opt. Eng. 20(1981), pp. 472-480
(43) Ostrem, J.S., „Homomorphic filtering of specular scenes," IEEE Trans. SMC-11(1981)5, pp.385-386
(44) Winter, H., Unger, S., and Osten, W.: „The application of adaptive and anisotropic filtering for the extraction of fringe patterns skeletons," In: Proc. Fringe `89, Akademie Verlag Berlin 1989, pp. 158-166
(45) Yu, Q., and Andresen, K., „New spin filters for interferometric fringe patterns and grating patterns," Appl. Opt. 33(1994)15, pp. 3705-3711
(46) Crimmins, T.R., „Geometric filter for reducing speckles," Opt. Eng. 25(1986)5, pp. 651-654
(47) Sonka, M., Hlavac, V., and Boyle, R., Image processing, Analysis and Machine Vision, Chapman & Hall Computing, London 1993
(48) Yu, Q., Andresen, K., Osten, W., and Jüptner, W.: „Analysis and removing of the systematic phase error in interferograms," Opt. Eng. 33(1994)5, 1630-1637
(49) Dorst, L., and Groen, F., „A system for quantitative analysis of interferograms," Proc. SPIE Vol. 599(1985), pp. 155-159
(50) Vrooman, H.A., and Maas, A.A.M., „Interferograms analysis using image processing techniques," Proc. SPIE Vol. 1121(1990), pp. 655-659

(51) Yu, Q., Andresen, K., Osten, W., and Jüptner, W.: Noise free normalized fringe patterns and local pixel transforms for strain extraction. Appl. Opt. 35(1996)20, pp. 3783-3790
(52) Robinson, D.W., „Automatic fringe analysis with a computer image processing system," Appl. Opt. 22(1983)14, 2169-2176
(53) Schemm, J.B., and Vest, C.M., „Fringe pattern recognition and interpolation using nonlinear regression analysis," Appl. Opt. 22(1983)18, 2850-2853
(54) Becker, F., and Yung, H., „Digital fringe reduction technique applied to the measurement of three-dimensional transonic flow fields," Opt. Eng. 24(1985),3, 429-434
(55) Joo, W., and Cha, S. S., „Automated interferogram analysis based on an integrated expert system," Appl. Opt. 34(1995)32, 7486-7496
(56) Mieth, U., and Osten, W.: „Three methods for the interpolation of phase values between fringe pattern skeleton," Proc. SPIE Vol. 1121, Interferometry'89, 1989, pp. 151-154
(57) Burton, D.R., and Lalor, M.J.,"Managing some problems of Fourier fringe analysis," Proc. SPIE Vol. 1163(1989), pp. 149-160
(58) Abramson, N., The making and evaluation of holograms, London, Academic Press, 1981
(59) Kreis, T., "Fourier-transform evaluation of holographic interference patterns," Proc. SPIE Vol. 814(1987), pp. 365-371
(60) Takeda, M., „Spatial-carrier fringe pattern analysis and its applications to precision interferometry and profilometry: An overview," Industrial Metrology 1(1990), pp. 79-99
(61) Womack, K.H., „Interferometric phase measurement using synchronous detection," Opt. Eng. 23(1984)4, pp. 391-395
(62) Mertz, L., „Real-time fringe pattern analysis," Appl. Opt. 22(1983), pp. 1535-1539
(63) Macy, W.W., „ Two-dimensional fringe-pattern analysis," Appl. Opt. 22(1983), pp. 3898-3901
(64) Ransom. P.L., and Kokal, J.V., „Interferogram analysis by modified sinusoid fitting technique," Appl. Opt. 25(1986), 4199-4204
(65) Morgan, C.J., „Least-squares estimation in phase-measurement interferometry," Optics Letters 7(1982)8, pp. 368-370
(66) Greifenkamp, J.E., „A generalized data reduction for heterodyne interferometry," Opt.Eng. 23(1984)4, pp. 350-352
(67) Kreis, T., Holographic Interferometry - Principles and Methods, Berlin, Akademie Verlag, 1996
(68) Wyant, J.C., and Creath, K., „Recent advances in interferometric optical testing," Laser Focus/Electro Optics, Nov. 1985, pp. 118-132

(69) Carré, P., „Installation et utilisation du comparateur photoelectrique et interferentiel du bureau international des poids et measures," Metrologia 2(1966)1, pp. 13-23

(70) Jüptner, W., Kreis, T., and Kreitlow, H., „Automatic evaluation of holographic interferograms by reference beam phase shifting," Proc. SPIE Vol. 398(1983), pp. 22-29

(71) Schwider, J., Burow, R., Elßner, K.-E., Grzanna, J., Spolaczyk, R. and, Merkel, K., „Digital wavefront measuring interferometry: Some systematic error sources," Appl. Opt. 22(1983)21, 3421-3432

(72) Hariharan, P., Oreb, B.F. and, Eiju, T., „Digital phase shifting interferometry: a simple erro-compensating phase calculation algorithm," Appl. Opt. 26(1987),pp. 2504-2505

(73) Koliopoulos, C., „Interferometric optical phase measurement techniques," Ph.D. Thesis, Optical Science Center Tucson, Univ. Arizona 1981

(74) Creath, K, „Comparison of phase-measurement algorithms," Proc. SPIE Vol. 680(1986), pp. 19-28

(75) Creath, K., „Phase-measurement interferometry techniques," In: E. Wolf (Ed.): Progr. in Optics, Vol. 26, Amsterdam, North-Holland, 1988, pp. 349-393

(76) Hunter, J.C., and Collins, M.W., „Holographic interferometry and digital fringe processing," J.Phys.D.: Appl. Phys. 20(1987), pp. 683-691

(77) Yu, Q., and Andresen, K., „Fringe orientation maps and 2D derivative-sign binary image methods for extraction of fringe skeletons," Appl. Opt. 33(1994)29, pp. 6873-6878

(78) Osten, W.; Nadeborn, W.; Andrä, P.: General hierarchical approach in absolute phase measurement. Proc. SPIE Vol. 2860(1996), 2-13

(79) Robinson, D.R, „Phase unwrapping methods," In: (10), pp. 194-229

(80) Idesawa, M, Yatagai, T., and Soma, T., Appl. Opt. 16(1977), pp. 2152-2162

(81) Varman, P.O., Optics and Lasers in Engineering 5(1984), pp. 41-58

(82) Colin, A., and Osten, W.: „Automatic support for consistent labeling of skeletonized fringe patterns," J. Mod. Opt. 42(1995)5, 945-954

(83) Ghiglia, D.C., Mastin, G.A., and Romero, L.A., „Cellular-automata method for phase unwrapping," J.O.S.A.(A) 4(1987), pp. 267-280

(84) Osten, W., and Höfling, R., „The inverse modulo process in automatic fringe analysis - problems and approaches," Proc.Int. Conf. on Hologram Interferometry and Speckle Metrology. Baltimore 1990, pp. 301-309

(85) Shough, D., "Beyond fringe analysis," Proc. SPIE Vol. 2003(1993), pp. 208-223

(86) Huntley, J.M., Buckland, J.R., „Characterization of sources of 2π-phase discontinuity in speckle interferograms," J.O.S.A. A 12(1995)9, 1990-1996
(87) Huntley, J.M., „Noise-immune phase unwrapping algorithm," Appl. Opt. 28(1989), 3268-3270
(88) Greivenkamp, J.E., „Sub-Nyquist interferometry," Appl. Opt. 26(1987), pp. 5245-5258
(89) Bone, D.B., „Fourier fringe analysis: the two-dimensional phase unwrapping problem," Appl. Opt. 30(1991)25, pp. 3627-3632
(90) Andrä, P., Mieth, U., and Osten, W., „Some strategies for unwrapping noisy interferograms in phase-sampling-interferometry," Proc. SPIE Vol.1508 (1991), pp. 50-60
(91) Judge, R.T., Quan, C., and Bryanston-Cross, P.J., „Holographic deformation measurements by Fourier transform technique with automatic phase unwrapping," Opt. Eng. 31(1992)3, 533-543
(92) Takeda, M., Nagatome, K., and Watanabe, Y., „Phase unwrapping by neural network," In: (7), pp. 137-141
(93) Huntley, J.M., Cusack, R., and Saldner, H., „New phase unwrapping algorithms," In: (7), pp. 148-153
(94) Huntley, J.M., and Saldner, H., „Temporal phase-unwrapping algorithm for automated interferogram analysis," Appl. Opt. 32(1993)17, pp. 3047-3052
(95) Ghiglia, D.C., and Romero, L.A., „Robust two-dimensional weighted and unweighted phase unwrapping that uses fast transforms and iterative methods," J.O.S.A. A 11(1994)1, pp. 107-117
(96) Judge, T.R., and Bryanston-Cross, P.J., „A review of phase unwrapping techniques in fringe analysis," Opt. Lasers Eng. 21(1994), pp. 199-239
(97) Takeda, M., " Current trends and future directions of fringe analysis," Proc. SPIE Vol. 2544 (1995), pp. 2-10
(98) Owner-Petersen, M., „Phase map unwrapping: A comparison of some traditional methods and a presentation of a new approach," Proc. SPIE Vol. 1508(1991), pp. 73-82
(99) Gierloff, , J.J., „Phase unwrapping by regions," Proc. SPIE Vol. 818(1987), pp. 267-278
(100) Ettemeyer, A., Neupert, U., Rottenkolber, H., and Winter, C., „Fast and robust analysis of fringe patterns - an important step towards the automation of holographic testing procedures," In: (5), pp. 23-31
(101) Lin, Q., Vesecky, J.F., and Zebker, H.A., „ Phase unwrapping through fringe-line detection in synthetic aperture radar interferometry," Appl. Opt. 33(1994)2, pp. 201-208

(102) Kreis, T.M., Biedermann, R., and Jüptner, W.P.O., „Evaluation of holographic interference patterns by artificial neural networks," Proc. SPIE Vol 2544(1995), pp. 11-24
(103) Marroquin, J.L.; Rivera, M., „Quadratic regularization functionals for phase unwrapping," J.O.S.A. A 12(1995), 2393-2400
(104) Marroquin, J.L; Servin, M.; Rodriguez-Vera, R., „Adaptive quadrature filters for multi-phase stepping images," Opt. Let. 23(1998)4, 238-240
(105) Marroquin, J.L; Rivera, M.; Botello, S.; Rodriguez-Vera, R; Servin, M., „Regularization methods for processing fringe pattern images," Proc. SPIE Vol. 3478(1998), 26-36
(106) Osten, W., „Active optical metrology - a definition by examples," Proc. SPIE Vol. 3478(1998), pp. 11-25
(107) B. Hofmann, „Ill-posedness and regularization of inverse problems – a review of mathematical methods," In: H. Lübbig, *The inverse problem.* Akademie Verlag Berlin 1995, pp. 45-66
(108) J. Hadamard, *Lectures on Cauchy's problem in partial differential equations.* Yale University Press, New Haven 1923
(109) A.N. Tikhonov, „Solution of incorrectly formulated problems and the regularization method," Sov. Math. Dokl. 4(1963), 1035-1038
(110) A.N. Tikhonov, V.Y. Arsenin, *Solution of ill-posed problems.* (Transl. From Russian) Winston and Sons, Washington 1977
(111) Ghiglia, D.C.; Pritt, M.D., *Two-dimensional phase unwrapping.* Wiley, New York 1998
(112) Stetson, K.A., „Use of sensitivity vector variations to determine absolute displacements in double exposure hologram interferometry," Appl. Opt 29(1990), pp. 502-504
(113) Osten, W.; Nadeborn, W.; and Andrä, P., „General hierarchical approach in absolute phase measurement," Proc. SPIE Vol. 2860(1996), 2-13
(114) Abramson, N, "The Holo-Diagram V: A Device for practical interpreting of hologram interference fringes," Appl. Opt. 11(1972), pp. 1143-1147
(115) Skudayski, U., and Jüptner, W., "Synthetic wavelength interferometry for the extension of the dynamic range," Proc. SPIE Vol. 1508(1991), pp. 68-72
(116) Yuk, K.C.,Jo, J. H. and Chang, S., "Determination of the absolute order of shadow moiré fringes by using two differently colored light sources," Appl. Opt. 33(1994)1, pp. 130-132
(117) Altschuler, M.D., Altschuler, B.R., and Taboda, J., "Measuring surfaces space-coded by laser-projected dot matrix," Proc. SPIE Vol. 182(1979), pp.163-172

(118) Wahl, F.M., "A coded light approach for depth map acquisition," G. Hartmann (Ed.): Mustererkennung 1986, Springer Verlag 1986, pp. 12-17
(119) Zumbrunn, R. " Automatic fast shape determination of diffuse reflecting objects at close range by means of structured light and digital phase measurement," ISPRS, Interlaken, Switzerland 1987, pp. 363-378
(120) Steinbichler, H., "Verfahren und Vorrichtung zur Bestimmung der Absolutkoordinaten eines Objektes," German Patent 4134546
(121) Nadeborn, W., Andrä, P, and Osten, W., „A robust procedure for absolute phase measurement," Opt. & Lasers in Eng. 24(1996), pp. 245-260
(122) Tiziani, H.: Optische Verfahren zur Abstands- und Topografiebestimmung. Informationstechnik 1(1991), 5-14
(123) Pfeifer, T.; Thiel, J.: Absolutinterferometrie mit durchstimmbaren Halbleiterlasern. Technisches Messen 60(1993)5, 185-191
(124) de Groot, P.J.: Extending the unambiguous range of two-color interferometers. Appl. Opt. 33(1994)25, 5948-5953
(125) Takeda, M.; Yamamoto, H.: Fourier-transform speckle profilometry: three-dimensional shape measurement of diffuse objects with large height steps and/or spatially isolated surfaces. Appl. Opt. 33(1994), 7829-7837
(126) Zou, Y.; Pedrini, G.; Tiziani, H.: Surface contouring in a video frame by changing the wavelength of a diode laser. Opt. Eng. 35(1996)4, 1074-1079
(127) Kuwamura, S.; Yamaguchi, I.: Wavelength scanning profilometry for real-time surface shape measurement. Appl. Opt. 36(1997)19, 4473-4482
(128) Huntley, J.M.; Saldner, H.: Temporal phase-unwrapping algorithm for automated interferogram analysis. Appl. Opt. 32(1993)17, 3047-3052
(129) Huntley, J.M.; Saldner, H.: Shape measurement by temporal phase unwrapping and spatial light modulator-based fringe projector. Proc. SPIE. Vol. 3100(1997), 185-192
(130) Osten, W., and Jüptner, W., „Measurement of displacement vector fields of extended objects," Opt. & Lasers in Eng. 24(1996), pp. 261-285
(131) Pryputniewicz, R. J., „Experiment and FEM modeling," In: W. Jüptner and W. Osten (Eds.): Fringe´93, Proc. of the 2nd Intern. Workshop on Automatic Processing of Fringe Patterns, " Bremen 1993, Akademie Verlag Berlin 1993, 257-275
(132) Abramson, N., „The holo-diagram: a practical device fo making and evaluating holograms," Appl. Opt. 8(1969), 1235-1240
(133) Abramson, N., „ Sandwich hologram interferometry. 4. Holographic studies of two milling machines," Appl. Opt. 18(1977), 2521
(134) Birnbaum, G.; Vest, C. M., „Holographic nondestructive evaluation: status and future," Int. Adv. in Nondestr. Test 9(1983), 257-282

(135) Sollid, J.E., and Stetson, K. A., „Strains from holographic data," Exp. Mech. 18(1978), 208-216
(136) Stoer, J., „Einführung in die numerische Mathematik I," Springer Verlag Berlin, Heidelberg, New York 1972
(137) Osten, W., „Theory and practice in optimization of holographic interferometers," Proc. SPIE VOL. 473(1988), 52-55
(138) Seebacher, S.; Osten, W., and Jüptner, W., „3D-Deformation analysis of microcomponents using digital holography," Proc. SPIE Vol.3098(1997), pp. 382-391
(139) Osten, W, and Häusler, F., „Zur Optimierung holografischer Interferometer," Preprint P-Mech-08/81
(140) Tiziani, H., „Optical techniques for shape measurement, In: W. Jüptner and W. Osten (Eds.): " Fringe´93," Proc. of the 2nd Intern. Workshop on Automatic Processing of Fringe Patterns, Bremen 1993, Akademie Verlag Berlin 1993, pp. 165-174
(141) Krattenthaler, W., Mayer, K.J., and Duwe, H.P., „3D-surface measurement with coded light approach," In: Österr. Arbeitsgem. Mustererkennung, Proc. ÖAGM 12(1993), 103-114
(142) Donges, A.; Noll, R.: Lasermeßtechnik. Hüthig Verlag Heidelberg 1993
(143) Nadeborn, W., Andrä, P., and Osten, W., "Model based identification of system parameters in optical shape measurement," In: W. Jüptner and W. Osten (Eds.): " Fringe´93," Proc. of the 2nd Intern. Workshop on Automatic Processing of Fringe Patterns, " Bremen 1993, Akademie Verlag Berlin 1993, 214-222
(144) Tsai, R. Y., "A versatile camera calibration technique for high-accuracy 3-D machine vision metrology using off-the-shelf TV cameras and lenses, " IEEE J. Robot. Automat. RA-3(4), pp. 323 - 344, 1987
(145) Peeck, A.: Bestimmung von Abweichungen nach Maß, Form und Lage mit Hilfe der holografischen Interferometrie. Dissertation, Technische Univ. Hannover 1974
(146) Wernicke, G.; Mente, L.; Osten, W.: Application of mismatch techniques in holographic interferometry. EAN 1982, Karlovy Vary, Conf. Proc. Vol. III, 31-38
(147) Pfeifer, T.; Mischo, H.; Koch, S.; Evertz, J.: Coded speckle-interferometrical formtesting. Proc. Fringe´97, Akademie Verlag Berlin 1997, pp. 171-178
(148) Schnars, U.: Direct phase determination in hologram interferometry with use of digitally recorded holograms. J.O.S.A A 11(1994)7, pp. 2011-2015
(149) Kreis, T., Jüptner, W., and geldmacher, J., „Principles of digital holographic interferometry," Proc. SPIE Vol. 3478(1998), 45-54

(150) Pryputniewicz, R.J., and Stetson, K. A., "Determination of sensitivity vectors in hologram interferometry from two known rotations of the object," Appl.Opt. 18(1980)13, 2201-2205

(151) Vogel, D., V. Großer, W. Osten, J. Vogel and R. Höfling, "Holographic 3D-measurement technique based on a digital image processing system. Proc. Fringe´89 , Akademie Verlag   Berlin 1989, pp. 33-41

(152) Schreiber, W., L. Wenke and W. Osten, " Bestimmung von Verschiebungsvektoren mittels Größen," Optica Acta 28(1981)9, 1163-1167

(153) Kohler, H. " Interferometric instead of geometric measurement of object points in holographic interferometric deformation analysis," Optica Acta 29(1982)3, 275-280

(154) Nobis, D., and Vest, C. M., "Statistical analysis of errors in holographic interferometry,"   Appl.Opt. 17(1987), 2198-2204

(155) W. Osten, "Some considerations on the statistical error analysis in holographic interferometry with application to an optimized interferometer," Optica Acta 32(1985)7, 827-838

(156) Ek, L., and Biedermann, K. "Analysis of a system for hologram interferometry with a continuously scanning reconstruction beam," Appl.Opt. 17(1977), 2535-2542

(157) Dhir, S.K., and Sikora, J.P.," An improved method for obtaining the general displacement field from a holographic interferogram," Exp. Mech. 12(1972),323-327

(158) Andrä, P.; Beeck, A.; Jüptner, W.; Nadeborn, W.; and Osten, W., „Combination of optically measured coordinates and displacements for quantitative investigation of complex objects," Proc. SPIE Vol. 2782(1996), pp.

(159) Jüptner, W.; Osten, W.; Andrä, P.; and Nadeborn, W., „Nondestructive quantitative 3D characterization of a car brake," Proc. SPIE Vol. 2861(1996), 170-179

(160) Birnbaum, G, and Vest, C. M., "Holographic nondestructive evaluation: status and future," Int. Adv. in Nondestr. Test 9(1983), pp. 257-282

(161) Osten, W., Saedler, J., and Wilhelmi, W., „Fast evaluation of fringe patterns with a digital image processing system," (in German), Laser Magazin 2(1987), pp.58-66

(162) Osten, W., Jüptner, W., and Mieth, U, „Knowledge assisted evaluation of fringe patterns for automatic fault detection," Proc. SPIE Vol. 2004(1993), pp. 256-268

(163) Jüptner, W., Kreis, T., Mieth, U., and Osten, W., „Application of neural networks and knowledge based systems for automatic identification of

fault indicating fringe patterns," Proc. SPIE Interferometry'94 Vol. 2342(1994), pp. 16-26
(164) Kreis, T., Jüptner, W., and Biedermann, R., „Neural network approach to nondestructive testing," Appl. Opt. 34(1995)8, pp. 1407-1415
(165) Bischof,Th., and Jüptner,W., „Determination of the adhesive load by holographic interferometry using the result of FEM-calculations," Proc. SPIE Vol. 1508(1991), pp. 90-95
(167) Osten, W.; Elandaloussi, F., and Jüptner, W., „Recognition by synthesis - a new approach for the recognition of material faults in HNDE," Proc. SPIE Vol. 2861(1996), 220-224
(168) Mieth, U.; Osten, W.; and Jüptner, W., „Numerical investigations on the appearance of material flaws in holographic interference patterns," Proc. International Symposium on Laser Application in Precision Measurement, Balatonfüred 1996, Akademie Verlag Berlin 1996, 218-225
(169) Jüptner, W., and Kreis, Th., „Holographic NDT and visual inspection in production line application," Proc. SPIE 604(1986), pp. 30-36
(170) Jüptner, W., „Nondestructive Testing with Interferometry," Proc. *"Fringe'93"*, Akademie Verlag, Berlin, 1993, pp. 315-324
(171) Hung, Y.,Y.: „Displacement and strain measurement," In: R.K. Erf (Ed.): Speckle Metrology. Academic Press, New York 1987, pp. 51-71
(172) Haykin, S., Neural Networks: A Comprehensive Foundation, MacMillan, New York, 1994
(173) Osten, W.; Elandaloussi, F., Jüptner, W.: Recognition by synthesis - a new approach for the recognition of material faults in HNDE. Proc. SPIE Vol. 2861(1996), 220-224
(174) Osten, W.; Elandaloussi; F.; Mieth, U.: Software brings automation to fringe-pattern processing. EuroPhotonics, February/March 1998, pp 34-35